Energy Storage, Grid Integration, Energy Economics, and the Environment

Nano and Energy Series

SERIES EDITOR

Sohail Anwar, Pennsylvania State University, Altoona College, USA

For more information about this series, please visit:
https://www.crcpress.com/Nano-and-Energy/book-series/NANANDENE

Energy Storage, Grid Integration, Energy Economics, and the Environment

Radian Belu

CRC Press
Taylor & Francis Group
Boca Raton London New York

CRC Press is an imprint of the
Taylor & Francis Group, an **informa** business

CRC Press
Taylor & Francis Group
6000 Broken Sound Parkway NW, Suite 300
Boca Raton, FL 33487-2742

First issued in paperback 2022

© 2020 by Taylor & Francis Group, LLC
CRC Press is an imprint of Taylor & Francis Group, an Informa business

No claim to original U.S. Government works

ISBN 13: 978-0-367-26140-5 (hbk)
ISBN-13: 978-1-03-233796-8 (pbk)
DOI: 10.1201/9780429322433

Contents

Preface

Two volumes cover main renewable energy sources, conversion and energy storage systems and technologies, grid integration, economic aspects and environmental implications of the renewable energy systems, from basic and fundamental knowledge, to resource assessment and design aspects. The second volume contains comprehensive descriptions and in-depth presentations of the main energy storage technologies, bioenergy, biomass and biofuels, a brief introduction of the hydrogen economy, microgrids and distributed generation, environmental impacts of renewable energy uses and applications, economic and cost analysis of renewable energy systems and projects, energy concentration and energy management. Chapter 1 of this volume introduces a comprehensive presentation of the major energy storage technologies, with the exception of the batteries and fuel cells, discussed in next volume chapter. This chapter discusses in detail the energy storage technology characteristics, functionalities, operation principles, performances and the most suitable applications into renewable energy and power systems. The reader will have a good understanding of the operation of each energy storage technology, functions and parameters that are critical in the energy storage technology selection for a specific application. Chapter 2 discusses in detail the batteries and fuel cells, while giving a brief discussion of the hydrogen economy. After an introduction of the electrochemistry basics, the chapter focuses on major battery types, including flow a special batteries, characteristics, operation, performances and equivalent circuits, followed by the presentation of the principles, operation and types of fuel cells, as well as the applications of batteries and fuel cells. Chapter 3 covers bioenergy importance and application, biomass uses and applications, biofuel production methods and potential, as well as further discussion on the hydrogen economy. Chapter 4 is reserved for utility and grid integration of the renewable energy systems, integration issues, challenges and methods. A chapter section gives a brief presentation of the main power electronic converters used in renewable energy applications and systems. Chapter 5 focuses on renewable energy economics and cost analysis methods and procedures, including life-cycle cost analysis and how they are applied to major renewable energy applications. Chapter 6 contains a good discussion of the distributed generation concepts and issues, microgrid architecture, structure, configuration, control and protection methods, including DC microgrids and nanogrids. Microgrids have gained increased attention over the last decades because they are viewed as a major component of the smart grids, and for their roles in the grid integration of renewables, improved grid resilience, service restoration and as an economic way to provide power in the areas where the grid is not available or too expensive to expand. Chapter 7 introduces a comprehensive discussion of the environmental impacts of the renewable energy uses and applications, life-cycle assessment analysis methods and procedures. The last section of this chapter discusses energy conservation issues and a short presentation of the energy management methods, procedures and importance. The materials in this volume originate from the courses that the author developed and taught in the areas of energy and power engineering, renewable energy systems, distributed generation, industrial energy systems and energy management, as well as from the research projects that the author was involved in the last 25 years. Similar to the first volume, this volume also assumes no specific knowledge of power and energy engineering, energy storage or renewable energy; it guides the reader through basic understanding from topic to topic and inside of each topic. Over 270 questions and problems are included in this volume. Each chapter also contains several solved examples to help the readers and instructors to understand and/or to teach the materials. A rich, comprehensive and up-to-date reference section is also included at the end of each of the volume chapters for professionals, engineers, students and interested readers in the renewable energy, storage, microgrids, grid integration or renewable energy topics and problems. The book is intended both as a textbook and as reference book for students, instructors, engineers and professionals interested in renewable energy systems, technologies, operation, use and design. The author is fully indebted to the students, colleagues and co-professionals for their feedback and suggestions over the years, and last but not least to the editorial technical staff for support and help.

Acknowledgment

To my wife Paulina Belu, my best friend and partner in life, for her patience, understanding and support. The book is also dedicated to my children, Alexandru, Ruxandra, Mirela and my grandchildren Stefan-Ovdiu and Ana-Victoria for their support and life's enjoys.

Author

Radian Belu, PhD, is an associate professor in the electrical engineering department at Southern University and A&M College, Baton Rouge, Louisiana. He has a PhD in power engineering as well as a PhD in physics. Before joining Southern University, Dr. Belu held faculty and research positions at universities and research institutes in Romania, Canada, and United States. He also worked for several years in the industry as a project manager, senior engineer, and consultant. His research focuses on energy conversion, renewable energy, microgrids, power electronics, climate, and extreme event impacts on power systems. He has taught and developed courses in power engineering, renewable energy, smart grids, control, electric machines, and environmental physics. His research interests include power systems, renewable energy systems, smart microgrids, power electronics and electric machines, energy management and engineering education. Dr. Belu has published 1 book, 15 book chapters, and over 200 papers in refereed journals and in conference proceedings. He has been PI or Co-PI for various research projects in the United States and abroad.

1

Energy Storage Systems

1.1 Introduction, Energy Storage Overview

Electricity distribution systems have entered in a period of considerable change, driven by several interconnected factors, aging network, distributed generation, renewable energy integration, pollutant reduction requirements, regulatory incentives, demand side management and new technologies. In this climate, the use of energy storage has emerged as an area of considerable interest. The end of this period of transition will be signaled by the successful establishment of the technology and practices that must go together to create what is termed the smart grid. A smart grid as part of an electricity power system can intelligently integrate the actions of all users connected to it, generators, consumers and those that do both, in order to efficiently deliver sustainable, economic and secure electricity supplies. The end state of this transition is not yet known. Moreover, the transmission capabilities are not expanding enough to meet the growing needs, causing serious constraint problems, while as a result of the electrical grid aging, outages cost billions of dollars annually in the U.S. economy and elsewhere. Energy storage, designed to provide support for both long-term applications and dynamic performance enhancement, can provide better balancing between the energy demand and supply, allowing the increased asset utilization, facilitating the renewable energy penetration, and improving the flexibility, reliability and the overall grid efficiency. On the other hand, most of the renewable energy sources are characterized by generation variability, intermittency and discontinuity; the generation cannot be controlled by the system operator, making it more difficult to integrate it into the generation pool when compared with conventional electricity generators.

The most important enabling technology for the renewable energy use on the utility scale is the energy storage. It has been in existence for a long time and has been utilized in many forms and applications—from a flashlight to spacecraft systems. Today energy storage is used to make the electric power systems more reliable and resilient or to broaden the renewable energy use a reality. Energy storage systems (ESSs) are critical in enabling the renewable energy integration and usage, by providing the means to convert non-dispatchable energy system into a dispatchable one. To match power demands, in the context of the renewable energy intermittent and the fairly predictable electrical demands, the energy storage is critical, allowing de-coupling of generation from consumption, reducing the needs for constant energy demand monitoring and prediction. Energy storage also provides economic benefits by allowing a reduction of plant energy production to meet average demands rather than peak power demands. Transmission lines and equipment can also be sized for average power demand, instead of peak demands. The energy storage is enabling the power plants to run at higher capacities, ensuring that electrical demands are met all times, reducing the needs for peaking power plants that have lowest efficiency, higher harmful emissions, and highest operating costs. Even the entire peaking power plant concept could be dismissed if adequate energy storage is utilized. Moreover energy storage helps the utilities to provide the required power quality and reliability by the increasingly complex and sensitive equipment, while maximizing the electrical capacity use. Overall it can improve system responsiveness, reliability, resilience and flexibility, reducing capital and operating costs. Suppliers can use energy storage for transmission line stabilization, spinning reserve, and voltage control, while customers receive improved power quality and more reliable supply. Technologies such as ultra-capacitors, flywheels, batteries, fuel cells and superconducting magnetic energy storage can be used for power quality and reliability purposes. These applications require a large power output over very short timescales, typically from tenths of a second to a few minutes, while not requiring large amount of energy to be stored, because of the operations' short timescales. ESSs can provide a wide array of solutions to key issues, affecting the power systems, such as spinning reserves, load leveling and shifting, load forecasting, frequency control, voltage regulation, relief of transmission line and system capacity, enhance power quality, more effective and efficient use of capital resources. Being able to store the excess available energy that has not been consumed not only helps with the variety of issues previously mentioned, but it also increases the overall power system efficiency.

There are several well-established electricity storage technologies, as well as a large number in process of development offering significant application potential. Economically viable storage of energy requires efficient conversion of electricity and storage in other energy form, which can be converted back to electricity when needed. All energy storage methods need to be feasible and environmentally safe. Energy storage technologies are separated into four major classes: *mechanical-, electrical-, thermal-* and *chemical*-based energy storage systems. Each class contains several types with specific characteristics and applications. Mechanical storage includes pumped hydro storage, compressed air energy storage and flywheels. Electrochemical storage includes all types of batteries, hydrogen-based energy storage and fuel cells. Here also is included thermochemical energy storage, such as solar hydrogen, solar metal, solar ammonia dissociation–recombination and solar methane dissociation–recombination. Electromagnetic energy storage includes super-capacitors and superconducting magnetic energy storage. Thermal energy storage includes two broad categories: (1) *low temperature energy storage*, such as aquiferous cold energy storage and cryogenic energy storage, and (2) *high temperature energy storage*, which is divided into sensible heat systems such as steam or hot water accumulators, graphite, hot rocks and concrete, and latent heat systems such as phase change materials. Figure 1.1 summarizes the most common energy storage technologies. Despite the opportunity offered by the energy storage systems to increased energy stability and reliability of the intermittent energy sources, there were only few energy storage technologies (batteries, pumped hydro energy storage, compressed air energy storage, and thermal storage) with a globally installed capacity exceeding 100 MW. There are several opportunities for significant improvement in energy storage, with the most appropriate technologies for application to power quality management, load shifting and energy management.

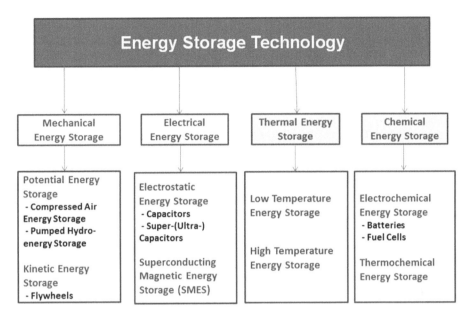

FIGURE 1.1
Classification of the major energy storage technologies.

In a weak power grid, the RES integration at remote connection points may generate unacceptable voltage variations due to power generation fluctuations. Upgrading the power transmission line to mitigate this problem is often very expensive, while the EES inclusion for power smoothing and voltage regulation at the connection point allows the weak grid operation, offering an economic alternative to transmission line upgrading. The current status shows that several drivers are emerging, spurring the growth in the energy storage demand, such as increase in the renewable energy usage, an increasingly strained transmission infrastructure as new lines lag well behind demand, the microgrid emergence as part of distributed grid architecture, and the increased needs for higher supply reliability and security. However, issues regarding the optimal active integration (operational, technical and market) of energy storage systems in the electric grid are still not fully developed, tested and standardized. The ESSs integration and further development of energy converting units including renewable energies must be based on the existing electric system infrastructure, requiring optimal integration of energy storage systems. Renewable energy systems with optimum energy storage can behave as conventional power plants, at least for short-time intervals, 1 hour to 1 day, depending on the energy storage system capacity. Renewable energy sources can rarely provide immediate response to demands as these sources do not deliver a supply easily adjustable to consumption, being in this regard low-inertia systems. Growth of decentralized power generation results in greater network load stability problems, requiring energy storage, as one of the potential solutions. Energy storage is crucial in the energy management from renewable energy sources, allowing energy to be released into the grid during peak hours when it is needed. In the conventional operation planning process of bulk power plants, normally a top-down strategy, coming from an energy consumption point of view down to a stepwise detailed description is used. In this strategy, the planning horizon is subdivided in long-, medium- and short-term planning. The time scale in each planning step is a compromise between accuracy and the number of technical and economical boundaries. Planning strategy is mainly driven by economic considerations in a unidirectional electrical power supply chain. In such planning strategies the detailed control functionality can be only figured out when the system planning model in the planning is detailed enough. This means the detailed model accuracy, inside of each planning stage (long-, medium-, short-term, quasi-stationary, or dynamic), defines the optimal layout of the conventional supply systems. In the distributed generation (DG) case the technical boundaries are critical for the planning process to get an economical optimal supply configuration under stable operation conditions. When this is not considered, a potential conflict may arise between the energy supplier and the operator of electricity distribution networks, especially in the case of renewable energy sources. Therefore, there are needs for planning and integration strategy that include clear system specifications and definitions, system modeling and structuring for optimal integration of energy storage and distributed energy resources (DERs) that include ESS and DER control functions, limits and boundaries. These requirements lead to bottom-up strategy starting from DG units up to the centralized conventional electrical supply system.

The most relevant renewable energy sources used in power generation are solar thermal energy, photovoltaic and wind energy systems. The main disadvantage of such energy sources is its generation discontinuity, their energy generation is not controlled by the system operator, making more difficult to integrate them into the conventional generation pools. Energy storage becomes a critical factor that can solve these issues. The major energy storage system applications are summarized in Figure 1.2, where the systems are classified according to their applications. Energy storage technologies for grid applications have achieved various levels of technical and economic maturity. For grid energy storage, challenges include roundtrip efficiencies that range from under 30% to over 90%. Efficiency losses represent a trade-off between the increased electricity

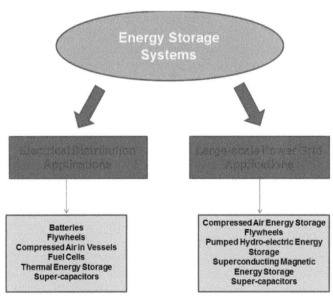

FIGURE 1.2
Applications of the principal energy storage systems.

cost, cycled through the energy storage system, and the increased value of greater dispatchability and other grid services. The capital cost of many grid storage technologies is highly relative to the conventional alternatives, such as gas-fired power plants, which can be constructed quickly and are perceived as a low-risk investment by both regulated utilities and independent power producers. Existing electricity market structures may also undervalue services that electricity storage can provide. However, substantial and extensive research and developments are underway in the United States and abroad to improve energy storage system performances. Changes to market structures and policies may be critical components of achieving competitiveness for electricity storage. Removing non-technical barriers may be as important as technology improvements in increasing energy storage adoption to improve grid performances. Understanding the energy storage potential in electric applications is complicated by several factors. First is the wide range of energy storage technologies either commercially available, in development, or being researched, making difficult to have a balanced understanding of the capabilities, costs or comparative advantages of the energy storage options. Second, there are different energy storage applications, each with distinct operational requirements. Certain energy storage technologies may suit certain applications better than others. Finally, there are many aspects of market structure and economic regulation that affect storage deployment. These factors make the development of energy storage research challenging. While there is general consensus that storage technology improvements are needed, there are multiple potential pathways to such improvements that cut across different scientific disciplines. A number of obstacles have hampered the energy storage usage, such as: (1) lack of design experience, operation of management, (2) inconclusive benefits, (3) high capital costs or (4) if utilities or RES developers should pay for energy storage. However, as renewable energy systems and power quality become increasingly important, costs and concerns regarding energy storage are expected to decline. This chapter was carried out to identify various types of energy storage currently available. The parameters used to describe an energy storage device and the applications they fulfill are explored and discussed, the analysis of each technology currently available are presented in details, indicating the operation, cost, applications, advantages, disadvantages and future developments.

1.2 Energy Storage System Functions

Electrical energy can be stored in *kinetic, potential, electrochemical* or *electromagnetic forms*, which can be transferred back to the electrical energy when required. The conversion of electrical energy to different forms and back to electrical energy is done by the power conversion systems. The energy generated during off-peak periods can be stored and used to meet the loads during peak periods when the energy is more expensive, improving the power system economics and reliability. Compared to conventional generators, many energy storage systems have a faster ramping rate which can quickly respond to the load fluctuations. Therefore, the energy storage systems can be a perfect spinning energy reserve which provides a fast load following and reduces the need for conventional spinning fossil fuel reserves of the conventional generation plants. Electrical energy storages were initially used only for load leveling applications, while they are considered now to improve the power system quality and stability, to ensure a reliable and more secure power supply, and to black-start the power system. Breakthroughs that are reducing the energy storage costs could drive revolutionary changes in the power system design and operation. Peak load problems can be reduced, stability can be improved, and power disturbances can be reduced or even eliminated. Storage can be applied at the power plant level, in support of the transmission system, at various points in power distribution system or on particular equipment on the customer meter side. Figure 1.3 shows the new electricity value chain, supported by the integration of energy storage systems. Energy storage systems in combination with advanced power electronics have a great technical role and lead to many financial benefits. Energy storage devices by their nature are typically suitable for a very particular application set, primarily due to their potential power and storage capacities. Therefore, in order to provide a fair comparison between the various energy storage technologies, they are grouped, based on their size of power and storage capacity, separated into four categories: devices with large power (>50 MW) and storage (>100 MWh) capacities; devices with medium power (1–50 MW) and storage capacities (5–100 MWh); devices with medium power or storage capacities but not both; and finally storage systems with power less than 1 MW and capacity less than 5 MWh. Energy storage technology is now a well-established concept, with pumped hydroelectric energy storage (PHES), in use since 1929 primarily to level the grid daily load. As the electricity sector is undergoing important changes, the energy storage is becoming a realistic option for: (1) restructuring the electricity market; (2) integrating renewable resources; (3) improving power quality; (4) aiding shift toward distributed energy; and (5) helping the network operate under more stringent environmental requirements.

ESSs are needed by the conventional electricity generating industry, because unlike any other commodities, the conventional electricity generating industries have little or no energy storage facility. The electricity transmission and distribution systems are operated for the simple one-way transportation from large power plants to consumers, meaning that electricity must always be used precisely when produced. However, the electricity demand varies considerably emergently, daily and seasonally, while the maximum demand may only last for very short periods, leading to inefficient, overdesigned and expensive power plants. ESS allows energy production to be de-coupled from its supply, self-generated or purchased. Having large-scale electricity storage capacity available, system planners would need to build only generating capacity to meet the average electricity demands rather than peak demands. This is particularly important to large utility generation units, e.g., nuclear power plants, which must operate near

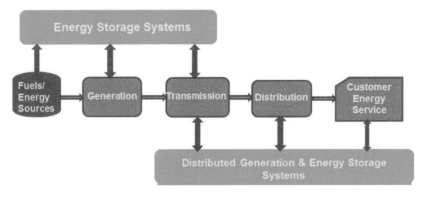

FIGURE 1.3
New chain of electricity, with energy storage as the sixth dimension.

full capacity for economic reasons. Therefore, ESS can provide substantial benefits, e.g., load following, peaking power and standby reserve, increasing the net efficiency of thermal power sources while reducing harmful emissions. Furthermore, ESS is regarded as an imperative technology for the distributed energy resource systems. The traditional electricity value chain has been considered to consist of five links: fuel/energy source, generation, transmission, distribution and customer-side energy service as shown in Figure 1.3. Supplying power when and where is needed, ESS is on the brink of becoming the *sixth link* by integrating the existing segments and creating a more responsive market. Storage technologies are numerous, covering the full spectrum ranging from larger scale, generation and transmission, to those primarily related to the distribution network and even "beyond the meter," into the end-user site.

ESS performances include cycle efficiency, cost per unit capacity, energy density, power capacity, life cycle, environmental effect including end-of-life disposal cost. An ideal ESS system is one exhibiting the best possible performances so that it will have the minimum amortized (dollar or environmental) cost during its whole lifetime. Unfortunately, no single ESS type can simultaneously fulfill all the desired characteristics of an ideal ESS system, and thus minimize the amortized lifetime cost of ESS. Capital cost, one of the most important criteria in the ESS design, can be represented in the form of cost per unit of delivered energy ($/kWh) or per unit of output power ($/kW). Capital cost is an especially important concern when constructing hybrid energy storage (HES) systems. Such systems often consist of several ESS elements with relatively low unit cost (e.g., lead-acid batteries) and ESS elements with relatively high unit cost (e.g., super-capacitors). The overall HES system cost must be minimized by allocating the appropriate amounts of low-cost vs. high-cost ESS elements while meeting some other constraints such as cycle efficiency, total storage capacity, or peak output power rate. The ESS cycle efficiency is defined by the "roundtrip" efficiency, i.e., the energy input and output efficiency for charging and then discharging. The cycle efficiency is the product of charging efficiency and discharging efficiency, where charging efficiency is the ratio of electrical energy stored in an ESS element to the total energy supplied to that element during the entire charging process, and discharging efficiency is the ratio of energy derived from an ESS element during the discharging process to the total energy stored in it. Charging/discharging efficiency can be significantly affected by the charging/discharging profiles and the ambient conditions. The ESS state of health (SOH) is a measure of its age, reflecting the ESS general condition and its ability to store and deliver energy compared to its initial state. During the ESS lifetime, its capacity or "health" tends to gradually deteriorate due to irreversible physical and chemical changes which taking place. Term "replacement" implies the ESS disposal and replace, meaning the use, until it is no longer usable (its end-of-life) and replace with the same or similar system. To indicate the rate at which SOH is deteriorating, the life cycle may be defined as the number of ESS cycles performed before its capacity drops to a specific threshold of its initial capacity. It is one of the key performance parameters and gives an indication of the expected ESS working lifetime. The life cycle is closely related to the replacement period and full ESS cost according to a life cycle assessment. The self-discharge rate is a measure of how quickly a storage element loses its energy when it simply sits on the shelf, being determined by the inner structure and chemistry, ambient temperature and humidity, and significantly affect the sustainable energy storage period of the given storage element.

Deferred from the conventional power system which has large, centralized units, DERs are usually installed at the distribution level, close to the place of utilization, and generate power typically in the small range of a few kW to a few MW. A DER is regarded as a sustainable, efficient, reliable and environmentally friendly alternative to the conventional generators. The energy system is undergoing the change to be a mixture of centralized and distributed subsystems with higher and higher DER penetration. However, more drastic load fluctuations and emergent voltage drops are anticipated, due to smaller capacity and higher line fault probability than in conventional power system. ESS is identified as a key solution to compensate the power flexibility and provide a secure power supply in cases of instantaneous voltage drop for such distributed energy networks. ESSs are also critically important to intermittent renewable energy supply. The penetration of renewable resources may displace significant amounts of energy produced by large conventional plants. However, intermittency and non-controllability are inherent characteristics of renewable energy-based electricity generation systems. A suitable ESS could provide an important (even crucial) approach to dealing with the RES intermittency and unpredictability of their output as the surplus could be stored during the periods when generation exceeds the demand and then be used to cover periods when the load is greater than the generation. Future development of renewable energy technologies will drive the cost down. Nonetheless, the widespread deployment of solar and wind power in the future will face the fundamental difficulty of intermittent supplies, requiring demand flexibility, backup power sources, and enough storage for hours to days and perhaps a week. For example, the EES applications to enhance wind generation were identified as: (i) transmission curtailment, mitigating the power delivery constraint imposed by insufficient transmission capacity; (ii) time-shifting, firming and shaping of wind-generated energy by storing it during the off-peak interval (supplemented by power purchased from the grid when wind generation is inadequate) and discharging during the peak interval; (iii) forecast hedge mitigating the errors (shortfalls) in wind energy bids into the market prior to required delivery, thus reducing volatility of spot prices and mitigating risk exposure of consumers to this volatility; (iv) grid frequency support through the energy storage during sudden, large decreases in wind generation over a short discharge interval; and (v) fluctuation suppression through stabilizing the wind farm generation frequency by suppressing fluctuations (absorbing and discharging energy during short duration variations in output). The key applications for storing the renewable electricity should be equally relevant to solar power generation as well as other intermittent renewables.

1. *Grid voltage support*, meaning that the power is provided to the electrical distribution grid to maintain voltages within the acceptable limits, which is a trade-off between the amount of "real" energy produced by generators and the amount of "reactive" power produced.

2. *Grid frequency support*, meaning that the real power is provided to the power distribution to reduce any sudden, large load-generation imbalance to keep the grid frequency within the permissible tolerance for periods up to 30 minutes.

3. *Grid angular (transient) stability*, meaning, the reduction in the power oscillations (due to rapid events) by injection and absorption of real power.

4. *Load leveling/peak shaving* represents the rescheduling of certain loads to cut electrical power demand, or the production of energy during off-peak periods for storage and use during peak demand periods. While peak shaving is reducing electric usage during peak periods or moving usage from the time of peak demand to off peak periods.

5. *Spinning reserve*, the amount of generation capacity that can be used to produce active power over a given period of time, which has not yet been committed to the ***energy*** production during this period.

6. *Power quality improvement* is basically related to the changes in magnitude and shape of voltage and current. This results in different issues, including Harmonics, Power Factor, Transients, Flicker, Sagand Swell, etc. DESSs can mitigate these problems.

7. *Power reliability*, defined as the percentage/ratio of interruption in delivery of electric power (may include exceeding the threshold and not only complete loss of power) versus total uptime. Distributed energy storage systems can help provide reliable electric service to consumers.

8. *Ride through support*, meaning the electric unit stays connected during system disturbance (voltage sag). ESSs have the potential of providing energy to ride-through.

9. *Unbalanced load compensation*, achieved in combination with four-wire inverters and also by injecting and absorbing power individually at each phase to supply unbalanced loads.

The advanced electric energy storage technologies, when utilized properly, would have an environmental, economic, and energy diversity advantages to the system, including:

1. *Matching electricity supply to load demand*: Electrical energy is stored during times when production (from power plants) exceeds consumption (at lower cost possible) and the stored energy is utilized at times when consumption exceeds production (again, assuming at higher cost level). In this way, electricity production need not be scaled up and down to meet momentary consumption—instead, production is maintained at a more constant and economic level. This has the advantage that fuel-based power plants (i.e., coal, oil, gas) can be operated more efficiently and easily at constant production levels. Also this technique would maintain a continuous power to the customer without fluctuation, and reducing the need for expensive, aging, and environmentally unfriendly fossil-fired generation plants.

2. *Reducing the risks of power blackouts*: Electricity storage technologies have the ability to provide power to the grid to smooth out short-term fluctuations caused by momentary interruptions and sudden load changes. If applied properly, real long-term energy storage technology can also provide power to the grid during the long blackouts such as losing a transmission line, or large generation unit.

3. *Enabling renewable technologies*: The sun and wind are the two largest renewable energy sources, considered for electric power production, but both are intermittent, and expected to produce a significant portion of electric energy (estimated at 20%–30%) when that energy has a low financial value (off-peak period). Electricity storage technologies can smooth out this variability, allowing the unused electricity to be dispatched at a later time (peak period), make RES more valuable, cost effective and reliable option.

4. *Power quality*: Power quality problem may cause poor operations or failures of end-user equipment. Distribution network, sensitive loads and critical operations suffer from outages and service interruptions which can cost financial losses to both utility and consumers. Energy storage, when properly engineered and implemented, can provide electricity to the customer without any secondary fluctuation or disruptions and overcome the power quality problems such as swells/sags, spikes or harmonics. Table 1.1 summarizes the energy storage applications and the system requirements for the above power system applications.

TABLE 1.1

Energy Storage Applications in Power Systems

Applications	Matching Supply & Demand	Providing Backup Power	Enabling Renewable Technology	Power Quality
Discharged power	1–100 MW	1–200 MW	20 kW–100 MW	1 kW–20 MW
Response time	<10 minutes	<10 ms (Quick) <10 minutes (Conventional)	<1 second	
Energy capacity	1–1000 MWh	1–1000 MWh	10 kWh–200 MWh	50–500 kWh
Efficiency needed	High	Medium	High	Low
Lifetime needed	High	High	High	Low

1.2.1 Summary of Benefits from Energy Storage

The electricity supply chain is deregulated, with clear divisions between generation companies, transmission system operators, distribution network operators and supply companies. Application of energy storage to distribution networks can benefit the customer, supply's company, and generation operator in several ways. Major areas where energy storage systems can be applied are summarized as:

1. *Voltage control*: Support a heavily loaded feeder, provide power factor correction, reducing the need to constrain DG, minimize on-load tap changer operations, mitigating flicker, sags and swells.
2. *Power flow management*: Redirect power flows, delay network reinforcement, reduce reverse power flows and minimize losses.
3. *Restoration*: Assist voltage control and power flow management in a post fault reconfigured network.
4. *Energy market*: Arbitrage, balancing market, reduce DG variability, increase DG yield from non-firm connections, replace spinning reserve.
5. *Commercial/regulatory aspects*: Assist in compliance with energy security standard, reducing customer minutes lost, while reducing generator curtailment.
6. *Network management*: Assist islanded networks, support black starts, switching ESS between alternative feeders at a normally open point.

It is evident that developing a compelling business case for installing EES at distribution level in today electricity market with present technology costs is difficult if value is accrued from only a single benefit. The importance of understanding the interactions between several objectives and quantifying the benefit brought to each of them is a critical activity in evaluating the ESS potential. EES systems can contribute significantly to meeting the needs of modern society for more efficient, environmentally benign energy use in buildings, industry, transportation, power and utility applications. Overall the use of EES systems often results in such significant benefits as: reduced energy costs and consumption, improved air and water quality, increased operation flexibility, reduced initial and maintenance costs, reduced equipment size, more efficient and effective utilization of equipment, fuel conservation, facilitating more efficient energy use and fuel substitution. EESs have an enormous potential to increase the effectiveness of energy-conversion equipment use and for facilitating large-scale fuel substitutions. EESs are complex and cannot be evaluated properly without a detailed understanding of energy supplies and end-use considerations.

1.3 Energy Storage Technologies

The energy storage technologies are classified in four categories depending on the type of energy stored—mechanical, electrical, thermal and chemical energy storage technologies, each offering different opportunities, but also consisting of own disadvantages. Each method of electricity storage is assessed and the characteristics of each technology, such as overall storage capacity, energy density (the amount of energy stored per kilogram), power density (the time rate of energy transfer per kilogram) and the round trip conversion efficiency are discussed. Energy storage is defined as the electric energy conversion from a power grid into a form that is stored until it is converted back to electrical energy. The electrical energy storage systems are currently characterized by: (a) disagreement on the role and design of energy storage systems, (b) common use of storage only by large pump hydro, small batteries or other energy storage devices, (c) new technologies are still under demonstration and illustration, (d) no overall recognized planning tools or models to aid the understanding of energy storage devices, (e) system integration including power electronics must be improved and (f) likely the small-scale storage will have greater importance in the future. Many energy storage systems are available today with different capabilities and applications. Detailed comparisons of energy storage technologies can be found in several papers or books published over the years. Most relevant ESS characteristics are summarized in Table 1.2.

1.3.1 Classification of Energy Storage Technologies

Energy storage technologies can be classified broadly into three categories: short-term (ranging from a few seconds to minutes), long-term (from minutes to hours) and very (real) long-term (ranging from several hours to days). This classification is basically a measure of the amount of MWh (energy) that storage system can provide. Both discharge duration (time) and the energy storage system capacity (MW) can be varied in order to design a suitable energy storage system for a specific application. Figures 1.1

TABLE 1.2

Major Energy Storage Technologies

Technology	Power	Back-Up Time	Response Time	Efficiency (%)
Pumped hydro	100 kW–2 GW	Hours	Minutes	70–80
CAES	100—200 MW	Hours	Minutes	85–95
BESS	100 W–100 MW	Hours	Seconds	60–80
Flywheels	5 kW–100 MW	Minutes	Minutes	85–95
SEMS	100 kW–100 MW	Seconds	Milliseconds	95
Super-capacitors	<1 MW	Seconds	Milliseconds	Above 95

and 1.2 are showing the energy storage classification and applications. The area (stored energy in MWh) can be the same or different depending on the discharge duration. Decreasing discharge rate can be applied to make the duration longer, while increasing discharge rate can make the time shorter. The energy storage time response classifications are defined here, such as:

1. *Short-term response energy storage*: Technologies with high power density (MW/m³ or MW/kg) and with the ability of short-time responses belongs, being usually applied to improve power quality, to maintain the voltage stability during transients (few seconds or minutes).

2. *Medium- or long-term response energy storage*: Long-term response energy storage technologies for power system applications can usually supply electrical energy for minutes to hours and are used for energy management, frequency regulation and grid congestion management.

3. *Real very long-term response energy storage*: Real long-term (days, weeks, or months) response energy storage technologies are usually applied to match demand over 24 hours or longer.

The EES elements are classified based on the power capacity (MW) and the discharge duration (time). They can be varied as needed to design a suitable energy storage technique for a particular application. The price range for each device has to do with a device capacity and efficiency, a larger device power capacity and higher efficiency typically means higher investment, maintenance and operation costs. Most relevant ESS technology characteristics are summarized in Table 1.2.

1.3.2 Pumped Hydroelectric Energy Storage

Pumped hydroelectric energy storage is the most mature and largest energy storage technique available. It consists of two large reservoirs located at different elevations and a number of pump/turbine units (see the schematic of Figure 1.4). During off-peak electrical demand, water is pumped from the lower reservoir to the higher reservoir where it is stored until it is needed. Once required the water in the upper reservoir is released through the turbine-generators to produce electricity. Therefore, during production a PHES facility operates similarly to a conventional hydroelectric system. The efficiency of modern pumped storage facilities is between 70% and 85%. However, variable speed machines are now being used to improve this. The efficiency is limited by the efficiency of the pump/turbine unit used in the facility. Until recently, PHES units have always used fresh water as the storage medium. A typical PHES facility has 300 m of hydraulic head (the vertical distance between the upper and lower reservoir). The power capacity (kW) is a function of the flow rate and the hydraulic head, whilst the energy stored (kWh) is a function of the reservoir volume and hydraulic head. To calculate the mass power output of a PHES facility, the following relationship can be used:

$$P_C = \rho g Q H \eta \tag{1.1}$$

where P_C is the power capacity in W, ρ is the mass density of water in kg/m³, g is the acceleration due to gravity in m/s², Q is the discharge through the turbines in m³/s, H is

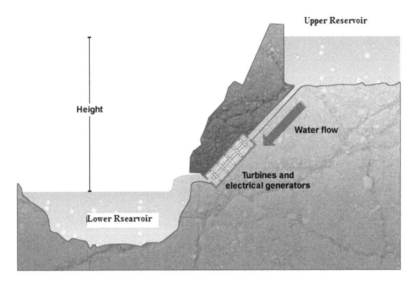

FIGURE 1.4
Typical pumped hydroelectric energy storage.

the effective head in m, and η is the efficiency. And to evaluate the storage capacity of the PHES the following must be used:

$$S_C = \frac{\rho g H V \eta}{3.6 \times 10^9} \tag{1.2}$$

where S_C is the storage capacity in megawatt-hours (MWh), V is the volume of water that is drained and filled each day in m^3. It is evident that the power and storage capacities are both dependent on the head and the volume of the reservoirs. However, facilities should be designed with the greatest hydraulic head possible rather than largest upper reservoir possible. It is much cheaper to construct a facility with a large hydraulic head and small reservoirs, than to construct a facility of equal capacity with a small hydraulic head and large reservoirs because:

1. Less material needs to be removed to create the reservoirs required.
2. Smaller piping is necessary, hence smaller boreholes during drilling.
3. The turbine is physically smaller.

> **Example 1.1:** A pumped storage facility has a head of 450 m and an efficiency of 93%. What is the flow rate needed to generate 100 MW for 3.5 hour per day? Assume the water density is 10^3 kg/m^3 and the acceleration due to gravity 9.80 m/s. What is the required working volume per day?
>
> **Solution:** From Equation (1.1) the flow rate is:
>
> $$Q = \frac{P_C}{\eta \rho g H} = \frac{10^8}{0.93 \times 9.806 \times 10^3 \times 450} = 24.383 \text{ m}^3/\text{s}$$

Working volume can be estimated as:

$$V \simeq Qt = 24.383 \times 4.5 \times 3600 = 394{,}996.7 \text{ m}^3 \simeq 3.95 \times 10^5 \text{ m}^3$$

Today, there is over 90 GW in more than 240 PHES facilities in the world, roughly 3% of the world's global generating capacity. Each individual facility can store from 30 to 4000 MW of electrical energy. Pumped storage has been commercially implemented for load balancing for over 80 years. The pumped energy storage is classified as real long-term response energy storage and typically used for applications needing to supply power for periods between hours and days (power outages). In the United States only, there are 38 pumped energy storage facilities, providing a total power capacity of about 19 GW.

Alternative configurations are underground pumped hydroelectric energy storage (UPHES), having the same operating principle as PHES system: two reservoirs with a large head between them. The only major difference is the reservoir locations in UPHES facilities have the upper reservoir at ground level and the lower reservoir deep below the Earth's surface. The depth depends on the amount of hydraulic head required for a specific application. UPHES has the same disadvantages as PHES (large-scale required, high capital costs, etc.), with one major exception. As stated previously, the most significant problem with PHES is their geological dependence. As the lower reservoir is obtained by drilling into the ground and the upper reservoir is at ground level, UPHES doesn't have such stringent geological dependences. The major disadvantage for UPHES is its commercial youth. To date there are a very few, if any, UPHES facilities in operation. Therefore, it is difficult to analyze and to trust the performance of this technology. However, the UPHES may have a bright future if cost-effective excavation techniques are identified for its construction. Its relatively large-scale storage capacity combined with location independence can provide storage with unique characteristics. However, besides the cost, a number of areas need to be investigated, such as its design, storage capacity and environmental impact to prove it is a viable option.

1.3.3 Compressed Air Energy Storage (CAES)

This is an established energy storage technology in the grid operation since 1970, being developed in conjunction to a Brayton power cycle. The energy stored mechanically by compressing the air, while the energy is released to the grid when the air is expanded. Compressed air energy storage is achieved at high pressures (40–70 bars), at near ambient temperatures. This means less volume and a smaller storage reservoir. Large caverns made of high-quality rock deep in the ground, ancient salt mines, or underground natural gas storage caves are the best options for CAES, as they benefit from geostatic pressure, which facilitates the containment of the air mass (see Figure 1.5). Studies have shown that the air could be compressed and stored in underground, high-pressure piping (20–100 bars). A CAES facility may consist of a power train motor that drives a compressor (to compress the air into the cavern), high pressure and low pressure turbines and a generator. In a gas turbine (GT), 66% of the energy is used is required to compress the air. Therefore, the CAES is pre-compressing the air using off-peak electrical power, taken from the grid to drive a motor (rather than using GT gas) and stores it in large storage reservoirs. An optimum CAES storage medium need to have the following characteristics: low thermal conductivity, in order to maintain higher air temperature; high permeability to have very low pressure drops, during charge-discharge cycle; lack of cracks and fissures into the reservoir walls to minimize the air mass losses; and higher wall elasticity to avoid wall damages under frequent loading and unloading.

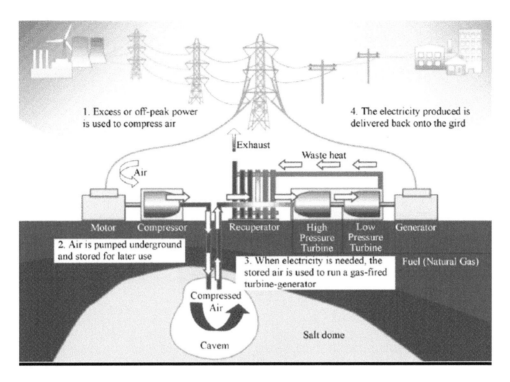

FIGURE 1.5
Typical compressed air energy storage.

CAES may use the peaks of energy generated by renewable energy plants to run a compressor that compresses the air into underground reservoirs or surface vessel/piping systems. The compressed air is used, combined with a variety of fuels in combustion turbines to generate electricity, during peak demand. The energy storage capacity depends on the compressed air volume and maximum storage pressure. When the grid is producing electricity during peak hours, the compressed air is released from the storage facility, so instead of using expensive gas to compress the air, cheaper off-peak electricity is used. However, when the air is released from the reservoir it must be mixed with small amounts of gas before entering the turbine, to avoid air temperature and pressure issues. If the pressure using air alone was high enough to achieve a significant power output, the temperature of the air would be far too low for the materials and connections to tolerate. The amount of gas required is very small so a GT working with CAES can produce three times more electricity than operating alone, using the same amount of natural gas. The reservoir can be man-made, an expensive choice, or CAES locations are usually decided by identifying suitable natural geological formations. These include salt-caverns, hard-rock caverns, depleted gas fields or an aquifer. Salt-caverns can be designed to suit specific requirements. Fresh water is pumped into the cavern and left until the salt dissolves and saturates the fresh water. The water is then processed at surface to remove salt, and the cycle is repeated until the required volume cavern is created. This process is expensive and can take up to two years. Hard-rock caverns are even more expensive, about 60% higher than salt-caverns. Finally, aquifers cannot store the air at higher pressures therefore have relatively lower energy capacities. CAES efficiency is difficult to estimate, because they use both electrical energy and natural gas. It is estimated that the efficiency based on the compression and expansion cycles is in the range of 68%–75%. Typical CAES capacities are between 50 and 300 MW. CAES is used for large- and medium-scale applications. The life of these facilities is far longer than existing gas turbines and the charge/discharge ratio is dependent on the compressor size used and the reservoir size and pressure. With assumption of ideal gas and isothermal process, the energy stored by compressing m amount of gas at constant temperature from initial pressure, P_i to final pressure, P_f is given by:

$$E = -\int_{P_i}^{P_f} V dP = mR_g \ln\left(\frac{P_i}{P_f}\right) = P_f V_f \ln\left(\frac{P_i}{P_f}\right) = P_f V_f \ln\left(\frac{V_i}{V_f}\right) \quad (1.3)$$

Here, air is assumed an ideal gas which specific heat is constant, R_g is the ideal gas constant (8.31447 J·K^{-1}mol^{-1}), V_i and V_f are the initial and final volumes of the compressed air. The compression power, P_C, depends on the air flow rate, Q (the volume per unit time) and the compression ratio (P_f/P_i), being expressed as:

$$P_C = \frac{\gamma}{\gamma-1} P_i \cdot Q \left[\left(\frac{P_f}{P_i}\right)^{\frac{\gamma}{\gamma-1}} - 1\right] \quad (1.4)$$

where γ is the ratio of air specific heat coefficients ($\gamma \approx 1.4$), P_i, P_f are the pressure of initial state and the final state, the atmospheric and compressed state. At Hundorf, Germany a 290-MW CAES (300,000 m^3 and 48,000 Pa) was built in 1978, and a 110-MW CAES (540,000 m^3 and 53,000 Pa) at McIntosh, Alabama, in 1991. CAES project is implemented or in process to be implemented in Canada, the United States and E.U. among others, using different technologies, geological structures and approaches at various power capacities.

Example 1.2: A CAES has a volume of 450,000 m^3, and the compressed air pressure range is from 75 bars to atmospheric pressure. Assuming isothermal process and an efficiency of 30% estimate the energy and average power for a 3-hour discharge period.

Solution: From Equation (1.3), assuming the atmospheric pressure 1 bar the energy is:

$$E = 45 \cdot 10^4 \times 75 \cdot 10^5 \ln\left(\frac{75}{1}\right) \approx 1457.2 \cdot 10^4 \text{ MJ}$$

The average power output is then

$$P = \frac{\eta \cdot E}{\Delta t} = \frac{0.3 \times 1457.2 \cdot 10^{10}}{2 \times 3600} \approx 404.76 \text{ MW}$$

CAES is the only very-large-scale commercially feasible energy storage technology other than PHES. CAES has a fast reaction time with plants usually able to go from 0% to 100% in less than 10 minutes, 10% to 100% in approximately 4 minutes and from 50% to 100% in less than 15 seconds. As a result, it is ideal for acting as a large bulk energy sink or supply, being also able to undertake frequent start-ups and shut-downs. CAES use compressed air so they do not suffer from the excessive heating effect of the conventional gas turbines, when operating on partial load. The CAES flexibilities mean that they can be used for ancillary services such as frequency regulation, load following and voltage control. As a result, CAES is a serious contender in the wind power energy storage market. A number of possibilities are being considered such as integrating a CAES facility with several wind farms within the same region. The excess off-peak power from the wind farms is stored a CAES facility and used when needed or there is no wind generation. Main CAES advantages are higher energy and power capacities, longer lifetime, while the major disadvantages are low efficiencies, some adverse environmental impact and difficulty of siting. CAES coupled with natural gas storage technology is based on the idea of coupling underground natural gas storage with electricity storage. The pressure difference between high-pressure gas storage (200 bars) in deep underground reservoirs (1500 m) and gas injected into the conduits with a maximum service pressure of 60–80 bars leads to the energy consumption of energy for compression, that can be released in the form of electricity during decompression. The liquefaction of natural gas or compressed air requires a large energy amounts. The idea is to use two storage reservoirs for liquefied natural gas and liquid air, regenerative heat exchangers, compressors and gas turbines. During the burning of natural gas to activate the turbine and generate electricity, the liquid air and the gas are vaporized and the cold created is conserved in the exchangers. During off-peak hours, the air is cooled by the already-stored air, compressed with an electric compressor, and is finally liquefied and store.

1.3.4 Flywheel Energy Storage (FES)

A flywheel energy storage system converts electrical energy supplied from DC or three-phase AC power source into kinetic energy of a spinning mass or converts kinetic energy of a spinning mass into electrical energy. Basically, a flywheel is a disk with a certain mass that can spin, holding kinetic energy. Modern high-tech flywheels are built with the disk attached to a rotor in upright position to prevent gravity influence, being charged or discharged by a reversible electric motor-generator system. When dealing with its efficiency however it gets more complicated, as stated by the physics principles, the friction during operation, being major factor, so to increase the efficiency is to minimize friction. This can be accomplished in two ways: the disk can spin in vacuum, less friction, and by spinning the rotor on electromagnetic bearings so it basically floats. The spinning speed for a modern flywheel reaches up to 16,000 rpm and offers a capacity up to 25 kWh, which can be absorbed and injected almost instantly. A complete flywheel energy storage system consists of the following components: high-speed rotor attached to a shaft supported by bearings with good lubrication or with magnetic suspension to minimize frication, an electromechanical energy converter, power electronic conditioning and control units.

Modern flywheel energy storage devices are comprised of a massive or composite flywheel coupled with a motor-generator and special brackets (often magnetic), set inside a housing at very low pressure to reduce self-discharge losses. They have a great cycling capacity (a few 10,000 to a few 100,000 cycles) determined by fatigue design, and to store energy in a power system, high-capacity flywheels are needed. Friction losses of a 200-tons flywheel are estimated at about 200 kW. However, for a flywheel with an instantaneous efficiency of 85%, the overall efficiency drops to 78% after 5 hours, and to 45% after 1 day. Long-term storage with this type of apparatus is therefore not foreseeable. An FES device is made up of a central shaft that holds a rotor and a flywheel, rotating on two magnetic bearings to reduce friction, all contained within a vacuum to reduce aerodynamic drag losses. Flywheels store energy by accelerating the rotor/flywheel to a very high speed and maintaining the energy in the system as kinetic energy. Flywheels release energy by reversing the charging process so that the motor is then used as a generator. As the flywheel discharges, the rotor/flywheel slows down until eventually coming to a complete stop. The rotor dictates the amount of energy that the flywheel is capable of storing. Due to their simplicity, flywheel energy storage systems have been widely used in commercial small units (about 3 kWh) in the range of 1 kW—3 hours to 100 kW—3 seconds. Energy is stored as kinetic energy using a rotor:

$$E = \frac{1}{2}J\omega^2 \tag{1.5}$$

where J is the momentum of inertia and ω is the angular velocity. The moment of inertia is given by the volume integral taken over the product of mass density ρ and squared distance r^2 of mass elements with respect to the axis of rotation:

$$J = \int \rho r^2 dV$$

However in the case of regular geometries, the momentum of inertia is given by:

$$J = kmr^2 \tag{1.6}$$

Here, m is the flywheel total mass, and r the outer radius of the disk. Equation (1.5) then becomes:

$$E = \frac{1}{2}kmr^2\omega^2 = \frac{1}{2}k(\rho\Delta V)r^2\omega^2 \tag{1.7}$$

Here, ΔV is the increment of the volume. The inertial constant depends on the shape of the rotating object. For thin rings, $k = 1$, while for a solid uniform disk, $k = 0.5$. Flywheel rotor is usually a hollow cylinder and has magnetic bearings to minimize the friction. The rotor is located in a vacuum pipe to decrease the friction even more. The rotor is integrated into a motor/generator machine that allows the energy flow in both directions. The energy storage capacity depends on the mass and shape of the rotor and on the maximum available angular velocity.

> **Example 1.3:** A flywheel is a uniform circular disc of diameter of 2.90 m, 1500 kg and is rotating at 5000 rpm. Calculate the flywheel kinetic energy.
>
> **Solution:** In Equation (1.7), $k = 0.5$, so:
>
> $$E = \frac{1}{2}kmr^2\omega^2 = \frac{1}{2}\times0.5\times1500\times\left(\frac{2.90}{2}\right)^2\times\left(\frac{2\pi 5000}{60}\right)^2 = 2.16142\times10^8\,J = 216.142\ MJ$$

There are two topologies, *slow* flywheels (with angular velocity below 6000 rpm) made of steel rotors, and *fast* flywheels (below 60,000 rpm) that use advanced materials (carbon fiber or glass fiber), having a higher energy and power density than steel rotors. The flywheel designs are modular and systems of 10 MW are feasible, the efficiency of 80%–85%, with a lifetime of over 20 years. The advances on the rotor technology are permitting high dynamic and durability of tenths of thousands of charge-discharge cycles. These characteristics make these systems suitable for power quality applications (frequency stability or temporary interruptions, voltage sags and voltage swells). For renewable energy applications the trend is to combine them with other energy storage technologies, such as micro-CAES or thermal energy storage. Flywheels store power in direct relation to the rotor mass and the square of its linear speed. At first sight from Equation (1.7), we are tempted to maximize the energy density by increasing the product of the three factors ω^2, ρ and the squared distance r^2 of the relevant mass elements with respect to the axis of rotation. However, the spatial energy densities cannot be extended excessively, due to the rotation centrifugal stresses, which are roughly increasing with the product of these three factors. However, the centrifugal stress is proportional to the square of outer tip speed, so the allowable maximum stress put the upper limit of the tip speed, and to avoid flywheel fragmentation, a certain material tensile stress level must not be exceeded. Consequently, the most efficient way to store energy in a flywheel is to make it spin faster, not by making it heavier. The energy density within a flywheel is defined as the energy per unit mass:

$$\frac{E_{KIN}}{m_{FW}} = 0.5 \cdot v_l^2 = \frac{\sigma}{\rho} \tag{1.8}$$

where E_{KIN} is the total kinetic energy in Joules, m_{FW} is the mass of the flywheel in kg, v_l is the linear velocity of the flywheel in m/s, σ is the specific strength of the material in Nm/kg, and ρ is the density of the material in kg/m³. For a rotating thin ring, therefore, the maximum energy density is dependent on the specific strength of the material and not on the mass. For these above reasons, the flywheel rotors in almost all practical designs are thin-rim configurations. For such rotors with inner radius R_1 and outer radius R_2, the maximum stored energy is given by:

$$E_{max} = Kv^2 \left[1 + \left(\frac{R_1}{R_2} \right)^2 \right] \tag{1.9}$$

Here, K is the proportionality constant, having values between 0.5 and 1, and v the linear velocity. The thin-rim flywheels with the ratio R_1/R_2 approaching unity have a higher specific energy for a given allowable stress limit. However, the higher the ultimate material strength the higher is the energy density. The energy density of a flywheel is normally the first criterion for the selection of a material. Regarding specific strength, composite materials have significant advantages compared to metallic materials. Table 1.3 lists some flywheel materials and their properties. The burst behavior is a deciding factor for choosing a flywheel material. For energy storage purposes, materials with higher strength and lower density are desirable. For this reason, composite materials having strength-to-density material ratios greater than 400 Nm/kg (or Wh/kg) are frequently used in advanced flywheels.

TABLE 1.3

Typical Materials Used for Flywheels and Their Properties

Material	Density (kg/m³)	Strength (MN/m²)	Specific Strength (MNm/kg)
Steel	7800	1800	0.22
Alloy (AlMnMg)	2700	600	0.22
Titanium	4500	1200	0.27
GRFP[a]	2000	1600	0.80
CFRP[a]	1500	2400	1.60

[a] GFRP, glass fiber-reinforced polymer; CFRP, carbon fiber-reinforced polymer.

Example 1.4: Compute the energy density for a steel flywheel.

Solution: Using Equation (1.8) and the density and specific strength values for steel (Table 1.3), the flywheel energy density is:

$$\frac{W_{\text{KIN}}}{m_{\text{FW}}} = \frac{0.22 \times 10^6}{7800} = 28.21 \text{ J/kg}$$

We have to keep in mind that not all energy stored in a flywheel system, during charging phase is returned during the discharge. The useful energy density that can be given back is:

$$\frac{W}{m} = \left(1 - s^2\right) \cdot k \cdot \frac{\sigma}{\rho} \tag{1.10}$$

where s is the ratio of maximum and minimum operating speeds, and k is an empirical constant, the rotor geometrical shape factor, having maximum value 1. The power and energy capacities are decoupled in flywheels. In order to obtain the required power capacity, you must optimize the motor/generator and the power electronics. These systems, so-called *low-speed flywheels*, usually have relatively low rotational speeds, approximately 10,000 rpm and a heavy rotor made from steel. They can provide up to 1650 kW, for very short times (up to 120 seconds). To optimize the flywheel storage capacities, the rotor speed must be increased. These high-speed flywheels, spin on a lighter rotor at much higher speeds, with prototype composite flywheels claiming to reach speeds in excess of 100,000 rpm. However, the fastest flywheels commercially available spin at about 80,000 rpm. They can provide energy up to an hour, with a maximum power of 750 kW. Over the past years, the flywheel efficiency has improved up to 80%, although some sources claim efficiencies as high as 90%. Flywheels have an extremely fast dynamic response, a long life (~20 years), requiring little maintenance, and are environmentally friendly. As the storage medium used in flywheels is mechanical, the unit can be discharged repeatedly and fully without any damage to the device. Flywheels are used for power quality enhancements, uninterruptable power supply, capturing waste energy that is very useful in electric vehicle applications and finally, to dampen frequency variation, making FES very useful to smooth the irregular electrical output from wind turbines.

A complete flywheel energy storage system consists of a high-speed rotor attached to a shaft, via a string hub, bearings with good lubrication or magnetic suspension to reduce friction, and electromechanical energy convertor, working as a motor during charging and as a generator while discharging the stored energy, power electronic conditioning and control units. Notice that the stored energy in flywheels has a significant destructive potential when released uncontrolled. Efforts are made to design rotors such that, in the case of a failure, many thin and long fragments, having little trans-lateral energy are released, so the rotor burst can be relatively benign. However, even with careful design, a composite rotor still can fail dangerously. The safety of a flywheel system is not related only to the rotor. The housing enclosure, and all components and materials within it, can influence the result of a burst significantly. To facilitate mechanical ESS by flywheel, low-loss and long-life bearings and suitable flywheel materials need to be developed. Some new materials are steel wire, vinyl-impregnated fiberglass and carbon fiber.

1.3.5 Superconducting Magnetic Energy Storage

Superconducting Magnetic Energy Storage (SMES) exploits advances in materials and power electronics technologies to achieve a novel means of energy storage based on three principles of physics: (a) some materials (superconductors) carry current with no resistive losses, (b) electric currents induce magnetic fields and (c) magnetic fields are a form of energy that can be stored. The combination of these principles provides the potential for the highly efficient storage of electrical energy in a superconducting coil. Operationally, SMES is different from other storage technologies in that a continuously circulating current within the superconducting coil produces the stored energy. In addition, the only conversion process in the SMES system is from AC to DC power conversion, i.e., there are no thermodynamic losses inherent in the conversion. Basically, SMESs store energy in the magnetic field created by the flow of direct current in a superconducting coil which has been cryogenically cooled to a temperature below its superconducting critical temperature. The idea is to store energy in the form of an electromagnetic field surrounding the coil, which operating at very low temperatures, to become superconducting, which made the system a superconductor. SMES makes use of this phenomenon and—in theory—stores energy without almost any energy loss (practically about 95% efficiency). SMES was originally proposed for large-scale, load leveling; however, because of its rapid discharge capabilities, it has been implemented on electric power systems for pulsed-power and system-stability applications.

The power and stored energy in an SMES system are determined by application and site-specific requirements. Once these values are set, a system can be designed with adequate margin to provide the required energy on demand. SMES units have been proposed over a wide range of power (1 to 1000 MW_{AC}) and energy storage ratings (0.3 kWh to 1000 MWh). Independent of size, all SMES systems include three parts: superconducting coil, power conditioning system (power electronics and control) and cryogenically cooled refrigerator. Once the superconducting coil is charged, the current is not decaying and magnetic energy can be stored indefinitely. The stored energy can be released back to the network by discharging the coil. The power conditioning system uses an inverter/rectifier to transform AC power to DC or convert DC back to AC. The inverter/rectifier accounts for about 2%–3% energy losses in each direction. SMES loses the least electricity amount in the energy storage process compared to any other methods of storing energy. There are several reasons for using superconducting magnetic energy storage instead of other energy storage methods. The most important advantage of SMES is that the time delay during charge and discharge is quite short. Power is available almost instantaneously and very high power output can be provided for a brief period of time. Other energy storage methods, such as pumped hydro or compressed air, have a substantial time delay associated with the energy conversion of stored mechanical energy back into electricity. Thus if demand is immediate, SMESs are a viable option. Another advantage is that the loss of power is less than other storage methods because electric currents encounter almost no resistance. Additionally the main parts in an SMES are motionless, which results in high reliability. The magnetic energy stored by a carrying current coil is given by:

$$E_{\text{SMES}} = \frac{1}{2}LI^2 \tag{1.11}$$

where E is the energy, expressed in J; L is the coil inductance measured in H; and I is the current. The total stored energy, or the level of charge, can be found from the above equation and the current in the coil. Alternatively, the SMES magnetic energy density per unit volume, for a magnetic flux density B (in T) and magnetic permeability of free space, μ_0 (=$4\pi \times 10^{-7}$ H/m), is expressed as:

$$E_{\text{SMES}} = \frac{B^2}{2\mu_0} \approx 4 \times 10^5 B^2 \text{ J} \cdot \text{m}^{-3} \tag{1.12}$$

Example 1.5: Find the energy density for an SMES that has a magnetic flux density of 4.5 T.

Solution: Applying Equation (1.12) the volume energy density is:

$$E_{SMES} \approx 4 \times 10^5 (4.5)^2 = 8.1 \text{ MJ} \cdot \text{m}^{-3}$$

The maximum practical stored energy, however, is determined by two factors: the size and geometry of the coil, which determine the inductance. The characteristics of the conductor determine the maximum current. Superconductors carry substantial currents in high magnetic fields. For example, at 5 T, which is 100,000 times greater than the earth's field, practical superconductors can carry currents of 300,000 A/cm². For example, for a cylindrical coil with conductors of a rectangular cross section, with mean radius of coil R, a and b are width and depth of the conductor, f, the called form function, determined by the coil shapes and geometries, ξ and δ are two parameters to characterize the dimensions of the stored energy is a function of coil dimensions, shape, geometry, number of turns and carrying current, given by:

$$E = \frac{1}{2} R N^2 I^2 f(\xi, \delta) \tag{1.13}$$

where I is the current, $f(\xi, \delta)$ is the form function (J/A-m), and N is the number of turns of coil.

Example 1.6: If the current flowing in an SMES coil having 1000 turns, a radius of 0.75 m is 250 A, compute the energy stored in this coil. Assume a form function of 1.35.

Solution: For the SEMS characteristics the stored energy is:

$$E = 0.5 \times 0.75 \times (10^3)^2 (250)^2 \times 1.35 = 31.64063 \times 10^9 \text{ J}$$

The superconductor is one of the major costs of a superconducting coil, so one design goal is to store the maximum amount of energy per quantity of superconductor. Many factors contribute to achieving this goal. One fundamental aspect, however, is to select a coil design that most effectively uses the material, minimizing the cost. Since not too many SMES coils have been constructed and installed, there is still little experience with a generic design. This is true even for the small or micro-SMES units for power-quality applications, where several different coil designs have been used. A primary consideration in the design of an SMES coil is the maximum allowable current in the conductor. It depends on: conductor size, the superconducting materials used, the resulting magnetic field and the operating temperature. The magnetic forces can be significant in large coils, a containment structure within or around the coil is needed. The superconducting SMES coil must be maintained at a temperature sufficiently low to sustain the conductor superconducting state, about 4.5 K (−269°C or −452°F). This thermal operating regime is maintained by a special cryogenic refrigerator that uses helium as the refrigerant, being the only material that is not a solid at these temperatures. Just as a conventional refrigerator requires power to operate, electricity is used to power the cryogenic refrigerator. Thermodynamic analyses have shown that power required for removing heat from the coil increases with decreasing temperature. Including inefficiencies within the refrigerator itself, 200–1000 W of electric power is required for each watt that is removed from the

4.5 K environment. As a result, design of SMES and other cryogenic systems places a high priority on reducing losses within the superconducting coils and minimizing the flow of heat into the cold environment. Both the power requirements and the physical dimensions of the refrigerator depend on the amount of heat that must be removed from the superconducting coil. However, small SMES coils and modern MRI magnets are designed to have such low losses that very small refrigerators are adequate.

Charging and discharging an SMES coil is different from that of other energy storage technologies, because it carries a current at any state of charge. Since the current always flows in one direction, the power conversion system (PCS) must produce a positive voltage across the coil when energy is to be stored, increasing the current to increase, while for discharge, the electronics is adjusted to make it appear as a load across the coil, producing a negative voltage causing the coil to discharge. The applied voltage times the instantaneous current determines the power. SMES manufacturers design their systems so that both the coil current and the allowable voltage include safety and performance margins. The PCS power capacity determines the rated SMES unit capacity. The control system establishes the link between grid power demands and power flow to and from the SMES coil. It receives dispatch signals from the power grid and status information from the SMES coil, and the integration of the dispatch request and charge level determines the response of the SMES unit. The control system also measures the condition of the SMES coil, the refrigerator, and other equipment, maintaining system safety and sends system status information to the operator. The power of an SMES system is established to meet the requirements of the application, e.g., power quality or power system stability. In general, the maximum power is the smaller of two quantities the PCS power rating and the product of the peak coil current and the maximum coil withstand voltage. The physical size of an SMES system is the combined sizes of the coil, the refrigerator and the PCS, each depending on a variety of factors. The overall efficiency of an SMES plant depends on many factors. In principle, it can be as high as 95% in very large systems. For small power systems, used in power quality applications, the overall system efficiency is lower. Fortunately, in these applications, efficiency is usually not a critical factor. The SMES coil stores energy with absolutely no loss while the current is constant. There are, however, losses associated with changing current during charging and discharging, and the resulting magnetic field changes. In general, these losses referred to as eddy current and hysteresis losses are also small. Major losses are in the PCS and especially in the refrigerator system. However, power quality and system stability applications do not require high efficiency because the cost of maintenance power is much less than the potential losses to the user due to a power outage.

1.3.6 Supercapacitors

Capacitors are devices in which two conducting plates are separated by an insulator. An example is shown in Figure 1.6. A DC voltage is connected across the capacitor, one plate being positive the other negative. The opposite charges on the plates attract and hence store energy. The electric charge $Q(C)$ stored in a capacitor of capacitance C (in F) at a voltage of V (V) is given by the equation:

$$Q = C \cdot V \tag{1.14}$$

Similar to the flywheels, the capacitors can provide large energy storage capabilities; they are used in small size configurations as components in electronic circuits and systems.

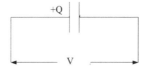

FIGURE 1.6
Capacitor symbol and diagram.

The large energy storing capacitors with very large plate areas are the so-called *super-capacitors* or *ultracapacitors*. The energy stored in a capacitor is given by the well-known equation:

$$E = 0.5C \cdot V^2 \tag{1.15}$$

where $E(J)$, and V the capacitor voltage. The capacitance C of a capacitor in Farads will be given by the equation:

$$C = \frac{\varepsilon A}{d} \tag{1.16}$$

Here ε is the electric permittivity of the material between the plates, A is the plate area and d is the separation of the plates, in standard units. The key technology to the modern super-capacitors is that the separation of the plates is so small, while the plate area is very large. The very large capacitances arise from the formation on the electrode surface of a layer of electrolytic ions (the double layer), having huge surface areas, leading the capacitances of tens, hundreds or even thousands of *Farads,* with the capacitor that can be fitted into a small container of the size of a beer can. However, the problem with this technology is that the voltage across the capacitor can only be very low, usually lower than 3 V. Equation (1.15) severely limits the energy that can be stored for a given capacity and so the voltage. In order to store charge at a reasonable voltage several capacitors are usually connected in series. This not only increases the cost, but putting capacitors in series the total capacitance is reduced, as well as the charge equalization problem. In capacitors in series the charge on each one should be the same, as the same current flows through any series circuit. However, the problem is that always there is a certain amount of self-discharge in each one, due to the fact that the insulation between the plates of the capacitors is not perfect. This self-discharge is not equal in all the capacitors, which is an issue that needs to be corrected, otherwise there may be a relative charge build-up on some of the capacitors, and this will result in a higher voltage on those capacitors. The solution to this issue, being essential in systems of more than about six capacitors in series, is to have charge equalization circuits. These are circuits connected to each pair of capacitors that continually monitor the voltage across adjacent capacitors, and move charge from one to the other in order to make sure that the voltage across the capacitors is the same. These charge equalization circuits increase the cost and size of a capacitor energy storage system, and they also consume some energy, though very efficient designs are available. In many ways the characteristics of supercapacitors are like those of flywheels. They have relatively high specific power and relatively low specific energy. They can be used as the energy storage for regenerative braking. Although they could be used alone on a vehicle, they would be better used in a hybrid as devices for giving out and receiving energy rapidly during braking and accelerating afterwards, e.g., at traffic lights. Supercapacitors are inherently safer than flywheels as they avoid the problems of mechanical breakdown and gyroscopic effects. Power electronics are needed to step voltages up and down as required.

1.3.7 Thermal Energy Storage

The thermal energy storage (TES) can be defined as the temporary storage of thermal energy at high or low temperatures. It is not a new concept and at has been used for centuries. Energy storage can reduce the time or rate mismatch between energy supply and energy demand, and it plays an important role in energy conservation. TES involves storing energy in a thermal reservoir so that it can be recovered at a later time. Thermal energy may be stored by changing the temperature of a substance (i.e., sensible heat), by changing the phase of a substance (i.e., its latent heat) or through a combination of both. TES systems are expected to have extended applications as new energy technologies are developed and implemented. The stored thermal energy may be used for space heating, process heating or for conversion to electricity. TES is the temporary storage of high- or low-temperature energy for later use. Examples include the storage of solar energy for overnight heating, or summer heat for winter use, winter ice for space cooling during the summer, or the heat or cool generated electrically during off-peak periods for use during peak demand hours. Solar energy, unlike fossil fuels, is not available at all times, and even cooling loads, that usually are coincident with solar radiation maximum, are quite often present after sunset. In thermal electrical generating systems, TES offers the possibility of storing energy before its conversion to electricity. Energy demands in the commercial, industrial, utility and residential sectors vary on a daily, weekly, seasonal or yearly basis. The use of TES in such sectors requires that they operate synergistically and to be carefully matched to each specific application. The use of TES for such thermal applications as space heating, hot water heating, cooling, or air-conditioning is important for energy conservation and savings, so new TES techniques has been developed over the past few decades in industrial countries with an enormous potential for making the use of thermal equipment more effective and for facilitating large-scale substitutions of fossil fuels. Sensible heat changes in a material are dependent on its specific heat capacity and the temperature changes. Latent heat changes associated with a phase change of a material and are occurring at constant temperature. Sensible heat storage systems commonly use rocks or water as the storage medium. Latent heat storage systems can utilize a variety of phase change materials, and usually store heat as the material changes from a solid to a liquid phase. One advantage of latent heat storage methods is that, if the storage material pressure is kept constant, the released energy is also at constant temperature, which can very important for some industrial processes or for power generation, where constant temperatures are desired. The latent heat storage is ideal for storing large thermal energy quantities for heating or for electricity generation.

TES systems are dealing with the storage of energy by material cooling, heating, melting, solidifying, or vaporizing, the thermal energy becomes available when the initial process is reversed. The materials that store heat are typically well insulated. Primary disadvantage of a thermal energy storage system, similar to other energy storage technologies is the large initial investments required to build the energy storage infrastructure. However, it has two primary advantages: (1) the energy-system efficiency is improved with the implementation of a thermal energy storage system (CHP has approximately 85%–90% efficient while conventional power plants are only 40% efficient), and (2) these techniques have already been implemented with good results. On the negative side, thermal energy storage does not improve flexibility within the transport sector like the hydrogen energy storage system, this not being a critical issue. Thermal energy storage does have a few disadvantages, but these are small compared to its advantages. Due to the efficiency improvements and maturity of technology, it is likely that they can become more prominent throughout the world. Not only do they enable the utilization of inter-mittent renewable energy (e.g., wind or solar energy), but also to maximize the fuel use within power plants. These systems are already put into practice with promising results. Therefore, it is evident that this technology can play a crucial role in future energy systems. TES operation and system characteristics are based on thermodynamics and heat transfer principles. There are two major TES types for storing thermal energy, sensible heat storage and latent heat storage. First consist of change the temperature of a liquid or solid, without changing its phase. Thermal energy quantities differ in temperature, and the energy required E to heat a volume V of a substance from a temperature T_1 to a temperature T_2 is expressed by the well-known relationship:

$$E = m \int_{T_1}^{T_2} C dT = mC(T_2 - T_1) = \rho VC(T_2 - T_1) \qquad (1.17)$$

where C is the specific heat of the substance, m is the mass, and ρ is its density. The energy released by a material as its temperature is reduced, or absorbed by a material as its temperature is increased, is called the sensible heat. Second type of energy storage implies the phase change. The ability to store sensible heat for a given material is strongly dependent of the value of ρC. For example, water has a high value, is unexpansive, however being a liquid must be contained in a better and more expensive container than the one for a sold material. For high-temperature sensible heat TES (i.e., in the range of hundred Celsius degrees), iron and iron oxide have very good characteristics comparable to water, low oxidization in high-temperature liquid or air flow, with moderate costs. Rocks are unexpansive sensible heat TES materials; however, the volumetric thermal capacity is half of the water. Some common TES materials and their characteristics are listed in Table 1.4. Latent heat is associated with the changes of material state or phase changes, for example from solid to liquid. The amount of energy stored (E) in this case depends upon the mass (m) and latent heat of fusion (λ) of the material:

$$E = m \cdot \lambda$$

TABLE 1.4

Thermal Capacities at 20°C for Some Common TES Materials

Material	Density (kg/m³)	Specific Heat (J/kg·K)
Aluminum	2710	896
Brick	2200	837
Clay	1460	879
Concrete	2000	880
Glass	2710	837
Iron	7900	452
Magnetite	5177	752
Sandstone	2200	712
Water	1000	4182
Wood	700	2390

The storage operates isothermally at the melting point of the material. If isothermal operation at the phase change temperature is difficult to achieve, the system operates over a range of temperatures T_1 to T_2 that include the melting point. The sensible heat contributions have to be considered in the top of latent heat, and the amount of energy stored is given by:

$$E = m \left[\int_{T_1}^{T_{melt}} C_{Sd} dT + \lambda + \int_{T_{melt}}^{T_2} C_{Lq} dT \right] \qquad (1.18)$$

Here, C_{Sd} and C_{Lq} represent the specific heats of the solid and liquid phases and T_{melt} is the melting point. It is relatively straightforward to determine the value of the sensible heat for solids and liquids, being more complicated for gases. If a gas restricted to a certain volume is heated, both the temperature and the pressure increase. The specific heat observed in this case is called the specific heat at constant volume, C_v. If instead the volume is allowed to vary and the pressure is fixed, the specific heat at constant pressure, C_p, is obtained. The ratio $\gamma = C_p/C_v$ and the fraction of the heat produced during compression can be saved, significantly affecting the energy storage system efficiency. TES specific applications determine the used method. Considerations include, among others, storage temperature range, storage capacity, having a significant effect on the system operation, storage heat losses, especially for long-term storage, charging and discharging rates, initial and operation costs. Other considerations include the suitability of materials used for the container, the means adopted for transferring the heat to and from the storage, and the power requirements for these purposes. A figure of merit that is used occasionally for describing the performance of a TES unit is its efficiency, which is defined by Equation (1.18). The time period over which this ratio is calculated would depend upon the nature of the storage unit. For a short-term storage unit, the time period would be a few days, while for a long-term storage unit it could be a few months or even 1 year. For a well-designed short-term storage unit, the value of the efficiency should generally exceed 80%.

$$\eta = \frac{T_{max} - T_{min}}{T_{charging} - T_{min}} \qquad (1.19)$$

where T_{min}, and T_{max} are the maximum and minimum temperatures of the storage during discharging respectively, and $T_{charging}$ is the maximum temperature at the end of the charging period. Heat losses to environment between the discharging end and the charging beginning periods, as well as during these processes, are usually neglected. Two particular problems of thermal energy storage systems are the heat exchanger design, and in the case of phase change materials, the method of encapsulation. The heat exchanger should be designed to operate with as low a temperature difference as possible to avoid inefficiencies. In the case of sensible heat storage systems, energy is stored or extracted by heating or cooling a liquid or a solid, which does not change its phase during this process. A variety of substances have been used in such systems, such as: (a) liquids (water, molten salt, liquid metals, or organic liquids) and (b) solids (metals, minerals, or ceramics). In the case of solids, the material is invariable in porous form and the heat is stored or extracted a gas or a liquid flowing through the pores or voids. For incompressible type of thermal storage, for example the ones using heavy oils or rocks, the maximum work that can be produced is given in terms of specific heat capacity, C, and mass:

$$W_{max} = m\left[C\left(T_{str} - T_{amb}\right) + CT_{amb}\ln\left(\frac{T_{str}}{T_{amb}}\right)\right] = \rho V\left[C\left(T_{str} - T_{amb}\right) + CT_{amb}\ln\left(\frac{T_{str}}{T_{amb}}\right)\right] \quad (1.20)$$

Here, T_{str} and T_{amb} are the storage material temperature and ambient temperature (in Kelvin degrees), respectively, and m is the storage material mass, ρ the storage material density and V the volume. The storage materials (water, steam, molten salt, heavy oil or solid rocks) are at a temperature that is significantly higher than the ambient one, so the heat is continuously lost from the thermal storage, regardless the insulation quality. Given enough time, the stored energy, if not used is dissipated. For this reason TES are very suitable for short-term or intermediate-period applications rather than long-term ones. The total rate of heat transfer, q, from a TES reservoir depends on the overall heat transfer coefficient, C_{trsf}, the reservoir instant temperature, T, the ambient temperature and the reservoir total surface, A_{tot}, expressed as:

$$q = C_{trsf} \cdot A_{tot} \cdot \left(T - T_{amb}\right) \quad (1.21)$$

A number of TES applications are used to provide heating and cooling including Aquifer Thermal Storage (ATS), and Duct Thermal Storage (DTS). However, these are heat-generation techniques rather than energy-storage techniques and therefore are discussed in detail here. An aquifer is a ground water reservoir, consisting of highly water permeable materials such as clay or rocks, having large volumes and high TES capacities. When heat extraction and charging performances are good, high heating and cooling powers can be achieved with such systems. The amount of energy that can be stored in an aquifer depends on the local conditions (allowable temperature changes, thermal conductivity and natural groundwater flows). An aquifer storage system can be used for storage periods ranging from short to medium, including daily, weekly, seasonal, or mixed cycles. In terms of storing energy, there are two primary thermal energy storage options. One option is a technology which is used to supplement air conditioning in buildings. The thermal energy storage system can also be used very effectively to increase the flexibility within an energy system. As mentioned previously in this chapter, by integrating various energy storage systems of an energy system, increased penetrations of renewable energy sources can be achieved due to the additional flexibility created. Unlike other energy storage systems, which enabled interactions among the electricity, heat and transport sectors, thermal energy storage only combines the electricity and heat sectors with one another. By introducing district heating into an energy system, then electricity and heat can be provided from the same facility to the energy system using CHP plants. This brings additional flexibility to the system, enabling larger renewable energy penetrations.

1.4 Energy Storage for Electric Grid and Renewable Energy Applications

The development and use of renewable energy has experienced rapid growth over the past few years. In the next 20–30 years all energy systems will be based on the rational use of conventional resources and greater renewable energy use. Decentralized and renewable energy-based electricity production yields a more assured supply for consumers with fewer environmental impacts. However, their unpredictable character requires that network provisioning and usage regulations be established for optimal system operation. *The criteria to identify the most suitable EES are the following for each: How it works, advantages, applications, cost, disadvantages and future.* A brief comparison indicates the broad range of operating characteristics available for energy storage technologies. For utilities and renewable energy integration, energy storage capacity, power output and life cycle are key performance criteria. The need for long life cycle has motivated the use of very reversible storage systems such as CAES or pumped hydro as an alternative to electrochemical batteries that present problems of aging and are difficult to recycle. In transportation applications, portability, scalability and energy and power density are key performance criteria. Therefore, due to their modularity and portability, and in spite of the numerous issues, including limited life, batteries are still considered the most viable transportation option. Energy storage is a well-established concept yet still relatively unexplored. Storage systems such as pumped hydroelectric energy storage (PHES) have been in use since 1929, primarily to level the daily loads. As the electricity sector is undergoing significant changes, energy storage is becoming a realistic option for: (1) restructuring the electricity market; (2) integrating renewable resources; (3) improving power quality; (4) aiding shift toward distributed energy; and (5) helping grid operate under stricter environmental requirements. It is possible to divide grid storage applications into two broad categories based on the length of time a storage device needs to provide service: high power applications where the device must respond rapidly and be able to discharge for only short-term periods (up to about 1 hour), and energy management-related applications where the device may respond more slowly but must be able to discharge for several hours or more. Ideally, all storage devices would be able to provide all services, but some technologies are technically restricted to provide only short-term services. However, many of these services have very high value in the grid, so short-term storage can still provide considerable benefits.

Energy storage can optimize the existing generation and transmission infrastructures whilst also preventing expensive upgrades. Power fluctuations from renewable resources will prevent their large-scale penetration into the network. However, energy storage devices can manage these irregularities and thus aid the amalgamation of renewable technologies. In relation to conventional power production energy storage devices can improve overall power quality and reliability, which is becoming more important for modern commercial applications. Finally, energy storage devices can reduce emissions by aiding the transition to newer, cleaner technologies such as renewable resources and the hydrogen economy. The concept of having energy storage for an electric grid provides all the benefits of conventional generation such as enhanced grid stability, optimized transmission infrastructure, high-power quality, excellent renewable energy penetration and increased wind farm capacity. However, energy storage technologies produce no carbon emissions and do not rely on imported fossil fuels. As a result, energy storage is an attractive option for increasing wind penetration onto the electric grid when it is needed. The rapid advances in energy storage technology have permitted such devices of reasonable size to be designed and commissioned successfully aiming at balancing any instantaneous mismatch in active power during abnormal operation of the power grid. Thus, fast-acting generation reserve is provided to the microgrid so that the dynamic security can be significantly enhanced. High power and energy density with outstanding conversion efficiency, and fast and independent power response in four quadrants make the selected DESs capable of providing significant benefits to many potential microgrid applications. Most of these advanced DES technology applications are described below.

Distributed energy storage: The selected advanced DES units could provide the potential for energy storage of up to several MWh with a high return efficiency (above 95% for SMES and SCES, while 85% for FES systems) and a rapid response time for dynamic change of power flow (millisecond for SMES and SCES devices, a few milliseconds for FES systems). This aspect makes them ideal for energy management with large variations in energy requirements, as well as for backup power supply in case of loss of the utility main power supply or as a replacement of major generating unit trips in the microgrid.

Spinning reserve: In case of contingencies, such as failures of generating units or other microgrid components, a certain amount of short-term generation must be kept unloaded as spinning reserve. This reserve must be appropriately activated by means of the primary frequency control (PFC). The selected DG unit(s) can be effectively used in order to store excess energy during off-peak periods for substituting the generation reserve during the action of the PFC, enhancing the microgrid dynamic security.

Load following: The selected advanced DES devices have the ability to follow system load changes almost instantaneously which allows for conventional generating units to operate at roughly constant or slowly changing output power.

Microgrid stability: The selected advanced DES units have the capability to damp out low frequency power oscillations and to stabilize the system frequency as a result of system transients. Since the considered advanced DES systems are capable of controlling both the active and reactive powers simultaneously, they can act as a good device in order to stabilize the microgrid with high level of penetration of renewable energy sources, such as photovoltaic or particularly wind generation.

Automatic Generation Control (AGC): The advanced DES systems can be used as a controlling function in an AGC system to support a minimum of area control error (ACE).

Tie line power flow stabilization and control: A schedule of power between various microgrids or control areas inside the same microgrid requires that actual net power matches closely with the scheduled power.

Unfortunately, generators with highly fluctuating active power profile in one microgrid produce an error in the actual power delivered respect to the scheduled one, which can result in inefficient use of generation and system components. Advanced DES systems can be designed with appropriate controls to provide power in order to nearly eliminate this error and to ensure that generation is efficiently used and power schedules are met. Power quality improvement: SMES can provide ride through capabilities and smooth out disturbances on the microgrid that would otherwise interrupt sensitive customer loads. All these devices have very fast response and can inject active power in less than one power cycle; thus providing premium power supply to critical customers and preventing from losing power.

Reactive power flow control and power factor correction: The selected advanced DES units are capable of controlling the generated reactive power simultaneously and independently of the active power, which enables the correction of the power factor.

Voltage control: The selected advanced DES systems have shown to be effective for providing voltage support and regulation by locally generating reactive power.

ESSs are increasing their impact on the utility grid as a solution to stability problems. The main advantage of a storage unit is to contribute to the quality of the grid by maintaining the power constant. The main role of the ESSs is to increase the RES penetration, to level load curve, to contribute to the frequency control, to upgrade the transmission line capabilities, to mitigate the voltage fluctuations and to increase the power quality and reliability.

1. *Increasing RES penetration*: Although RESs are environmentally beneficial, the intermittent nature of two fast growing energies, wind and solar, causes voltage and frequency fluctuations on the grid. That represents a significant barrier to widespread penetration and replacement of fossil-fuel source base-load generation, because integrating renewable sources introduces some new issues on the operation of the power system, such as potential unbalancing between generation and demand. However, the intermittent RESs, such as solar and wind, need to be supported with other conventional utility power plants. It is estimated that, for every 10% wind penetration, a balancing power from other generation sources equivalent to 2%–4% of the installed wind capacity is required for a stable power system operation. Thus, with more penetration of intermittent renewable energy like wind power, the system operation will be more complex, and it will require additional balancing power. This is critical in countries with a large penetration of solar and wind systems, as Denmark or Spain, where it is estimated that approximately 20% and 10% of the electricity generation come from wind power, respectively. A large storage capacity allows higher percentage of wind, photovoltaic and other DG units in the electrical mix contributing to fulfill the objectives for a more sustainable future. In order to integrate RESs, it is necessary to propose a suitable storage system, offering capacities of several hours and power level from up to 100 MW. Nowadays, high-temperature thermosolar power plants are including a TEES, and it is expected that other storage systems to be included in the new RES generation and DG units in the future. Recently, the concept of vehicle-to-grid (V2G) has been introduced. It describes a system in which electric or plugin hybrid vehicles communicate with the power grid to sell demand respond services by either delivering electricity into the grid or throttling their charging rate. When coupled to an electricity network, EVs can act as a controllable load and energy storage in power systems with high penetration of RESs. The reliability of the renewable electricity will be enhanced with the vast untapped storage of EV fleets when connected to the grid. The market for the V2G systems represents 1 million vehicles with 20–50-kWh capacity, where 10% of this capacity is available for utility applications, including integration of RES (Belu, 2014, Huggins, 2016). The benefits of energy storage applications in RESs have been deeply studied in the bibliography.

2. *Load leveling*: Load leveling refers to the use of electricity stored during times of low demand to supply peak electricity demand, which reduces the need to draw on electricity from peaking power plants or increase the grid infrastructure. To deliver more power to the load, there are two possibilities, increase the

infrastructure and the generator capacity or install an ESS. The ESS allows one to postpone a large infrastructure investment in transmission and distribution network. New technologies, which are not restricted by their geographic limitations, have been proposed as more suitable for load leveling such as TESS and BSS. Rechargeable battery technologies like sodium sulfur (NaS) technology are attractive candidates for use in many utility scale energy storage applications. These advanced battery systems can be utilized with existing infrastructure, helping energy providers to meet peak demands and critical load.

3. *Energy arbitrage*: The term energy arbitrage refers to earning a profit by charging ESS with cheap electricity when the demand is low and selling the stored energy at a higher price when the demand is high (Eyer, and Corey, 2010, Belu, 2014). This activity can also be used to influence in the demand side, such as using higher peak prices to induce a reduction in peak demand through demand charges, real-time pricing, or other market measures. This function has been traditionally performed by pumped hydro storage (PHS). PHS is appropriate for energy arbitrage because it can be constructed at capacities over 100-MW and discharged over periods of time from 100 to 1000 minutes. These installations allow storage when the demand is low and the energy is cheap. This ESS is the most widely used energy storage technology at utility scale (100 GW installed worldwide). CAES is also appropriate for energy arbitrage because it can be constructed in capacities of a few hundreds of megawatts and can be discharged over long periods of time. A new trend for this application is to use the ancillary services that offer the battery of electrical V2G. The large quantity of this V2G expected in a next future could contribute to a new concept of the energy marker.

4. *Primary frequency regulation*: The technical application of ESS includes transient and permanent grid frequency stability support. To contribute to the frequency stabilization during transient, called grid angular stability (GAS), low- and medium-capacity ESSs are needed. This low-energy-storage requirement is because GAS operation consists of injection and absorption of real power during short periods of time, 1–2 seconds. Modern variable-speed wind turbines and large photovoltaic power plants connected to the utility grid do not contribute to the frequency stability as the synchronous generators of the conventional gas or steam turbine do. This creates a new application of ESS that is to be used to emulate the inertia of these steam turbine generators to complement this angular stability deficit. Another solution is to use the power electronic converter of variable-speed wind turbines to emulate the steam turbine inertia using the inertial energy storage of the rotors of these wind turbines. SMESs are getting increasing acceptance in variation applications of damping frequency oscillations because of their higher efficiency and faster response. EDLC, FESS and BSS are also very suitable for this application.

5. *End-user peak shaving*: There are several undesired grid voltage effects at the end-user level, depending on the duration and variability. Typical voltage effects are long-period interruptions (blackouts), short-period interruptions (voltage sags), voltage peaks and variable fluctuation (flicker).

To perform peak shaving and to mitigate blackouts, the typical approach involves installing UPSs. If an online UPS is installed in series, this isolates the load from the grid, and fluctuations produced by the utility have no effect on the users. However, this solution may not be optimal for all applications.

One solution, presented as grid voltage stability, involves the mitigating against degraded voltage by providing additional reactive power and injecting real power for durations of up to 2 seconds. The energy storage needed to protect the load against this voltage degradation is low. The energy storage demanded is even lower in applications with ride-through capability, where the electric load or the generator stays connected during the system disturbance, because part of the energy obtained from the grid during the under-voltage period. The voltage flicker is caused by rapid changes of RES, industrial or domestic loads, such as electric arc furnaces, rolling mills, welding equipment and pumps. An ESS can help to reduce voltage fluctuations at the point of common coupling produced by these transitory generators and loads.

1.4.1 Parameters of an Energy Storage Device

- *Power capacity*: is the maximum instantaneous output that an energy storage device can provide, usually measured in kilowatts (kW) or megawatts (MW).
- *Energy storage capacity*: is the amount of electrical energy the device can store usually measured in kilowatt-hours (kWh) or megawatt-hours (MWh).
- *Efficiency*: indicates the quantity of electricity which can be recovered as a percentage of the electricity used to charge the device.
- *Response time*: is the length of time it takes the storage device to start releasing power.
- *Round-trip efficiency*: indicates the quantity of electricity which can be recovered as a percentage of the electricity used to charge and discharge the device.

1.5 Summary

Developing efficient and inexpensive energy storage devices is as important as developing new sources of energy. Energy storage systems are the key enabling technologies for transportation and utility applications. In particular, the proliferation of energy storage will enable the integration and dispatch of renewable generation and will facilitate the emergence of smarter grids with less reliance on inefficient peak-power plants. There is a range of options available to store intermittent energy until it is needed for electricity production. The benefits obtained in transport and utility go from technical aspects to economic objectives. For instance, a reduction of CO_2 emissions can be achieved by the use of electric vehicles or increase of profits can be obtained with a load-leveling application in a power transmission line. Anyway, the continuous development of the storage technologies and the evolution of their applications will motivate further research to solve the existing issues and improve the energy storage systems. As the percentage of renewable energy sources in the grid continues to increase, the storage methods will become critical to the provision of secure and uninterrupted power. With further use of storage, prices and efficiencies will become more favorable, such that coupled renewable and storage energy systems will be economical. The choice of storage system will depend on the individual needs. However, it will commonly be necessary to incorporate more than one energy storage system in the power systems providing large amounts of energy. In this way, both short and longer term power interruptions can be compensated from stored energy. Continuing investment and research in this area will help to mark the way for greater use of renewable energy sources in future electricity supply grids.

ESSs are the key enabling technologies for transport and utility applications. In particular, the proliferation of energy storage will enable the integration and dispatch of renewable generation and will facilitate the emergence of smarter grids with less reliance on inefficient peak power plants. In the transportation sector, the emergence of viable onboard electric energy storage devices such as high-power and high-energy lithium ion batteries will enable the widespread adoption of plug-in electric and HEVs which will also interact with the smart grids of the future. Mature storage technologies can be used in several applications, but in other situations, these technologies cannot fulfill with the application requirements. Thus, new storage systems have appeared, opening new challenges that have to be solved by the research community. Transport and utility applications operate with a wide range of time versus power storage requirements. The benefits obtained in transport and utility go from technical aspects to economic objectives. For instance, a reduction of CO_2 emissions can be achieved by the use of EVs or increase of profits can be obtained with a load leveling application in a power transmission line. Anyway, the continuous development of the storage technologies and the evolution of their applications will motivate further research to solve the existing issues and improve the ESSs.

Questions and Problems

1. What are the benefits of the energy storage systems?
2. What are the major problems for using: (a) pumped water energy storage, (b) compressed air energy storage and (c) the flywheels?
3. List some of the potential applications of energy storage.
4. Which of the following is not a method to store energy: (a) battery, (b) motor, (c) compressed air, (d) flywheel and (e) all of the above store energy?
5. List the benefits of electricity energy storage for power grid operation.
6. Classify and explain each compressed-air storage systems.
7. List the essential criteria for comparing energy storage systems.
8. Sensible heat storage can be performed by using: liquid, solid, liquid-solid media or by all of these media.
9. List a few of the potential applications of energy storage options.
10. How much energy can be delivered from a pumped storage facility of 8 million m^3, a head of 500-m and the overall efficiency of 85%?
11. List the essential criteria for comparing energy storage methods.
12. How much energy is needed to heat a volume of $10 m^3$ of water from 20°C to 81°C? Take the density of water as 1000 kg/m^3 and the specific heat as: 4.185 kJ/kg·°C.
13. List the various TES processes and explain each. Describe the operation of an aquifer-based TES.
14. What are the major benefits of TES systems?

15. Estimate the heat that can be stored in 10 m³ of water if the temperature increases with 10°C.

16. What energy storage options are available for solar energy applications?

17. List the key merit measures for any energy storage system.

18. A sensible heat energy storage system for solar energy depends on which: (a) material mass; (b) material specific heat; (c) temperature difference; or (d) all of the above factors.

19. How much kinetic energy does a flywheel (steel disk of diameter 6 in, thickness 1 in, speed of rotation 30,000 rpm) have?

20. A water-pumped energy storage facility has a level difference (head) of 500 m and a working volume of 081 km³. Estimate how much power is generated if it is required to operate 1.5 hours per day. The overall facility efficiency is 83% and the water density is 1000 kg/m³.

21. Estimate the energy of the CAES facilities at Hundorf, Germany, and McIntosh, Tennessee, USA.

22. A CAES has a volume of 500,000 m³, and the compressed air pressure range is from 80 bars to 1 bar. Assuming isothermal process and an efficiency of 33%, estimate the energy and power for a 3-hour discharge period.

23. If the power generated by CAES of 300,000 m³, in 2-hour discharge period from 66 bars to the atmospheric pressure is 360 MW, what is the system efficiency?

24. An old salt mine having a 15,000 m³ storage capacity has been selected for pressurized air storage at 33 bar, if the temperature during the filling/charging phase is 175°C, assuming an isothermal process and an efficiency of 35% estimate the energy and average power for a 3-hour discharge period.

25. Assuming a flow rate of 36 m³/s, what is the power capacity of the CAES in Example 1.2?

26. A thin-rim flywheel energy storage system has the R_1/R_2 ratio 0.85, estimate the maximum stored energy, assuming $K = 0.9$ and the linear speed 20, 50 and 60 m/s.

27. A flywheel has a weight of 20 kg, 8.5 m diameter, an angular velocity of 1200 rad/s, a density of 3200 kg/m³, and a volume change of 0.75 m³. Evaluate for $k = 0.5$, $k = 0.85$ and $k = 1$ how much energy this flywheel can store.

28. Very high-speed flywheels are made of composite materials. If a flywheel has the following characteristics: (a) a ring with radius of 2.5 m, mass of 100 kg, and speed of 25,000 rpm; and (b) a sold uniform disk, with same mass, radius and running at the same speed of rotation. How much energy is stored in each system?

29. Estimate the energy stored in a CAES with capacity of 250,000 m³, if the reservoir air presses is 60-bar, and the overall system efficiency is 33%. Estimate also the total generated electricity if the air is discharged over a period of 90 minutes.

30. Compute the kinetic energy density of flywheels made of alloy, GRFP and CRFP.

31. A flywheel is constructed in a toroidal shape, resembling a bicycle wheel, has a mass of 300 kg. Assuming all mass is concentrated at 1.8 m, what is its rpm to provide an 800 kW for 1 minute? In your opinion, is this system physically possible?

32. Calculate the required time for a heat storage temperature to decrease from 36°C to 8°C, if there is no load heat removal from, and the ambient temperature is 12°C,

the reservoir storage capacity is 2.0 m^3, the heat storage area is 30 m^2, and the over-all heat transfer reservoir liquid-environment is 6.5 W/m^2·°C.

33. Estimate the heat that can be absorbed by an ice cooling system having a volume of 2.4 m^3, through the ice melting, and by increasing the resulted water temperature by 10°C. Neglect any losses.

34. Calculate the theoretical efficiency of a TES operating between 122°C and 25°C, assuming a charging temperature of 150°C.

35. Assuming a cylindrical reservoir, made of bricks, with 10 m diameter and height of 12 m, compute the heat transfer rate, assuming the ambient temperature of 20°C and the storage material temperature of 105°C.

36. Calculate the required storage medium volume for the following thermal energy storage systems in order to store 1 MWh$_{th}$ of thermal energy for the temperature range from 350°C to 20°C, assuming the sold rocks and heavy oil storage materials.

References and Further Readings

1. I. Dincer and M. Rosen, *Thermal Energy Storage: Systems and Applications*, Wiley, New York, 2002.
2. B. Sorensen, *Renewable Energy—Its Physics, Engineering, Environmental Impacts, Economics & Planning* (3rd ed.), Elsevier/Academic Press, Boston, MA, 2004.
3. F.A. Farret and M.G. Simões, *Integration of Alternative Sources of Energy*, Wiley-Interscience, New York, 2006.
4. J. Andrews and N. Jelley, *Energy Science, Principles, Technologies, and Impacts*, Oxford University Press, Oxford, UK, 2007.
5. M. Kaltschmitt, W. Stteicher, and A. Wiese, *Renewable Energy: Technology, Economics and Environment*, Springer, Berlin, Germany, 2007.
6. H.O. Paksoy (ed.), *Thermal Energy Storage for Sustainable Energy Consumption: Fundamentals, Case Studies and Design*, Springer, Dordrecht, the Netherlands, 2007.
7. H. Ibrahim, A. Ilinca, and J. Perron, Energy storage systems—Characteristics and comparisons, *Renewable and Sustainable Energy Reviews*, 2008, Vol. 12, pp. 1221–1250.
8. A. Ter-Gazarian, *Energy Storage for Power Systems*, The IET Press, London, UK, 2008.
9. H. Chen, T.N. Cong, W. Yang, C. Tan, Y. Li, and Y. Ding, Progress in electrical energy storage system: A critical review, *Progress in Natural Science*, 2009, Vol. 19, pp. 291–312.
10. A. Vieira da Rosa, *Fundamentals of Renewable Energy Processes* (2nd ed.), Academic Press, Amsterdam, the Netherlands, 2009.
11. M. Beaudin, H. Zareipour, A. Schellenberglabe, and W. Rosehart, Energy storage for mitigating the variability of renewable electricity sources: An updated review, *Energy for Sustainable Development*, 2010, Vol. 14, pp. 302–314.
12. S. Vazquez, S.M. Lukic, E. Galvan, L.G. Franquelo, and J.M. Carrasco, Energy storage systems for transport and grid applications, *IEEE Transactions on Industrial Electronics*, 2010, Vol. 57, pp. 3881–3895.
13. R. Zito, *Energy Storage: A New Approach*, Wiley, New York, 2010.
14. J. Eyer and G. Corey, Energy storage for the electricity grid: Benefits and market potential assessment guide, Sandia National Laboratory, Report No. SAND2010-0815, February 2010.
15. F.S. Barnes and J.G. Levine (eds.), *Large Energy Storage Systems Handbook*, CRC Press, Boca Raton, FL, 2011.
16. E.E. Michaelides, *Alternative Energy Sources*, Springer, Berlin, Germany, 2012.

17. R. Belu, Renewable energy: Energy storage systems, in *Encyclopedia of Energy Engineering & Technology* (Online) (eds. Sohail Anwar, R. Belu et al.), CRC Press/Taylor & Francis Group, Vol. 2, 2014.
18. R.A. Dunlap, *Sustainable Energy*, Cengage Learning, Stamford, CT, 2015.
19. V. Nelsson and K. Starcher, *Introduction to Renewable Energy* (2nd ed.), CRC Press, Boca Raton, FL, 2016.
20. R.A. Huggins, *Energy Storage Fundamentals, Materials and Applications* (2nd ed.), Springer, Cham, Switzerland, 2016.
21. R. Belu, *Industrial Power Systems with Distributed and Embedded Generation*, The IET Press, Stevenage, UK, 2018.

2

Batteries, Fuel Cells and Hydrogen Energy

Energy storage systems are playing and will play critical roles in present and future electricity systems and in the extend uses of the intermittent renewable energy sources. Energy storage can take a number of forms, with one of the most commonly used are the electrochemical energy storage systems, such as batteries and fuel cells. Electrochemical energy conversion and storage represent basically the direct energy conversion of chemical energy (what is defined in thermodynamics as free energy) into electrical energy and vice versa. Storage media that can take and release energy in the form of electricity have the most universal values, because the electricity can efficiently converted into mechanical or thermal energy forms, whereas other energy conversion processes are less efficient. The focuses of this chapter are to look, to understand and to study principles of operation and characteristics of the most common types of electrochemical energy storage systems. The batteries are at the present the most practical and widely used electricity storage systems. The terms *battery* and *cell* are often interchanged, although strictly a battery consists of a group of cells connected together and built as a single unit. Batteries are classified into primary (non-rechargeable) and secondary (rechargeable) types. A *fuel cell* is an energy conversion device, closely related to a battery, both being electrochemical devices that are converting the chemical energy to electrical energy. Fuel cells are devices that utilize electrochemical reactions to generate electric power. They are believed to give a significant impact on the future energy system. Especially, when the hydrogen is generated from renewable energy systems, it is certain that the fuel cell will play a significant role. However, in a battery the chemical energy is stored internally, whereas in fuel cells the chemical energy is supplied externally and can be continuously replenished. Fuel cells are electrochemical engines that produce electricity from paired *oxidation-reduction* (*redox*) *reactions*. We can think of the fuel cell systems as batteries with flows of reactants in and products out. In contrast, the battery has a fixed supply of reactants that transform into products without being steadily replaced. The chapter is intended to provide solid background information on the operation of electrochemical energy storage devices that are sharing common principles and characteristics. In this chapter the focus is on the most common types of batteries and fuel cells, as well as their characteristics, parameters, performances, operation modes and applications. The mechanisms, operation and applications related to other electrical energy storage devices and systems, such as compressed air storage, flywheels, pumped hydroelectric, supercapacitors and superconducting magnetic energy storage, are reserved in the next book chapter.

2.1 Introduction, Electrochemistry Basics

Energy storage technology has been in existence for a long time and has been utilized in many forms and applications from a flashlight to spacecraft systems. Today energy storage is used to make the electric power systems more reliable, more robust, to improve the system operation and power quality, as well as making the broader renewable energy use and applications a reality. Energy storage systems (ESS) become critical in enabling renewable energy applications, providing the means to make the non-dispatchable resources into dispatchable ones. To match power demand, in the context of the renewable energy intermittent and the fairly predictable electrical demand behavior, the energy storage is critical, allowing de-coupling of generation from consumption, reducing the need for constant monitoring and prediction of energy demands. Energy storage also provides economic benefits by allowing a plant energy generation reduction to meet average demands rather than peak power demands. The current status shows that several drivers are emerging and are spurring growth in the energy storage demands, such as the growth of the renewable energy, an increasingly strained grid infrastructure as new lines lag well behind demand, the microgrid emergence as part of smart grid architecture, the increased need for higher reliability and security in electricity supply, the energy storage needed to improve and optimize building and industrial process energy usage and efficiency. However, issues regarding the optimal active integration (operational, technical and market) of energy storage into the electric grid are still not fully developed, tested and standardized. Systems and devices for electrochemical energy storage include batteries, fuel cells and electrochemical capacitors (discussed in a later boo chapter). Their energy storage and conversion mechanisms are different, however, there significant electro chemical similarities, such as that the energy-generating processes are taking place at the phase boundary of the electrode-electrolyte interface and the electron and ion charge carriers) transport are separated.

Chemical energy storage can be further classified into electrochemical and thermo-chemical energy storage. The electrochemical energy storage refers to conventional batteries, such as lead-acid (LA), nickel–metal hydride, and lithium-ion (Li-ion), and flow batteries, such as zinc/bromine (Zn/Br) and vanadium redox and metal air batteries. Electrochemical energy storage is also achieved in fuel cells (FCs), most commonly hydrogen fuel cells, but also include direct-methanol, molten carbonate and solid oxide fuel cells. Batteries and fuel cells are the most common energy storage devices used in power systems and in several other applications. For example, to date the only portable and rechargeable electrochemical energy storage systems (with the exception of the capacitors, which are electrical energy storage devices) are the batteries. Basic configuration of a battery or a fuel cell consists of two electrodes and an electrolyte, placed together in a special container and connected to an external device (a source or a load). Usually a battery consists of two or more electric cells joined together, in series, parallel or series-parallel configurations in order to achieve the desire voltage, current and power. The cells consisting of positive and negative electrodes joined by an electrolyte are converting chemical energy to electrical energy. It is the chemical reaction between the electrodes and the electrolyte which generates DC electricity. In the case of secondary or rechargeable batteries, the chemical reaction is reversed by reversing the current and the battery returned to a charged state.

To understand how a battery, a flow battery or a fuel cell functions and how they store or how they generate electric energy, it is necessary to know some generalities and have some basic understanding of the electrochemistry. From the invention of the lead-acid battery by Gaston Planté in 1859, electrochemistry has progressed steadily over the last 150 years, while the electrochemistry is defined as the branch of chemistry that examines the phenomena resulting from combined chemical and electrical effects. The electrochemistry field covers basically two areas: (a) *electrolytic processes*, where reactions in which chemical changes occur due to the passage of electrical currents, and (b) *galvanic or voltaic processes*, where chemical reactions are resulting in the production of electrical energy. The latest electrochemical processes are the subject of this chapter. With the miniaturization of electronic components, proliferating over the last half century, the demand for smaller and lightweight portable electronic equipment has dramatically increased the need for research on battery technologies and methods for the optimal battery power management. Battery types come in two different forms, *disposable* or *primary batteries* and *rechargeable* or *secondary batteries*. Mature rechargeable battery chemistries are: (a) lead-acid (LA), (b) nickel-cadmium (Ni-Cd), (c) nickel-metal hydride (Ni-MH), (d) lithium-ion (Li-ion), (e) lithium-polymer/lithium metal (Li-polymer), (f) sodium-sulfur (Na-S), (g) sodium-nickel chloride and (h) lithium-iron phosphate. Different types are the flow batteries, discussed later. With the increased demand growing from electric vehicles and portable consumer products, significant research funds are spent by companies on new battery technologies, such as zinc-based chemistries and silicon as a material for improving battery properties and performances. Higher energy density, better life cycle, environmental friendliness and safer operation are among the general design research targets for secondary batteries. Primary batteries are a reasonably mature technology, in terms of chemistry range, but still there is research to increase the energy density, reduce self-discharge rate, increase the battery life, or to improve the usable temperature range. To complement these developments, many semiconductor manufacturers continue to introduce new integrated circuits for battery power management. Other chemical energy storage options, such as the thermochemical energy storage, although promising, are still at an infant development stages and are not included in this entry.

Chemical energy storage is usually achieved through accumulators or batteries, characterized by a double function of storage and release of electricity by alternating the charge—discharge phases. They can transform chemical energy generated by electrochemical reactions into electrical energy and vice versa, without almost any harmful emissions or noise, and require little maintenance. There is a wide range of battery technologies in use, and their main assets are their energy densities (up to 2000 Wh/kg for lithium) and technological maturity. Chemical energy storage devices (batteries) and electrochemical capacitors (ECs) are among the leading EES technologies today. Both technologies are based on electrochemistry, and the fundamental difference between them is that batteries store energy in chemical reactants capable of generating charges, whereas electrochemical capacitors store energy directly as charges. Although the electrochemical capacitor is a promising technology for electrical energy storage, especially considering its high power capability, its energy density is too low to be considered for large-scale energy storage.

The most common rechargeable battery technologies are lead-acid, sodium-sulfur, vanadium redox and lithium ion types. A battery comprises one or more electrochemical cells, with each cell comprising a liquid, paste or solid electrolyte, positive and negative electrodes. During discharge, electro chemical reactions at the two electrodes generate an electron flow through an external circuit. Vanadium redox batteries (VRBs) have good prospects because they can be scaled up to much larger storage capacities and are showing great potential for longer lifetimes and lower per-cycle costs than conventional batteries requiring refurbishment of electrodes. Lithium ion batteries are also displaying very high potential for large-scale energy storage. A battery consists of one or more electrochemical cells, connected in series, in parallel or series-parallel configuration in order to provide the desired voltage, current and power. The anode is the electronegative electrode from which electrons are generated to do external work. The cathode is the electropositive electrode to which positive ions migrate inside the cell and electrons migrate through the battery external electrical circuit. The electrolyte allows the flow of ions and electrons, from one electrode to another, being commonly a liquid solution containing a salt dissolved in a solvent, and must be stable in the presence of both electrodes. The current collectors allow the transport of electrons to and from the electrodes, typically made of metals and must not react with the electrode or electrolyte materials. The cell voltage is determined by the chemical reaction energy occurring inside the cell. The anode and cathode are, in practice, complex composites, containing, besides the active material, polymeric binders to hold together the powder structure and conductive diluents such as carbon black to give the whole structure electronic conductivity so that electrons can be transported to the active material. In addition, these components are combined to ensure sufficient porosity to allow the liquid electrolyte to penetrate the powder structure and permit the ions to reach the reacting sites. During the charging process, the electrochemical reactions are reversed via the application of an external voltage across the electrodes.

Modern battery technologies are ranging from the mature and long-established lead-acid system through to various more recent and emerging systems and technologies. Newest technologies are attracting an increased interest for possible use in power systems, having achieved market acceptance and uptake in consumer electronics in the so-called 3Cs sector (cameras, cell phones and computers). Batteries have the potential to span a broad range of energy storage applications, in part to their portability, ease of use, large power storage capacity (100 W up to 20 MW), and easy to connected in series-parallel combinations to increase their power capacity for specific applications. Major battery advantages include: stand-alone operation, no need to be connected to an electrical system, easily to expand, while typical disadvantages are: expensive, limited life cycle and maintenance. These systems could be located in any place and be rapidly installed, less environment impacts of other ESS technologies and can be located in a building (or similar facility) near the demand point. Power system connected BESS uses an inverter to convert the battery DC voltage into AC grid-compatible voltage. These units present fast dynamics with response times near 20 ms and efficiencies ranging from 60% to 80%. The battery temperature change during charge and discharge cycles must be controlled because it affects its life expectancy. Depending on how the battery and cycle are, the BESS can require multiple charges and discharges per day. The battery cycle is normal while the discharge depth is small, but if the discharge depth is high the battery cycle duration could be degraded. The expected useful life of a Ni-Cd battery is 20,000 cycles if the discharge depth is limited to 15%. Example of large-scale BESSs installed today are: 10 MW (40 MWh) Chion system, California and 20 MW (5 MWh) Puerto Rico. Their main inconvenient however is their relatively low durability for large-amplitude cycling. They are often used in emergency back-up, renewable-energy system storage, etc. The minimum discharge period of the electrochemical accumulators rarely reaches less than 15 minutes. However, for some applications, power up to 100 W/kg, even a few kW/kg, can be reached within a few seconds or minutes. As opposed to capacitors, their voltage remains stable as a function of charge level. Nevertheless, between a high-power recharging operation at near-maximum charge level and nearing full discharge, the voltage can easily vary by a ratio of two.

Fuel cells are electrochemical devices sometimes compared to conventional batteries. However, there is a fundamental difference in that fuel and oxidizer are supplied to fuel cells continually so that they can generate continuous electric energy. Though research and development activities are still required, the fuel cell technology is one of the most important technologies that allow us to draw the environment friendly society in the twenty-first century. Therefore, a fuel cell is important as an energy vector, combining with the stored fuel as a potential energy source. Even today, several types of fuel cells have been already used in practical applications such as combined heat and power generation applications, portable electronics, transportation and space vehicle applications. A fuel cell is an electricity generation system, which is utilizing electrochemical reactions. It can produce electric power by inducing both a reaction to oxidize hydrogen obtained by reforming natural gas or other fuels, and a reaction to reduce oxygen in the air, each occurring at separate electrodes connected to an external circuit. The principal reaction is often described as the opposite of the water electrolysis, because the hydrogen and the oxygen undergo an electrochemical reaction to produce electric power and water. In fact, in the case of the fuel cells used in space vehicles, water produced in electricity generation is used for routine daily activities within the space vehicle. Whereas fuel cells are based on the electrochemical reaction of hydrogen or other fuels, in ordinary batteries the reactants are either consumed or must be regenerated through electric recharge, as in lead-acid batteries. The hydrogen combines with the oxygen inside a combustion-free process, releasing electric energy in a chemical reaction that is very sensitive to the operating temperature and other environmental conditions. However, the principle of electric power generation in fuel cells is entirely the same as the one for batteries. On the other hand, a fuel cell can be considered as a "cross-over" of a battery and a thermal engine. It resembles a thermal engine because theoretically it can operate as long as fuel is fed to it. It is worth to remember, that in a conventional power generation system, the fuel is combusted to generate heat and then heat is converted to mechanical energy before it can be used to produce electrical energy. The maximum efficiency of a thermal engine is when it operates at the Carnot cycle, and the cycle efficiency is determined by the ratio of the heat source and heat sink absolute temperatures. On the other hand, fuel cell operation is based on electrochemical reactions and not fuel combustion. By passing the conversion of the chemical energy to the thermal and then mechanical energy enables fuel cells can achieve efficiency potentially much higher than that of conventional thermal power generation technologies. However, similar to a battery, the fuel cell operation is based on electrochemical reactions. This combination of processes provides significant advantages for the fuel cells. In principle, fuel cells resemble more rechargeable battery, and from a thermodynamic point of view the differences between fuel cells and batteries are based on the definition of closed systems and control volume. A battery is a thermodynamically closed system, while a fuel cell is a thermodynamically control volume, in which the fuel and the oxidizer can flow across the system's boundaries.

Hydrogen is an important energy carrier, and when used as a fuel, can be considered as an alternative to the fossil fuels, such as coal, crude oil and natural gas, and their derivatives. It has the potential to be a clean, reliable and affordable energy source, having the major advantage that the by-product of its combustion with oxygen is water, rather than CO and CO_2, which contain carbon and are considered greenhouse gases. It is expected that hydrogen to play a major role in future energy systems. Hydrogen, used in fuel cells, is also versatile energy storage medium due to its conversion to useful work, and eventually into electricity in fuel cells or in internal combustion engines. Hydrogen, manufactured in other energy sources, such as renewable energy conversion systems can be developed as an electrical power storage medium that may solve part of the challenges related to grid integration and operation of such renewable energy conversion systems. Hydrogen can be used directly in internal reciprocating combustion engines, requiring relatively minor modifications, if it is raised to a moderately high pressure, as well as in gas turbines and process heaters. It can also be used in hydrogen/oxygen fuel cells to directly produce electricity with the only by-products are water heat. The energy efficiency of fuel cells can be as high as 60%, compared to the fossil fuel systems, having typically efficiencies about 34% efficient.

When high temperature fuel cells are used, it is possible to obtain electricity and also to use the heat generated in the fuel cell, related to its inefficiency, for heating, cooling or power generation purposes in combined-heat, cooling and power generation systems. These systems are called cogeneration, and it is possible to obtain total energy (overall) efficiencies to about 80% or even higher in such systems.

2.2 Battery Types and Characteristics

Batteries convert the chemical energy contained in its active materials into electrical energy through an electro chemical oxidation-reduction reversible reaction. As we mentioned before, batteries and fuel cells, while structurally and functionally distinct, are based on similar electrochemical principles and technologies. For that reason we are introduce them together to emphasize their common scientific foundations. They are, however, easily distinguishable, because batteries are devices for the storage of electrical energy, while fuel cells are energy converters with no inherent energy storage capability. A battery contains a finite amount of chemicals that spontaneously react to produce a flow of electrons when a conducting path is connected to its terminals, as is shown in Figure 2.1. Fuel cells behave similarly, except that chemical reactants are continuously supplied from outside the cell and products are eliminated in continuous, steady streams. The chemical reactions in batteries and fuel cells are *oxidation-reduction reactions*. A common terminology exists for both types of cells. Both contain electrode pairs in contact with an intervening electrolyte, the charge-carrying medium, as is shown in Figure 2.1. The electrolyte is a solution through which positively charged ions pass from the anode to the cathode. The negative electrode, called the *anode*, is where an oxidation reaction takes place, delivering electrons to the external circuit. The positive electrode, where reduction occurs, is called the *cathode*. Electrons flow through the external load from the negative to the positive electrode, while the conventional current flows in the opposite direction from high to low potential. At the cathode, electrons arriving from the external

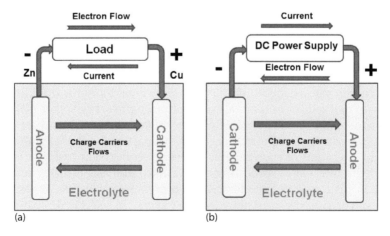

FIGURE 2.1
Electrochemical operation of a galvanic cell: (a) during discharging and (b) charging.

load are neutralized by reaction with positive ions from the electrolyte. Battery fundamental principles and operation, regardless the battery type, can be explained by using the so-called galvanic element or electrochemical cell. In summary a galvanic cell, as one shown in Figure 2.1, consists of three main components: the *anode*, the *cathode* and the *electrolyte*, while the electrons, needed for electrical conduction are produced or generated by a chemical oxidation-reduction reaction.

The process or mechanism by which a cell, as one shown in Figure 2.1 is able to convert input electrical energy (during the cell charging phase) and vice versa (during cell discharging phase) is the redox reaction. Through oxidation and reduction, the electrons are transferred from one substance to another, the electrons being lost through oxidation, while trough reduction the electrons are gain. The cell operation during discharging and charging phases are shown schematically in Figure 2.1. When the galvanic cell is connected to an external load, electrons flow from anode, which is oxidized, through external load to the cathode, where the electrons are accepted and the cathode material is reduced. During recharging (charging phase), the current is reversed and the oxidation takes place at the positive electrode and the reduction at the negative one. In the cell example of Figure 2.1, we are assuming a metal (Zn for example) as the anode material and chlorine (Cl_2) as a cathode material. The overall reactions during the discharging and charging phases are as follows:

$$\text{Discharge Reaction:} \quad Zn + Cl_2 \Rightarrow Zn^{2+} + 2Cl^- \left(ZnCl_2 \right) \tag{2.1a}$$

and, respectively

$$\text{Charge Reaction:} \quad Zn^{2+} + 2Cl^- \Rightarrow Zn + Cl_2 \tag{2.1b}$$

The cell standard potential is determined by its active materials, computed from experimental free energy data, under standard experimental conditions, as the ones listed in Table 2.1. The cell standard potential is computed from the electrode potentials, taking the oxidation potential as negative value of the reduction potential, as shown in Equation (2.2).

$$\text{Anode (Oxidation Potential)} + \text{Cathode (Reduction Potential)}$$

$$= \text{Cell Standard Potential} \tag{2.2}$$

TABLE 2.1

Electrode Material Voltaic Series

Metal Electrode	Li	K	Na	Mg	Zn	Fe	Cd	Al	Ni	PbO$_2$
$\Delta\phi_{metal\text{-}electrolyte}$ (V)	−3.02	−2.92	−2.71	−2.35	−0.762	−0.44	−0.402	−1.66	−0.25	1.69
Electrochemical equivalent (Ah/g)	3.86		1.16	2.20	0.82	0.96	0.48	2.98		0.224
Metal electrode	Pb	H$_2$	Cu	Ag	Hg	Au	Cl$_2$	O$_2$	Br$_2$	NiOOH
$\Delta\phi_{metal\text{-}eletrolyte}$ (V)	−0.126	0.00	0.345	0.80	0.86	1.50	1.36	1.23	1.07	0.49
Electrochemical equivalent (Ah/g)	0.26	26.59					0.756	3.35	0.335	0.292

Example 2.1: Estimate the cell standard potential for the cell with redox reaction of Equation (2.1a).

Solution: For the reaction of Equation (2.1a), the cell standard potential, from Equation (2.2) is:

$$Zn \rightarrow Zn^{2+} + 2e, \quad -(-0.76) \text{ V}$$

$$Cl_2 \rightarrow 2Cl^- - 2e, \quad 1.36 \text{ V}$$

$$\text{Cell Potential: } E^0 = 0.76 + 1.36 = 2.12 \text{ V}$$

From thermodynamics we know that the work done by the electrochemical cell comes at a cost, which in turns implies that the chemical reactions taking place within the cell must lead to a decrease in free energy. In fact for a reversible process at constant temperature and pressure the maximum work done by the system, W, is equal the free energy (Gibbs free energy) change $-\Delta G$. The work performed when transporting an electric charge e (in C) through a potential difference E (in V) is simply the product of eE. Here of interests is to express this work in a per mole basis. The total charge carried by one mole of positively charged ions of valence +1 is 96,487 C and this number, denoted by F, the Faraday's constant. Thus, the work produced by the electrochemical cell is:

$$W = -\Delta G = \xi F \cdot E_{max} \Rightarrow E_{max} = \frac{-\Delta G^0}{\xi F} \tag{2.3}$$

where ξ is the valence of the ions produced in the chemical reaction. The electric potential difference across the cell electrodes, E, is the electromotive force, or EMF, of the galvanic cell. It is clear from Equation (2.3) that in order to have higher work and potential, we need to find reactions with the highest driving force, the free energy change $-\Delta G$. On the other hand, if we consider an electrochemical cell in which the activity of species are different in the two electrodes, $a_1(-)$ in the negative electrode and $a_2(+)$ in the positive, the electromotive forces (EMF) of such a cell can be calculated with the help of the Nernst equation:

$$E_{cell} = \left(\frac{RT}{n\mathbf{F}}\right)\ln\left(\frac{a_1}{a_2}\right) \tag{2.4}$$

where n is the number of electrons needed to get one atom or molecule of into its ionic form in the electrolyte, and a_1 and a_2 are the activities at electrodes 1 and 2, respectively. This relation, often called the Nernst equation, is very useful, for it relates the measurable cell voltage to the chemical difference across an electrochemical galvanic cell. That is, it transduces between the chemical and electrical driving forces. If the activity of species i in one of the electrodes is a standard reference value, the Nernst equation provides the relative electrical potential of the other electrode. We can make use of such a cell in the following ways: (i) If $a_1 > a_2$ and the fuel is continually added on the left, and removed on the right, we have a source of energy, a fuel cell; and (ii) if we are applying a greater voltage than E in the opposite sense we can drive the species from one side to the other, hence we have an ion pump or an electrolyzer. So a fuel cell is an electrochemical cell which can continuously change the chemical energy of a fuel and oxidant to electrical energy with high efficiency, it is not surprising that a variety of synthetic fuels have been proposed, such as hydrogen, methanol, ammonia and methane.

Example 2.2: Estimate the maximum output voltage that a Zn-Cu electrochemical cell can generate.

Solution: The Zn-Cu reaction taking place in this electrochemical cell is:

$$Zn + Cu^{2+} \rightarrow Cu + Zn^{2+}$$

By using Equation (2.3) and from the data tables of the book appendixes (for Zn-Cu cell ξ is equal to 2, and ΔG^0 is equal to 216,160 kJ/kmole), the maximum voltage is:

$$E_{max} = \frac{-\Delta G^0}{\xi F} = \frac{216,160 \text{ kJ/kmole}}{2 \times 96,500,.0} = 1.12 \text{ V}$$

Batteries, as mentioned in the previous sections, convert the chemical energy contained in its active materials into electrical energy through an electrochemical oxidation-reduction reversible reaction. The operation and battery fundamental principles, regardless the battery type is explained by using the galvanic element or cell model. If, for example a galvanic element, made of a zinc (Zn) electrode, a copper (Cu) electrode and an electrolyte ($CuSO_4$, for example) the two electrodes are electrically connected (Figure 2.2), through electrolyte, the Zn^{2+} ions flow from Zn electrode, while the electrons migrate from Zn, and eventually combine with Cu^{2+} ions, residing into the electrolyte and from copper atoms, increasing the Cu electrode volume. The direction of electrons is determined by the potential difference, between metal and electrolyte, $\Delta \phi_{metal-electrolyte}$, the electron transport is in such way that the metal (Zn, in this case) separate the electrons and ions, because it has a lower metal-electrolyte potential difference than that of the Cu electrode against electrolyte. The metal (electrode) with higher potential is serving as positive electrode (and cell terminal) of a galvanic element. The metals can be arranged in voltaic sequences, in a way that all metals (e.g., Fe, Cd, Ni, Pb, or Cu) in the right side of a certain metal (e.g., Zn) to for a positive pole in a combination of electrodes, chosen based of the metal-electrolyte potential difference values of Table 2.1. For example Zn is a negative pole with respect to Fe, Cd, Ag, or Au electrodes, while Li electrode forms a negative pole with respect to K, Na, Mg, Zn or Fe electrodes. The potential difference (external voltage) between the cell (galvanic element) terminals is the voltage difference, for this galvanic element, existing between Zn electrode and electrolyte ($CuSO_4$) and between Cu electrode and electrolyte. Notice that is not possible to directly measure the potential difference between the metal electrode and electrolyte, only the metal-metal potential difference can be measured.

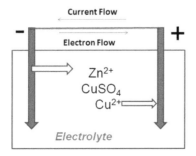

FIGURE 2.2
Galvanic element diagram (consisting of Zn and Cu electrodes, and $CuSO_4$ electrolyte). Current flows are also shown here.

Example 2.3: Determine the potential differences between Li-Au, Li-Ag and Pb-H$_2$, using the values in Table 2.1.

Solution: The required potential differences are:

$$\Delta\phi_{Li-Au} = -3.02 - 1.50 = -4.52 \text{ V}$$

$$\Delta\phi_{Li-Au} = -3.02 - 0.86 = -3.88 \text{ V}$$

And

$$\Delta\phi_{Ni-Cu} = -0.25 - 0.345 = -0.595 \text{ V}$$

Cell theoretical capacity, a critical parameter characterizing electrochemical energy storage devices and systems, is determined by the amount of cell active materials, and is expressing the total electricity quantity in the electrochemical reaction in terms of ampere-hour (Ah, in fact coulombs). It is directly associated with electricity obtained from the cell active materials. As specified in the electrochemistry, one gram-equivalent weight of substance (a gram-equivalent weight is the ratio of the atomic or molecular weight of the active material in grams over number of electrons involved in reaction) delivers 96,487 C or 26.8 Ah. The theoretical capacity of an electrochemical cell, based only on active materials involved in the cell electrochemical reaction, is computed from the equivalent weight of the reactants.

Example 2.4: Determine the Zn-Cl$_2$ cell theoretical capacity.

Solution: Using the appropriate values from Table 2.1, the cell theoretical capacity is:

$$Zn + Cl_2 \rightarrow ZnCl_2$$

$$1.22 \text{ g/Ah} + 1.32 \text{ g/Ah} = 2.54 \text{ g/Ah or } 0.394 \text{ Ah/g}$$

In a same way the volumetric ampere-hour capacity can be computed. Notice that water, electrolyte or any other materials that may be involved in the galvanic cell reaction are not included in the cell theoretical capacity or voltage calculations.

2.2.1 Battery Types

Lead-acid batteries (Pb-acid/LA) are the most common electrical energy storage device used at the present, especially in transportation, renewable energy and stand alone hybrid power systems. Its success is due to its maturity (research has been ongoing for about 150 years), low cost, long lifespan, fast response and low self-discharge rate. They are used for both short-term (seconds) and long-term applications (up to 8 hours). A lead-acid battery works somewhat differently than series galvanic elements, employing the polarization of the positive and negative charges. Two lead plates of a cell are immersed in a dilute sulfuric acid solution, each covered with a PbSO$_4$ layer, basically consisting of spongy lead anode and lead acid cathode, with lead as the current collector. The lead-acid batteries for electric vehicles are using a gel rather than a liquid electrolyte, in order to withstand deep cycle that is required here, for that reason being more expensive than regular ones. The sulfuric acid combines with the lead and lead oxide to produce lead sulfate and water, and the electrical energy is released during the process. The overall lead-acid battery reaction is:

$$Pb + PbO_2 + 2H_2SO_4 \rightleftarrows 2PbSO_4 + 2H_2O \tag{2.5}$$

This reaction proceeds from left to right, during battery discharge and in opposite direction during the battery charging process. During the discharging process, the electrons flow from the cathode to anode (current is flowing in opposite direction), until the chemical process is reversed and $PbSO_4$ forms on the anode and cathode, while the anode is covered with PbO_2 which gets reduced as the discharging process progresses. The voltage across of the two lead plates is $e_0 = 2.02$ V. If six sets for lead plates are connected in series, then the total terminal voltage is $E^0 = 6 \cdot e_0 = 12.12$ V. The efficiency of a lead-acid battery is from 80% to 90% of the charged energy. During the discharging the electrolyte gradually is loses the sulfuric acid and becomes more dilute, while during charging process, the electrolyte revert to lead and lead dioxide, and also recovering the its sulfuric acid and concentration increases.

There are two lead-acid battery types: (a) flooded lead-acid (FLA), and (b) valve-regulated lead-acid (VRLA). From application point of view, the lead-acid batteries split into stationary, traction and car batteries. Stationary batteries ensure uninterruptible electric energy supply, in case of the power system failure, undergoing usually only few cycles, so have about 20 years life time. Traction batteries, used for power supply of electric vehicles, electro-mobiles or industrial tracks, etc. are working in deep cycle charging-discharging regime with the lifespan of about 5 years (about 1000 charge-discharge cycles). Automotive (car) batteries, used to crank car engine, being able to supply short and intense discharge current, and also to support car electrical devices, when the engine is not running. FLA batteries, described above, are made up of two electrodes (lead plates), immersed in a mixture of water (65%) and sulfuric acid (35%). VRLA batteries have the same operating principle as FLA batteries, with the difference that they are sealed with pressure-regulating valves, to prevent the air entering the cells and hydrogen venting. VRLA batteries are operating with the help of an internal oxygen cycle, emitted in the later charging stages and during overcharging on the positive electrode, and it travels through a gas separator to the cathode where it is reduced to water. This oxygen cycle is making the cathode potential less negative, so the rate of hydrogen evolution decreases. Part of the generated electricity is used by the internal recombination oxygen cycle and is converted to heat. VRLA batteries have lower maintenance costs, less weight and space, but these advantages are coupled with higher initial costs and shorter lifetime. During discharge, lead sulfate is the product on both electrodes. Sulfate crystals become larger and difficult to break up during recharging, if the battery is over-discharged or kept discharged for a prolonged time period. Hydrogen is produced during charging, leading to water losses if overcharged. Their popularity is result of their wide availability, robustness and reasonably low cost. The disadvantages of the lead-acid batteries are their weight, low specific energy and specific power, short cycle life (100–1000 cycles), high maintenance requirements, hazards associated with lead and sulfuric acid during production and disposal, and capacity drop at low temperatures. FLA batteries have two primary applications: starting and ignition, short bursts of strong power, e.g., car engine batteries and deep cycle, low steady power over a long time applications. VRLA batteries are very popular for backup power, standby power supplies in telecommunication centers and for uninterruptible power supply.

Among the major reasons for failure of the lead-acid batteries include: positive plate expansion, positive mass fractioning, water loss, acid stratification, incomplete charging causing active mass sulfating, positive grid corrosion and negative active mass sulfating.

Construction of a lead-acid battery depends on the application. Electrodes consist of grid (usually made from lead alloys, being bearing structure needs to be mechanically and corrosion proofed) and active mass is made from lead oxides. Separators, another LA battery component, separating positive and negative electrodes, are providing electric insulation between positive and negative plates (preventing short-circuits), act as mechanical spacers, retaining active materials in close contact with the grid, and permit both the free diffusion of electrolyte and ion migration. Materials used for separators are wood veneers, paper or synthetic materials. Electrolytes are aqueous solutions of H_2SO_4 (sulfuric acid) with density 1.22–1.28 g/cm^3, liquid, gel or completely absorbed into separators. Lead-acid batteries are affected one or more of the following failure mechanisms, such as positive plate expansion, positive active mass fracturing, water losses, acid stratification, incomplete charging causing active mass sulfation, positive grid corrosion and negative active mass sulfation. Care must be taken during the battery maintenance.

Nickel-cadmium (Ni–Cd) *and nickel metal hydride* (Ni–MH) *batteries* use nickel oxyhydroxide for the cathode and metallic cadmium as the anode with a potassium hydroxide as an electrolyte. Other possible systems are Ni-Fe, Ni-Zn and Ni-H$_2$. These types of batteries were very popular before 1990 but have been largely superseded by Ni-MH due to their inferior cycle life, memory effect, energy density and toxicity of the cadmium in Ni-Cd, compared to Ni-MH batteries. Ni-MH also have the advantage of improved high rate capability (due to the endothermic nature of the discharge reaction), and higher over-discharge tolerance. Ni-MH use nickel oxyhydroxide for the cathode, with a potassium hydroxide electrolyte and a hydrogen-absorbing alloy usually made of lanthanum and rare earths serving as a solid source of reduced hydrogen which can be oxidized to form protons. Ni-Cd and Ni-MH batteries are mature technologies, relatively rugged with higher energy density, lower maintenance requirements and higher cycle life than LA batteries. They are, however, more expensive, with limitations on the long-term cost reduction potential due to material costs. In addition, Ni-Cd batteries contain toxic components making them environmentally challenging. With a higher energy density, longer cycle life and lower maintenance requirements, Ni-Cd batteries have a potential edge over LA batteries. The drawbacks of this technology include the cadmium toxicity and higher self-discharge rates than LA batteries. Also, Ni-Cd batteries may cost up to ten times more than a lead-acid battery, making it quite a costly alternative.

Ni-Cd batteries are produced in a wide commercial variety from sealed free-maintenance cells (capacities of 10 mAh to 20 Ah) to vented standby power units (capacities of 1000 Ah or higher), longer cycle life, overcharge capacity, high charge-discharge rates, almost constant discharge voltage, and they can operate at low temperature. Depending on construction, Ni-Cd batteries have energy densities in the range of 40–60 Wh/kg, and cycle life above several hundred for sealed cells to several thousand for vented cells. There are two main construction types of Ni-Cd batteries. There to main construction types of Ni-Cd batteries. First construction type are using pocket plate electrodes, in vented cells, where the active material is found in pockets of finely perforated nickel plated sheet steel, positive and negative plates are separated by plastic pins or ladders and plate edge insulators. Second type are using sintered, bonded or fiber plate electrodes, for both vented and sealed cells. Active material is distributed within the pores. The electrolyte is an aqueous solution of KOH, with 20%–28% weight concentration and density of 1.18–1.27 g/cm^3 at 25°C, with a 1%–2% of LiOH usually added to minimize the coagulation of the NiOOH electrode during the charge-discharge cycling. An aqueous NaOH electrolyte may be used in cells operating at higher temperatures. The overall Ni-Cd cell discharge reaction is:

$$2NiOOH + Cd + 2H_2O \rightarrow 2Ni(OH)_2 + Cd(OH)_2 \qquad (2.6)$$

The Ni-Cd cell voltage is $E^0 = +1.30$ V. The Ni-Cd batteries are designed as positive limited oxygen cycle use, amount of water decreases during discharge. The oxygen evolved at anode during charging diffuses to the cathode are reacts with cadmium, forming $Cd(OH)_2$, and carbon dioxide from air an react with KOH in electrolyte to form K_2CO_3, and $CdCO_3$ at cathode, increasing internal resistance and reducing capacity of the Ni-Cd batteries. Sealed Ni-MH cell has an active negative material hydrogen absorbed in a metal alloy that increases its energy density and making it more environmentally friendly compared with Ni-Cd cells. However, these cells have higher self-discharge rates and are less tolerant to overcharge than Ni-Cd cells. The hydrogen absorption alloys are made tow metals (first metal absorbs hydrogen exothermically and the second endothermically), and is serving also as catalyst for the absorption of the hydrogen into the ally lattice. A hydrolytic polypropylene separator is used in Ni-MH cells. The electrolyte, used in Ni-MH cells is potassium hydroxide, and the cell voltages range from 1.32 to 1.35 V. Their energy density is 80 Wh/kg (25% higher than that of Ni-Cd cells), life cycle over 1000 cycles, and a high self-discharge rate of 4%–5% per day. Ni-MH batteries are used in electric vehicles, consumer electronics, phones, medical instruments and other high-rate and longer life-cycle applications. The overall Ni-MH cell reaction during discharge cycle is:

$$NiOOH + MH \rightarrow Ni(OH)_2 + M \qquad (2.7)$$

Ni-Dc and Ni-MH cells suffer, both from memory effects, a temporary capacity reduction, but reversible caused by the shallow charge-discharge cycling. After a shallow cycling there is a voltage step during discharge, mining that the cell remembers the shallow cycling depth. The voltage reduction depends on the number of previous shallow cycles and the discharge current. However, its capacity is not affected, if the cell is fully discharged and then charged, then a deep discharge shows a normal discharge curve.

Lithium ion (*Li-ion*) *batteries* are largely cobalt or phosphate based. In both embodiments lithium ions flow between the anode and cathode to generated current flow. Li-ion batteries have a high energy to weight ratio (density), no memory effect, light weight, high reduction potential, low internal resistance and low self-discharge rates. Prices may be high and increasing penetration may push prices higher as limited lithium resources are depleted. Lithium ion battery technology has progressed from developmental and special-purpose status to a global mass-market product in less than 20 years. The technology is especially attractive for low-power and portable applications because Li-ion batteries offer high-power densities, typically 150–200 Wh/kg and generally acceptable cycle life, about 500 cycles, very low self-discharge rate (less than 10% per month) and high voltage, about 3.6 V. The operation of the lithium ion cells involves the reversible transfer of lithium ions, between electrodes, during charge and discharge. During charging, lithium ions move out (de-intercalate) from the lithium metal oxide cathode and intercalate into the graphite-based anode, with the reverse happening during the discharge reaction. The non-aqueous ionic conducting electrolyte takes no part in the reaction except for conducting the lithium ions during the charge and discharge cycles. Notwithstanding their significant advantages, lithium ion systems must be maintained within well-defined operating limits to avoid permanent cell damage or failure. The technology also possesses no natural ability to equalize the charge amount in its component cells. This, and the closely defined operational envelope of lithium ion batteries, essentially dictates the use of relatively sophisticated management systems. Most of commercial Li-ion cells have anodes of cobalt oxide, or manganese oxide. Negative electrode is made of carbon, in graphite form (light weight a low price) or an amorphous material with a high surface-area. Electrolyte is made of an organic liquid (ether) and a dissolved slat, and the positive and negative active mass is applied to both sides of thin metal foils, aluminum on the positive side and copper on the negative. A micro-porous polymer is used as separator between positive and negative electrode. The positive electrode reaction in a Li-ion cell is:

$$LiCoO_2 \rightarrow Li_{1-x}CoO_2 + xLi^+ + xe^- \tag{2.8a}$$

And the negative electrode reaction is:

$$xLi^+ + xe^- + C_6 \rightarrow Li_xC_6 \tag{2.8b}$$

Here the index, x is from 0 to 1 on the negative electrode, and from 0 to 0.45 on the positive one. Li-ion batteries are manufactured in coin, cylindrical and prismatic shapes. On the other hand, the application of the technology to larger-scale systems is relatively limited to date, although various developments are in hand in relation to the automotive, power utility, submersible and marine sectors. Lithium-polymer (Li-pol) batteries, a newer technology, are employing the property of the polymers, containing a hetero-atom (i.e., oxygen or sulfur) to dissolve lithium salts in high concentrations, leading at higher temperatures to a good electric conductivity (0.1 S/m at 100°C), allowing such polymers to be used as electrolyte for lithium batteries. Polymer electrolyte is safer than the liquid one, not being flammable.

2.2.2 Battery Fundamentals, Parameters and Electric Circuit Models

Battery or cell capacity ($CAP(t)$) is defined as the integral of current, $i(t)$ over a defined time period as:

$$CAP(t) = \int_0^t i(t)dt \tag{2.9}$$

The above relationship applies to either battery charge or discharge, meaning the capacity added or capacity removed from a battery or cell, respectively. The capacity of a battery or cell is measured in milliampere-hours (mAh) or ampere-hours (Ah). This basic definition is simple and straight; however, several different forms of capacity relationship are used in the battery industry. The distinctions between them reflect differences in the conditions under which the battery capacity is measured. Standard capacity measures the total capacity that a relatively new, but stabilized production cell or battery can store and discharge under a defined standard set of application condition, assuming that the cell or battery is fully formed, that it is charged at standard temperature at the specification rate, and that it is discharged at the same standard temperature at a specified standard discharge rate to a standard end-of-discharge voltage (EODV). The standard EODV is subject to variation depending on discharge rate. When the application conditions differ from standard ones, the cell or battery capacity changes, so the term actual capacity includes all nonstandard conditions that alter the amount of capacity the fully charged new cell or battery is capable of delivering when fully discharged to a standard EODV. Examples of such situations might include subjecting the cell or battery to a cold discharge or a high-rate discharge. That portion of actual capacity that is delivered by the fully charged new cell or battery to some nonstandard EODV is called available capacity. Thus, if the standard EODV is 1.6 V/cell, the available capacity to an EODV of 1.8 V/cell would be less than the actual capacity. Rated capacity is defined as the minimum expected capacity when a new, fully formed, cell is measured under standard conditions. This is the basis for C rate (defined later) and depends on the standard conditions used which may vary depending on the manufacturers and the battery types. If a battery is stored for a period of time following a full charge, some of its charge will dissipate. The capacity which remains that can be discharged is called retained capacity. In most of the practical engineering applications, the battery capacity, C, for a constant discharge rate of I (in A) as:

$$C = I \times t \tag{2.10}$$

The capacity of a battery, sometimes referred to as C_{load} or simply C, is an inaccurate measure of how much charge a battery can deliver to a load. It is an imprecise value because it depends on temperature, age of the cells, state of the charge and on the rate of discharge. It has been observed that two identical, fully charged batteries, under the same circumstances, are delivering different charges to a load depending on the current drawn by the load. In other words C is not constant and the value of C is for a fully charged battery is not an adequate description of the characteristic of the battery unless it is accompanied by an additional information, *rated time of discharge* with the assumption that the discharge occurs under a constant current regime. Usually, lead-acid batteries are selected as energy storage for the building DC microgrids, because of relatively low cost and mature technology. Moreover, the capacity of a battery is reduced if the current is drawn more quickly, being important to predict the effect of current on the battery capacity, both for application design and for battery measurements. One of the most used is the Peukert model (a slightly different variant of Equation 2.9), which is modeling very well the battery behavior at higher currents, while not very accurate at lower currents. Starting point of this model is the Peukert capacity, C_P, a constant, given by the following relationship:

$$C_P = I^k \times T \tag{2.11}$$

where k is a constant, Peukert coefficient, with typically value of about 1.2 for lead-acid batteries, and T is the discharge time in hours. The total charge removed from the battery, CR_{N+1}, after N steps, each time step, δt (in seconds) is given by:

$$CR_{N+1} = CR_N + \frac{\delta t \times I^k}{3600} \text{ Ah} \qquad (2.12)$$

Example 2.5: Peukert coefficient for a battery was found to be 1.107, find the Puekert capacity for a 4.2-A current and T equal to 10 hours.

Solution: From Equation (2.10), the Puekert capacity is:

$$C_P = 4.2^{1.107} \times 10 = 49 \text{ Ah}$$

The energy storage systems are usually operated by current closed-loop control, while the storage power is controlled by supervision unit which calculates the corresponding power reference. The storage state of charge (SOC) must be respected to its upper (maximum) and lower (minimum) SOC limits, SOC_{max} and SOC_{min}, respectively, to protect the battery from over-charging and over-discharging, as given in Equation (2.12). SOC is calculated with Equation (2.14), where SOC_0 is the initial SOC at t_0 (initial time), CREF is the storage nominal capacity (Ah) and V_S is the storage voltage.

$$SOC_{min} \leq SOC(t) \leq SOC_{max} \qquad (2.13)$$

and

$$SOC(t) = SOC_0 + \frac{1}{3600 \times V_S \times CREF} \int_{t_0}^{t} \left(P_{SC} - P_{SD}(t) \right) dt \qquad (2.14)$$

Accurate battery models are required for the simulation, analysis and design of energy consumption of electric vehicles, portable devices, or renewable energy and power system applications. The major challenge in modeling a battery are the nonlinear characteristics of the equivalent circuit parameters, which depend on the battery state of charge, and are requiring complete and complex experimental and/or numerical procedures. The battery itself has internal parameters, which need to be taken care of for modeling purposes, such as internal voltage and resistance. All electric cells in a battery have nominal voltages which gives the approximate voltage when the cell is delivering electrical power. The cells can be connected in series to give the overall voltage required by a specific application.

Example 2.6: A battery bank consists of several cells, connected in series. Assuming that the cell internal resistance is 0.012 Ω and the cell electrochemical voltage is 1.25 V. If the battery bank needs to deliver 12.5 A at 120 V to a load, determine the number of cells.

Solution: With cells in series, the number of cell can be estimated as:

$$N(\text{cells}) = \frac{V_L}{V_{OC}(\text{cell})} = \frac{120}{1.25 - 12.5 \times 0.012} = 109.09$$

We are choosing 110 cells, round off to upper integer, meaning a higher terminal voltage than 120 V. However, the terminal voltage is decreasing, over the battery bank lifetime.

There are three basic battery models, most used in engineering applications: *the ideal, linear and Thévenin models.* The battery ideal model, a very simple one is made up only by

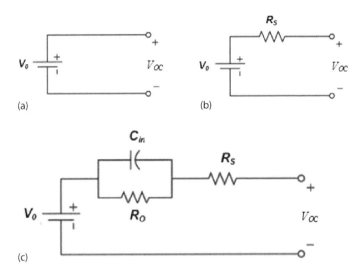

FIGURE 2.3
Battery electric diagrams, steady-state conditions: (a) battery ideal model, (b) battery linear model and (c) Thévenin battery model.

a voltage source (Figure 2.3a), and ignores the internal parameters and, hence. The battery linear model (Figure 2.3b), a widely used battery model in applications, and consists of an ideal battery with open-circuit voltage, V_0, and an equivalent series resistance, R_S, while V_{Out} represents the battery terminal (output) voltage. This terminal voltage is obtained from the open-circuit tests as well as from load tests conducted on a fully charged battery. Although this model is quite widely used, it still does not consider the varying characteristics of the internal impedance of the battery with the varying state of charge (SOC) and electrolyte concentration. The Thévenin model (Figure 2.3c), a more accurate and complex model, consists in addition to the open-circuit voltage and internal resistance, an internal capacitance, C_{in}, and the overvoltage resistance, R_O. The capacitor C_{in} accounts for the capacitance of the parallel plates and resistor R_O accounts the nonlinear resistance by the plate to the electrolyte. All the elements in this model are assumed to be constants. However, in reality, they depend on the battery conditions. Thus, this model is also not the most accurate, but is by far the most widely used. In this view, a new approach to evaluate batteries is introduced.

Any battery, regardless the type in the first approximation (for steady-state operation) works as constant voltage's source with an internal (source/battery) resistance, by considering battery linear model as shown in Figure 2.3b. These parameters, the internal voltage (V_{OC}) and resistance (R_S), are dependent of the discharged energy (Ah) as:

$$V_{OC} = V_0 - k_1 \times \text{DoD} \tag{2.15a}$$

and

$$R_S = R_0 + k_2 \times \text{DoD} \tag{2.15b}$$

Here, V_{OC}, known also as the open-circuit or electrochemical voltage decreases linear with depth of discharge, DoD, while the internal resistance, R_S increases linear with DoD. V_0 and R_0 are the values of the electrochemical (internal) voltage and resistance, respectively, when the battery is fully charged (DoD = 0). The constants k_1 and k_2 are determined from the battery test data, through curve fitting or other numerical procedures. The depth of discharge (DoD) is defined from the battery state of charge (SOC) as:

$$\text{DoD} = \frac{\text{Ah drained from battery}}{\text{Battery rated Ah capacity}} = 1 - SOC \tag{2.16}$$

While SOC (as defined in Equation 2.13) is computed, for practical application by a much simpler relationship that can be estimated from the battery monitoring (test data), as:

$$\text{SOC} = 1 - \text{DoD} = \frac{\text{Ah remaining in the battery}}{\text{Rated Ah battery capacity}} \tag{2.17}$$

Notice the battery terminal voltage is lower and the internal resistance is higher in a partially battery discharge state (i.e., any time when DoD > 0). The terminal voltage, V_L, of a partially discharged battery, with notation of Figure 2.4 is expressed as:

$$V_L = V_0 - I \cdot R_S = V_0 - k_1 \times \text{DoD} - I \cdot R_S \tag{2.18}$$

Then the power delivered to the load is I^2R_L, the battery internal loss is I^2R_S, dissipated as heat inside the battery. In consequence, as the battery discharge internal resistance, R_L increases, so more and more heat is generated.

> **Example 2.7:** A 12-V lead-acid car battery has a measured voltage of 11.2-V when delivers 40 A to a load. What are the load and the internal battery resistances? Determine its instantaneous power and the rate of sulfuric acid consumption.
>
> **Solution:** For the battery equivalent circuit of Figure 2.4, the voltage across the load is:
>
> $$V_L = V_0 - I \cdot R_S$$
>
> And the internal battery and load resistances are:
>
> $$R_S = \frac{12.0 - 11.2}{45} = 0.02 \ \Omega$$
>
> $$R_L = \frac{V_0}{I} - R_S = \frac{12}{40} - 0.02 = 0.3 - 0.02 = 0.28 \ \Omega$$
>
> The power delivered by the battery is:
>
> $$P = V_L \times I = 11.2 \times 40 = 448 \ \text{W}$$

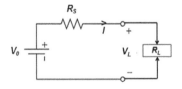

FIGURE 2.4
Electric diagram of battery-load, by using the steady-state battery linear model.

To estimate the sulfuric acid consumption rate, we suppose that the battery consumes N_{Rct} moles of reactant per second, and the released electrons in reaction flow through electrodes and external circuit proportional to the reaction rate, jN_{Rct}, which is the rate of flow of electrons from the cell (in mole per second), while j ($j = 2$ in the case of lead-acid batteries) is the number of moles of electrons released per mole of reactant. The electron current from the cell is then:

$$I = jN_{Rct}F \text{ (A)} \tag{2.19}$$

F is the Faraday constant equal to the Avogadro number times the electron charge:

$$F = 6.023 \times 10^{23} \cdot 1.609 \times 10^{-19} = 96{,}488 \text{ C/gram-mole}$$

The instantaneous power delivered by the cell is then:

$$P = jN_{Rct} \cdot F \cdot V \tag{2.20}$$

For the battery in this example the sulfuric acid consumption rate is:

$$N_{Rct} = \frac{I}{jF} = \frac{40}{2 \times 96488} = 2.0728 \times 10^{-4} \text{ gram-mole/s}$$

Because the sulfuric acid is consumed at the same rate at both anode and cathode, the above value is multiplied by 2, and it can be expressed in kg/hr, by using the sulfuric acid molecular weight

$$N_{Rct} = 2 \times 2.0728 \times 10^{-4} \cdot 10^{-3} \cdot 98 \times 3600 = 0.1463 \text{ kg/hr}$$

2.2.3 Summary of Battery Parameters

Battery performances are characterized and evaluated in terms of, among other parameters: charge-discharge (C/D) voltages, charge-discharge Ah ratio, self-discharge (trickle charge) rates, round-trip energy efficiency and number of charge-discharge cycles.

Cell and battery voltage: All battery cells have a nominal voltage which gives the approximate voltage when delivering power. Cells are connected in series to give the overall battery voltage required. Figure 2.4 is showing one of the equivalent battery electric circuits. However, the battery or cell voltage vs. time diagrams, during both discharging and charging phases are represented by quite non linear curves, usually with some sort of a middle time rage plateaus. The average voltage values of the plateau diagrams are the discharge voltage and charge voltage, respectively. The C/D voltages are dependent on the temperature and how fast the battery is charged or discharged.

Battery charge capacity (Ah): The most critical parameter is the electric charge that a battery can supply, and is expressed in Ampere-hour (Ah). For example if the capacity of a battery is 100 Ah, the battery can supply 1 A for 100 hours.

Stored energy: The energy stored in a battery depends on the *battery voltage*, and the *charge* stored, being another way to characterize a cell or a battery. This theoretical energy value is the maximum value that can be delivered by a specific electrochemical system, here a cell or a battery. The SI unit for energy is Joule (J); however, this is an inconveniently small unit, and in practical application Watt-hour (Wh) is used instead. The energy is expressed in Wh as:

$$\text{Energy (Wh)} = V \times Ah \tag{2.21}$$

In the Zn-Cl$_2$ cell, for the standard potential 2.12 V and the theoretical capacity per gram of active materials, 0.394 Ah/g, the theoretical mass energy density is 0.835 Wh/g or 835 Wh/kg.

Specific energy: Specific energy is the amount of electrical energy stored in a battery per unit battery mass (kg), and in practical applications is expressed in Wh/kg.

Energy density: Energy density is the amount of electrical energy per unit of battery volume, expressed in practical applications and engineering in Wh/m^3.

Charge (Ah) *efficiency*: An ideal battery will return the entire stored charge to the load, so its charge efficiency is 100%. However, the charge efficiency of actual batteries is less than 100%, and depends on the battery type, temperature, rate of charge, and varies with the battery state of charge. The *C/D ratio* (*charge efficiency*) is defined as the ratio of *Ah charged* over *Ah discharged* with no net change in the state of charge, being always greater than one, depending on the temperature and on the charge and discharge rates. The *charge (Columbic) efficiency* is the inverse of the C/D ratio.

Energy efficiency: It is another important parameter, being defined as the ratio of the electric energy supplied by the battery to the electric energy required to return the battery at its state before discharge. The energy efficiency in one round-trip (*round-trip energy efficiency*) of a cell or a battery, defined as the ratio *energy output/energy input* at the terminals, is expressed by:

$$\eta_{\text{Energy}} = \frac{\text{Average Discharge Voltage} \times Ah}{\text{Average Charge Voltage} \times C/D \text{ Ratio} \times Ah} \tag{2.22}$$

Example 2.8: Estimate the round-trip efficacy at the terminals for a cell with an average voltage 1.35 V, average charge voltage 1.50 V, and C/D ratio 1.15.

Solution: The round-trip efficiency, given by Equation (2.22) is:

$$\eta_{\text{Energy}} = \frac{1.35 \times C}{1.50 \times 1.15 \times C} = 0.753 \text{ or } 75.3\%$$

Self-discharge rate: Most of the battery types are discharging, when left unused (the self-discharge), an important battery characteristic. This means that some batteries cannot be left for longer periods without recharging. Batteries slowly discharge, even in open-circuit at a rate typically less than 1% of Ah per day, and to maintain full charge, the battery must be continuously trickle-charged to counter the discharge rate. Notice that after a battery is fully charge, the stored energy stops increasing, and any additional charge is converted into heat. If overcharged at a higher rate than the self-discharge rate for extended periods, the battery overheats and may pose safety hazards of potential explosion. The self-discharge rate varies with battery type, temperature and storage conditions.

Battery temperature, heating and cooling: Most of the battery types are running at ambient temperatures; however, there are battery types that need heating at start and then cooling when in use. For some battery types performances vary with temperature. The operating temperature influences the battery performances through charge storage capacity, the battery internal resistance and charge efficiency are all decreasing with temperature, while self-discharge rate is increasing with temperature. The temperature effects, cooling and heating are important parameters that designers need to take in consideration.

Battery life and number of deep cycles: Most types of rechargeable batteries have limited number of deep cycles of 20% of the battery charge, in the range of hundreds or thousands cycles. The number of deep cycles depends on the battery type, its electrochemistry, discharge depth, temperature, design details, and the way and conditions that battery is used. This is a very important parameter in battery specifications, reflecting the battery lifetime, so the application cost. Battery life is basically determined by the weakest cell in the battery stack. The battery cell can fail short-circuit, losing the voltage or open-circuit, losing the current, which is disabling the entire battery string. Charging, for example a battery may result in heat-related damages of the battery or charger. However, even without a random failure, the cell electrodes eventually wear-out and fail due to repeated C/D cycles. Notice that the lower DoD is the longer the battery cycle life is, being expressed by the following approximate relationship:

$$\text{Number of Cycles until Failure} \times \text{DoD} = \text{Constant } K \qquad (2.23)$$

This equation holds for most of the electro chemistries, and the constant K decreases with temperature. Life consideration is a key parameter in system design and battery selection.

2.3 Flow Batteries and Special Battery Types

Flow batteries (redox flow cells) are a two-electrolyte system in which the chemical compounds used for energy storage are in liquid state, in solution with the electrolyte, overcoming the limitations of standard electrochemical batteries (lead-acid or nickel-cadmium) in which the electrochemical reactions create solid compounds that are stored directly on the electrodes on which they form, limiting the capacity of standard batteries. In same way, a flow battery is a type of rechargeable secondary battery in which energy is stored chemically in liquid electrolytes. They all operate in a similar fashion, two charged electrolytes are pumped to the cell stack where a chemical reaction occurs, allowing current to be obtained from the device when required. On the other hand, secondary batteries are using the electrodes as an interface for collecting or depositing electrons and as a storage site for the products or reactants associated with the battery's reactions. Consequently, their energy and power densities are set by the size and shape of the electrodes. Flow batteries, on the other hand are storing and releasing the electrical energy by means of reversible electrochemical reactions of the two liquid electrolytes, so in contrast with conventional batteries, their power and energy ratings are independent variables, being determined by the active area of the cell stack assembly and their storage capacity by the electrolyte quantity. Their electrolytes contain dissolved electro-active species that flow through a power cell that converts chemical energy to electricity. Simply stated, flow batteries are fuel cells that can be recharged. From a practical view, storage of the reactants is very important. The storage of gaseous fuels in fuel cells requires large, high pressure tanks or cryogenic storage that is prone to thermal self-discharge. There are some advantages of the flow battery over a conventional secondary battery, such as the system capacity is scalable by simply increasing the amount of solution, leading to cheaper installation costs as the systems get larger, the battery can be fully discharged with no ill effects and has little loss of electrolyte over time. Over the past 20 years, development and demonstration activities have centered around four principal electro chemistry combinations for flow batteries: vanadium-vanadium, zinc-bromine, polysulfide-bromide and zinc-cerium, although others are under development. Installations to date have principally used the vanadium redox and zinc-bromine. The polysulfide-bromide system was developed for grid-connected, utility-scale storage applications at power ratings from 5 MW upwards. This development program was stopped in 2003. Production of flow cell-based energy storage systems proceeds at a slow pace, via the activities of a relatively small number of developers and suppliers. There are two costs associated with flow batteries: the power cost (kW) and the energy cost (kWh), as they are independent of each other. Their major advantages are: high power capacity, long life time, while the main disadvantages are: low energy density and efficiency. Flow batteries can be used for all energy storage requirements including load leveling, peak shaving, and integration of renewable resources. However, PSB batteries have a very fast response time; it can react within 20 ms if electrolyte is retained charged in the stacks (of cells). Under normal conditions, flow batteries can charge or discharge power very fast, making them useful for frequency response and voltage control.

Figure 2.5 depicts a flow cell energy storage system, having two compartments, one for each electrolyte, physically separated by an ion exchange membrane, basically two flow loops. Electrolytes flow into and out of the cell through separate manifolds and undergo chemical reaction inside the cell, with ion (proton) exchange through the cell membrane and electron exchange through the external electric circuit. The electrolytes' chemical energy is turned into electrical energy and vice versa. All flow batteries work in the same way but are varying in the chemistry of electrolytes. Because the electrolytes are stored

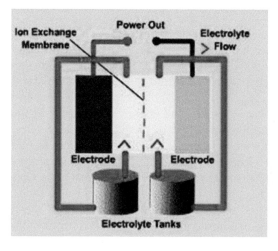

FIGURE 2.5
Flow battery schematic diagram.

separately and in large containers (low surface area to volume ratio), flow batteries show promise to have some of the lowest self-discharge rates of any energy storage technology available. The three types of flow batteries that are close to commercialization are: *vanadium redox, polysulfide bromide,* and *zinc bromide.* Production of flow cell-based energy storage systems proceeds at a slow pace, via the activities of a relatively small number of developers and suppliers. However, installations to date have principally used the vanadium redox and zinc-bromine. The polysulfide-bromide system was developed for grid-connected, utility-scale storage applications at power ratings from 5 MW upwards.

Zinc-bromine (Zn-Br) *batteries* are a type of redox flow battery, operating by using a pump system which circulates reactants through the battery. One manufacturer, ZBB Energy, builds Zn-Br batteries in 50 kWh modules made of three parallel connected 60-cell battery stacks. The battery modules are rated to discharge at 150 A at an average voltage of 96 V for 4 hours. The battery stack design allows for individual stacks to be replaced instead of the entire module. Zinc-bromine battery life time is rated at 2500 cycles. The electrochemical charging and discharging reaction is characterized as reversible and non-destructive, meaning it is capable of 100% depth of discharge. The Zn-Br battery uses less toxic electrolytes when compared to lead-acid batteries making them a more environmentally friendly choice. Zinc-bromine flow batteries lack technological maturity, so there are only a few examples of real installations. The overall chemical reaction during discharge:

$$Zn + Br_2 \rightarrow ZnBr_2, \quad \text{and with the theoretical voltage: } E_0 = 1.85 \text{ V} \tag{2.24}$$

During discharge product of reaction, the soluble zinc bromide is stored, along with the rest of the electrolyte, in the two loops and external tanks. During charge, bromine is liberated on the positive electrode and zinc is deposited on the negative electrode. Bromine is then mixed with an organic agent to form a dense, oily liquid poly-bromide complex system. It is produced as droplets and these are separated from the aqueous electrolyte on the bottom of the tank in positive electrode loop. During discharge, bromine in positive electrode loop is again returned to the cell electrolyte in the form of a dispersion of the poly-bromide oil. Vanadium redox batteries have the advantage of being able to store electrolyte solutions in non-pressurized vessels at room temperature. Vanadium-based redox flow batteries hold great promise for storing electric energy on a large scale (theoretically, they have infinite capacity).

Vanadium redox batteryflow battery (VRB) was pioneered at the University of New South Wales, Australia, and has shown potentials for long cycle life and energy efficiencies of over 80% in large installations. A vanadium redox battery is another type of a flow battery in which electrolytes in two loops are separated by a proton exchange membrane (PEM). The VRB uses compounds of the element vanadium in both electrolyte tanks. The electrolyte is prepared by dissolving of vanadium pentoxide (V_2O_5) in sulfuric acid (H_2SO_4). The electrolyte in the positive electrolyte loop contains $(VO_2)^+ - (V^{5+})$ and $(VO)^{2+} - (V^{4+})$ ions, the electrolyte in the negative electrolyte loop, V^{3+} and V^{2+} ions. Chemical reactions proceed on the carbon electrodes, while the reaction chemistry at the positive electrode is:

$$VO_2^+ + 2H^+ + e^- \leftrightarrow VO^{2+} + H_2O, \quad \text{with } E_0 = +1.00 \text{ V} \tag{2.25a}$$

And at the negative electrode, the reaction is:

$$V^{2+} \leftrightarrow V^{3+} + e^-, \text{ with } E_0 = -0.26 \text{ V} \tag{2.25b}$$

Under the actual VRB cell conditions, an open circuit voltage of 1.4 V is observed at 50% state of charge, while a fully charged cell produces over 1.6 V at open-circuit, and a fully discharged cell of 1.0 V. The extremely large capacities possible for VRB batteries make them well suited in large grid applications, where they could to average out the production of highly variable wind and/or solar power sources. Their extremely rapid response times make them suitable for uninterruptible power system applications, where they can be used to replace lead-acid batteries. Disadvantages of vanadium redox batteries are low energy density of about 25 Wh/kg of electrolyte, low charge efficiency (need to use pumps) and a high price.

Polysulfide-bromide (PSB) battery was developed in Canada for utility-scale storage applications at power ratings from 5 MW upwards. A PSB battery utilizes two salt solution electrolytes, sodium bromide (NaBr) and sodium polysulfide (Na_2S_x). PSB electrolytes are separated in the battery cell by a polymer membrane that only passes positive sodium ions. The chemical reaction at the positive electrode is:

$$NaBr_3 + 2Na^+ + 2e^- \leftrightarrow 3NaBr \tag{2.26a}$$

And at the negative electrode the reaction is:

$$2Na_2S_2 \leftrightarrow Na_2S_4 + 2Na^+ + 2e^- \tag{2.26b}$$

This technology is expected to attain energy efficiencies of approximately 75%. Although the salt solutions themselves are only mildly toxic, a catastrophic failure by one of the tanks could release highly toxic bromine gas. Nevertheless, the Tennessee Valley Authority released a finding of no significant impact for a proposed 430-GJ facility and deemed it safe. PSB batteries have a very fast response time; it can react within 20 ms if electrolyte is retained charged in the stacks (of cells).

Sodium-Sulfur (Na-S) battery was originally developed by the Ford Motor Company in the 1960s. The Na-S battery is one of the most advanced high temperature battery concepts. It contains sulfur at the positive electrode and sodium at the negative electrode. These electrodes are separated by a solid beta alumina ceramic. All Na-S batteries use a tube of Na-conducting beta-alumina ceramic as an electrode which is able to conduct sodium ions with molten Na on one side, and a sodium polysulfide melt on the other contained in a carbon felt for current collection. Through an electrochemical reaction, electrical energy is stored and released on demand. The cell voltage, about 2.08 V is derived from the chemical reaction between sodium and sulfur to produce sodium poly-sulfide. The theoretical energy density of these batteries is about 2.7×10^6 J/kg, much higher than the 0.61×10^6 J/kg of the lead-acid battery. Na-S batteries have an operating temperature between 300°C and 360°C. The primary manufacturer of Na-S batteries, NGK Insulators builds the batteries in 50 kW modules which are combined to make MW power battery systems. Na-S batteries exhibit higher power and higher energy density, higher columbic efficiency, good temperature stability, long cycle life, and low material costs. Their energy density is approximately three times that of traditional lead-acid batteries, with a high DC conversion efficiency of approximately 85%, making them an ideal candidate for the use in future DC distribution networks. Na-S batteries can be used for a wide variety of applications including peak shaving, renewable energy grid integration, power quality management, and emergency power units. They have the ability to discharge above their rated power, which makes them ideal to operate in both a peak shaving and power quality management environment. Lifetime is greater than 1000 cycles and is expected to target 2000 cycles for utility applications. The Na-S batteries have durability, cost and safety problems (sodium, being a very reactive metal), and while high operation temperature is another disadvantage. The reaction is given by:

$$\left. \begin{array}{l} Na \rightarrow Na^+ + e^- \\ \\ S + 2e^- \rightarrow S^= \end{array} \right\} \Rightarrow xNa + yS \rightarrow Na_xS_y \ (2.08 \ V) \tag{2.27}$$

Despite some quite significant progress, much remains to be done, especially in the areas of ceramic production technology, safety and durability but if these problems can be overcome the Na-S batteries could be very attractive for power system applications.

2.4 Fuel Cells

Fuel cells have been of interest for over 150 years as a potentially more efficient and less polluting technology for converting hydrogen and carbonaceous or fossil fuels into electricity compared to conventional thermal engines. A well-known and the first commercial application of the fuel cell has been the use of the hydrogen-oxygen fuel cell, using cryogenic fuels, in space vehicles. Use of the fuel cell in terrestrial applications has been developing slowly, but recent advances has revitalized interest in air-breathing systems for a variety of applications, including power system, load leveling, dispersed, distributed or on-site electric generation, transportation and electric vehicles. More recently, fuel cell technology has moved slowly toward portable applications, historically the domain of batteries, with power levels from less than 1 W to about 100 W, blurring the distinction between batteries and fuel cells. Fuel cell technology can be classified into two categories: (a) direct systems where fuels, such as hydrogen, methanol and hydrazine, can react directly in the fuel cell, and (b) indirect systems in which the fuel, such as natural gas or other fossil fuel, is first converted by reforming to a hydrogen-rich gas which is then fed into the fuel cell. Fuel cell systems can take a number of configurations depending on the combinations of fuel and oxidant, the type of electrolyte, the temperature of operation, the application, etc. There are different types of fuel cells classified according to their operation at low or high temperatures. Low-temperature fuel cells (up to about 100°C) are available and ready for mass production. They are used in applications where high-temperature fuel cell systems are not suitable, such as in commercial and residential power sources as well as in electric vehicles. The most compact systems are appropriate for powered electric vehicles and are available in the range of 5–100 kW. High temperatures (up to or about 1000°C) are typically used for industrial and large commercial applications and operate as decentralized stationary units of electric power generation. Because of such high operating temperatures, large heat dissipation is expected to be recovered and the integration with microturbines or as part of a micro-combined heat and power generation systems are very successful, with overall efficiencies of 70% or higher in such applications, representing a great contrast with respect to the energy generation in coal power plants, whose efficiency is up to 40%.

2.4.1 Fuel Cell Principles and Operation

Fuel cells are electrochemical devices that are producing electricity from paired oxidation or reduction reactions, being in some way batteries with flows/supplies of reactants in and products out. Fuel cells are hardly a new idea, being invented in about 1840, but they are really making their mark as a power source for electric vehicles, space applications and consumer electronics, only in the second part of twentieth century and the time is about to come. The fuel cells are distinguished from the secondary rechargeable batteries by their external fuel storage and extended life time. Basic energy conversion in fuel cells is summarized in the diagram of Figure 2.6, where chemical energy of the fuel is converted into electrical energy and heat energy (as one of the by-product, water being the other one), while the input energy is that produced during reactions at the electrodes.

Fuel cells have the advantages of high efficiency, low emissions, quite operations, good reliability and fewer moving parts (only pumps and fans to circulate coolant and reactant gases) over other energy generation systems. Their generation efficiency is very high in fuel cells, higher power density, lower vibration characteristics. A fuel cell is a DC voltage source, operating at about 1 V level or so. However, this might be set to change over the next 20 or 30 years. Certainly most of the major motor companies are spending very large sums of money developing fuel cell powered vehicles. The basic principle of the fuel cell is that it uses hydrogen fuel to produce electricity in a battery-like device, as discussed earlier. Figure 2.7 shows an example of fuel cell, based on the free energy, $\Delta G = -7.90 \times 10^{-19}$ J, for the reaction. The basic chemical reaction of this fuel cell is:

$$2H_2 + O_2 \rightarrow 2H_2O + Energy \tag{2.28}$$

The reaction products are thus water and energy. Because the types of fuel cell likely to be used in vehicles and mobile application work at quite lower temperatures (~85°C) there is no nitrous oxide produced by reactions between the components of the air used in the cell. A fuel cell vehicle could thus be described as zero-emission. Furthermore, because they

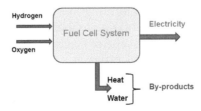

FIGURE 2.6
Fuel cell operation diagram.

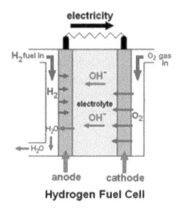

electricity

H_2 fuel in

H_2

OH^-

electrolyte

OH^-

O_2 gas in

O_2

H_2O

H_2O

anode   cathode

Hydrogen Fuel Cell

FIGURE 2.7
Schematic diagram of a hydrogen-oxygen fuel cell.

run on a fairly normal chemical fuel (hydrogen), very reasonable energies can be stored, and the range of fuel cell based vehicles is potentially quite satisfactory. Fuel cells are thus offering a real prospect of a silent zero-emission vehicle with a range and performance broadly comparable with IC engine vehicles. It is not surprising then that there have, for many years, been those who have seen fuel cells as a technology that shows great promise, and could even make serious inroads into the domination of the internal combustion engine. There are quite a few problems and challenges for fuel cells to overcome before they become a commercial reality as a vehicle power source or other power applications. The main problems and issues are listed here. Fuel cells are currently far more expensive than IC engines, and even hybrid IC/electric systems, so the cost reduction is a priority. Water management is another important and difficult issue with automotive fuel cells. The thermal management of fuel cells is actually rather more difficult than for IC engines. Hydrogen is the preferred fuel for fuel cells, but hydrogen is very difficult to store and transport. There is also the vital question of "where does the hydrogen come from" these issues are so difficult and important, with so many rival solutions. However, there is great hope that these problems can be overcome, and fuel cells can be the basis of less environmentally damaging transport.

We have seen that the basic principle of the fuel cell is the release of energy following a chemical reaction between hydrogen and oxygen. The key difference between this and simply burning the gas is that the energy is released as an electric current, rather that heat. How is this electric current produced?

To understand this we need to consider the separate reactions taking place at each electrode. These important details vary for different types of fuel cell, but if we start with a cell based on an acid electrolyte, we are considering the simplest and the most common type. At the anode of an acid electrolyte fuel cell the hydrogen gas ionizes, releasing electrons and creating H^+ ions (or protons):

$$2H_2 \rightarrow 4H^+ + 4e^- \tag{2.29}$$

During this reaction energy is released. At the cathode, oxygen reacts with electrons taken from the electrode, and H^+ ions from the electrolyte, to form water.

$$O_2 + 4H^+ + 4e^- \rightarrow 2H_2O \tag{2.30}$$

This reaction is built up with only two components, such as oxygen firs picking up electrons or first associating with hydrogen ions. Notice that at positive electrode, the reaction rate can be stimulated by a catalyst as usually in any fuel cell. Clearly, for both these reactions to proceed continuously, electrons produced at the anode must pass through an electrical circuit to the cathode, and H^+ ions must pass through the electrolyte. An acid is a fluid with free H^+ ions, and so serves very well this purpose. Certain polymers can also be made to contain mobile H^+ ions. The reactions given above may seem simple enough, but they do not proceed rapidly in normal circumstances. Also, the fact that hydrogen has to be used as a fuel can be a disadvantage. To solve these and other problems many different fuel cell types have been tried. The different types are usually distinguished by the used electrolyte, though there are always other important differences as well. The energy drop in free energy (Equation 2.3) is usually associated with reaction 2.28, G being expressed in terms of chemical potential of the H^+ ions dissolved in the electrolyte. According to the reactions of Equations (2.28) and (2.29), the chemical free energy is converted into electricity, and using Equation (2.3), this cell maximum electromotive force is:

$$E^0 = \frac{\Delta G}{nF} = 1.229 \text{ V} \tag{2.31}$$

In fuel cells, similar to batteries, the electrode reactions are surface phenomena, occurring at a liquid-solid or gas-solid interface and therefore proceed at a rate proportional to the exposed solid areas. For this reason porous electrode materials are used, often porous carbon impregnated or coated with a catalyst to speed the reactions. Thus, because of microscopic pores, the effective area of the electrodes is very large. Any phenomenon that prevents the gas from entering the pores or deactivates the catalyst must be avoided in order that a fuel cell to function effectively over longer periods. Because of the reaction rate-area relation, fuel cell current and power output increase with increased cell area. The surface power density (W/m^2) is an important parameter in comparing fuel cell designs, and the fuel cell power output can be scaled up by increasing its surface area. The electrolyte acts as an ion transport medium between electrodes. The passage rate of the positive charge through the electrolyte must match the rate of electron arrival at the opposite electrode to satisfy the physical requirement of electrical neutrality of the discharge fluids. Impediments to the ion transport rate through the electrolyte can limit current flow and the power output. Thus care must be taken in design to minimize the length of ion travel path and other factors that are retarding the ion transport. The theoretical maximum energy of an isothermal fuel cell (or other isothermal reversible control volume) is the difference in free energy (Gibbs) functions of the cell reactants and the products. The drop in free energy is mainly associated with Equation (2.30), expressing free energy of the chemical potential, e.g., of the H^+ ions dissolved in the electrolyte, which related to the maximum (open-circuit) voltage of a fuel cell through Equation (2.3). The maximum voltage of a fuel cell is about 1.23 V, as estimated in the example below. The maximum work output of which a fuel cell is able to perform is given by the decrease in its free energy, ΔG. The fuel cell conversion efficiency, η_{fc}, is defined as the electrical energy output per unit mass (or mole) of fuel to the corresponding heating value of the fuel consumed, the total energy drawn from the fuel, given by the maximum energy available from the fuel in an adiabatic steady-flow process, which the difference in the inlet and exit enthalpies, ΔH. The thermal fuel cell efficiency is expressed as:

$$\eta_{fc}(\text{thermal}) = \frac{\Delta G}{\Delta H} \tag{2.32}$$

Example 2.9: Determine the open-circuit voltage, maximum work and the thermal efficiency for a direct hydrogen-oxygen fuel cell at standard reference conditions (see Table 3, Appendix B). Consider two cases when the product is liquid water and vapor.

Solution: The maximum work of the fuel cell is the difference of the free energy of the products and the free energy of the reactants. The maximum work for the liquid-water product and the water-vapor product are, respectively:

$$W_{max}\left(\text{liquid}\right) = G_r - G_p(l) = 0 - (-237{,}141) = 237{,}141 \text{ kJ/kg} \cdot \text{mole}$$

$$W_{max}\left(\text{vapor}\right) = G_r - G_p(v) = 0 - (-228{,}582) = 228{,}582 \text{ kJ/kg} \cdot \text{mole}$$

The theoretical ideal one-circuit voltages, in the two cases are computed with Equation (2.3), as:

$$V_{max}\left(\text{liquid}\right) = \frac{W_{max}(l)}{nF} = \frac{237{,}141 \times 10^3}{2 \times 96400 \times 10^3} = 1.2299 \approx 1.23 \text{ V}$$

$$V_{max}\left(\text{vapor}\right) = \frac{W_{max}(v)}{nF} = \frac{228{,}582 \times 10^3}{2 \times 96400 \times 10^3} = 1.1855 \approx 1.19 \text{ V}$$

The change in the enthalpies for water-liquid product and waver-vapor product (Table 3 of Appendix C) is: 285,830 kJ/kg·mole and 241,826 kJ/kg·mole, respectively. The thermal maximum efficiencies in the two case, computed with Equation (2.32), are:

$$\eta_{fc}(\text{liquid}) = \frac{\Delta G}{\Delta H} = \frac{237{,}141}{285{,}830} = 0.83 \text{ or } 83\%$$

$$\eta_{fc}(\text{vapor}) = \frac{\Delta G}{\Delta H} = \frac{228{,}582}{241{,}826} = 0.95 \text{ or } 95\%$$

However, the actual efficiencies are lower than in this example, while the actual open-circuit voltage (potentials) of 0.7–0.9 V is typical for most of the fuel cell regardless the type, and losses or inefficiencies in fuel cells, often called polarizations, are reflected in a cell voltage drop when the fuel cell is under load. Three major types of polarization are: (a) the *ohmic polarization* is due to the internal resistance to the motion of electrons through electrodes and of ions through the electrolyte; (b) the *concentration polarization* due to the mass transport effects relating to diffusion of gases through porous electrodes and to the solution and dissolution of reactants and products; and (c) the *activation polarization*, which is related to the activation energy barriers for the various steps in the oxidation-reduction reactions at the electrodes. The net effect of these and other polarizations is a decline in terminal voltage with increasing current drawn by the load. This voltage drop is reflected in a peak in the fuel cell power-current characteristic. The high thermal efficiencies of the fuel cells are based on the operation of the fuel cell as an isothermal, steady-flow process. Although heat must be rejected, the transformation of chemical energy directly into electron flow does not rely on heat rejection in a cyclic process to a sink at low temperature as in a heat engine, which is why the Carnot efficiency limit does not apply. However, the inefficiency associated with the various polarizations and the incomplete fuel utilization tends to reduce overall cell conversion efficiencies to well below the thermal efficiencies predicted in the example above. Fuel cells, like other energy conversion systems, do not function in exact conformance with simplistic models or without inefficiencies.

Defining the voltage efficiency, η_V, as the ratio of the terminal voltage to the theoretical EMF, V_T/EMF, and the current efficiency, η_I, as the ratio of the cell electrical current to the theoretical charge flow associated with the fuel consumption, the overall fuel cell conversion efficiency is related to the thermal efficiency by this relationship:

$$\eta_{fc} = \frac{V_T}{EMF} \times \frac{I}{jN_{Rct}F} \times \eta_{th} = \eta_V \cdot \eta_I \cdot \eta_{th} \tag{2.33}$$

Example 2.10: The actual DC output voltage of hydrogen-oxygen fuel cell is 0.80 V. Assuming its current efficiency 1.0, what is the fuel cell actual efficiency?

Solution: From previous example the fuel cell thermal efficiency is 0.83, and the voltage efficiency is equal to $0.80/1.23 = 0.65$, then the actual fuel cell efficiency, Equation (2.31) is:

$$\eta_{fc} = \eta_V \cdot \eta_I \cdot \eta_{th} = 0.65 \times 1.0 \times 0.83 = 0.54 \text{ or } 54\%$$

In actual a number of factors and loss mechanisms tend to diminish the output voltage and the fuel cell efficiency. However, if the heat rejected in high-temperature fuel cells is recovered and not wasted, the overall fuel cell stack efficiencies as high as 80% are attainable.

2.4.2 Fuel Cell Types and Applications

The type of a fuel cell is primarily determined by the electrolyte that it employs, which, in turn determines the operating temperature, varying widely between fuel cell types. High temperature fuel cells operate above 600°C (or 1100°F), which are permitting spontaneous internal reforming of light hydrocarbon fuels, such as methane into hydrogen and carbon in presence of water. The reaction is taking place at anode over a nickel catalyst provided that adequate heat is all the time available. The internal reforming is eliminating the need of a separate fuel processor, allowing the use of other fuels than pure hydrogen. These features are leading to an overall efficiency increase of about 15%. Moreover, high temperature fuel cells also generate high-grade waste heat that can be used in downstream processes for cogeneration, further increasing the overall system efficiency. High-temperature fuel cells react easily and efficiently without the need of expensive noble metal catalysts (e.g., Platinum). However, the released energy amount by the electrochemical reaction decreases with temperature. High-temperature fuel cells also suffer of materials problems, because few materials can work for extended time periods in a chemical environment at higher temperatures without degradation. Furthermore the high-temperature operation does not lend itself easily to large-scale operations and is not suitable where quick start-up is required. As a result, high-temperature fuel cells are focusing on stationary applications where the efficiencies of internal reforming and cogeneration capabilities outweigh the disadvantages of material breakdown and slow start-up. Low temperature fuel cells operate below 250°C (480°F), which are preventing the internal reforming, and therefore require an external source of hydrogen, which reduces the system efficiency, while increasing the system cost. On the other hand, they exhibit quick start-up, suffer fewer materials problems and are easier to handle in transportation or building applications.

An important element of fuel cell design is that, similar to the large battery systems, are built from a large number of identical units or cells. Each has an open-circuit voltage on the order of 1 V, depending on the oxidation-reduction reactions taking place in the cell.

The fuel cells are usually built in sandwich-style assemblies called *stacks*, while the fuel and oxidant cross-flow through a portion of the stack. Electrically conducting bipolar separator plates serve as direct current transmission paths between successive cells. This modular type of construction allows research and development of individual cells and engineering of fuel cell systems to proceed in parallel. A fuel cell stack can be configured with many groups of cells in series and parallel connections to further tailor the voltage, current and power produced. The number of individual cells contained within one stack is typically greater than 50 and varies significantly with stack design. A *fuel cell system* consists of one or multiple stacks connected in series and/or parallel, and the auxiliary components whose types and structure depend on the fuel cell type and primary fuels used. The major accessories include thermal management (cooling) subsystem, fuel supply, storage and processing subsystems, and oxidant (typically air) supply, control and conditioning subsystems. The basic fuel cell stack components include the electrodes and electrolyte with additional components required for electrical connections and insulation and the flow of fuel and oxidant through the stack. These key components include current collectors and separator plates. The current collectors conduct electrons from the anode to the separator plate. The separator plates provide the electrical series connections between cells and physically separate the oxidant flow of one cell from the fuel flow of an adjacent cell. The channels in the current collectors serve as the distribution pathways for the fuel and oxidant. Often, the two current collectors and the separator plate are combined into a single unit called a bipolar plate. Electrically conducting bipolar separator plates serve as direct current transmission paths between successive stack cells. This modular structure allows research and development of individual cells and engineering of fuel cell systems to proceed in parallel. The preferred fuel for most fuel cell types is hydrogen. Hydrogen is not readily available, however, but the infrastructure for the reliable extraction, transport or distribution, refining and/or purification of hydrocarbon fuels is well established. Thus, fuel cell systems that have been developed for practical applications to date have been designed to operate on hydrocarbon fuels. In addition to the fuel cell system requirement of a fuel processor for operation on hydrocarbon fuels, a power conditioning, and for grid-connection of to supply an AC load an inverter is also needed. There are six major types of fuel cells being known or used in the stationary and mobile applications. All these fuel cell types have the same basic design as stated above, but with different electrolytes. These fuel cells are:

1. Alkaline Fuel Cells (AFCs);
2. Phosphoric Acid Fuel Cells (PAFCs);
3. Molten Carbonate Fuel Cells (MCFCs);
4. Solid Oxide Fuel Cells (SOFCs);
5. Direct Methanol Fuel Cells (DMFCs); and
6. Proton Exchange Membrane Fuel Cells (PEMFCs).

All these fuel cells, regardless the types require fairly pure hydrogen fuel to run. However, large amount of hydrogen gas is still difficult to produce transport and store. Therefore, a reformer is usually equipped inside the fuel cell stacks to generate hydrogen gas from liquid fuels such as gasoline or methanol.

Among these five types of fuel cell, PEMFC has the highest potential for widespread use. PEMFC is getting cheaper to manufacture and easier to handle. It operates at relatively low temperature when compared with other types of fuel cell. AFC systems have the highest efficiency and are therefore being used to generate electricity in spacecraft systems, however, it requires very pure hydrogen and oxygen to operate and thus the running cost is very expensive. As a result, AFCs are unlikely to be used extensively for general purposes, such as in vehicles and in buildings. In contrast, the MCFC and the SOFC are specially designed to be used in power stations to generate electricity in large-scale power systems. Nevertheless, there are still a lot of technical and safety problems associated with the use of these fuel cells (MCFC and SOFC) in the long-term applications. Apart from these fuel cell types, a new type of fuel cell, the direct methanol fuel cell (DMFC), being under vigorous on-going research is coming on the market. This type of fuel cell has the same operating mechanism as PEMFC, but instead of using pure hydrogen, it is able to use methanol directly as the basic fuel. A reformer is therefore not essential in this fuel cell system to reform complex hydrocarbons into pure hydrogen. Several companies around the world are presently working on DMFC to power electronic equipment. The DMFC appears to be the most promising alternative to replace the battery used in portable electronics. Although the development is fast, the fuel cells have not yet reached its potential level of commercial success due to high material costs (Pt electrode) and market barriers. To tackle the electrode problem, platinum or platinum-based nanoparticles are coated on the surface of carbon black. This porous electrode used in fuel cells can give higher current densities than geometric plate electrode. More importantly, it can reduce the quantity of platinum used.

AFCs, first type of fuel cells used in applications, are one of the most efficient fuel cells, giving the best performances among all fuel cell types under the same or similar operating conditions when running on pure hydrogen and oxygen. They are employing potassium hydroxide solutions (KOH) as electrolytes, immobilized in asbestos matrix. As a result, the AFCs operate at high pressure to prevent the boiling and depletion of the liquid electrolyte, so severe operating conditions of high temperature and high pressure are imposing extremely strict cell material requirements (electrodes made of gold, platinum or silver either these precious metals are not necessary for the electrochemical reactions) to withstand the extreme corrosive oxidizing and reducing environments of the cathode and anode. Other problems associated to AFCs are the concerns for safety and reliability, the extreme intolerance to CO_2, their complex water management requirement and short lifetime. However, the AFCs, besides the low temperature operation, need little or no expensive platinum catalyst, are relative easy to operate, fast start-up time, low weight and volume. They are operating at temperatures between 100°C and 250°C (211°F to 482°F). Higher temperature AFCs are using higher concentrated electrolyte solutions, while lower temperatures are using diluted electrolyte solutions. Alkaline fuel cells operate using pure hydrogen, free of carbon oxides, having the following overall cell anode and cathode reactions:

$$H_2 + 2OH^- \Rightarrow 2H_2O + 2e^- \text{ (anode)}$$

$$\frac{1}{2}O_2 + H_2O + 2e^- \Rightarrow 2OH^- \text{ (cathode)}$$

(2.34)

Phosphoric acid fuel cell technology has been in commercial use for longest time among all other fuel cell technologies, unlike AFCs which were targeted for terrestrial applications. PAFC employs phosphoric acid (often in silicon carbide ceramic matrices). Phosphoric acid fuel cells use an electrolyte that conducts hydrogen ions (H^+) from the anode to the cathode. Their electrolyte is composed of liquid phosphoric acid within a silicon carbide matrix material. PAFCs run at temperatures at about 150°C–205°C (300°F–400°F), which is increasing the catalyst activity and also reducing and preventing the oxygen to adsorbed by the phosphate anions, which can reduces the catalytic performance. PAFCs accept as fuel either direct hydrogen or a mixture of hydrogen, carbon dioxide, water, carbon monoxide and possible nitrogen, produced from conversion of fossil fuels into hydrogen and carbon monoxide. The PAFC basic components are the electrodes consisting of finely dispersed platinum catalyst or carbon paper, SiC matrix holding the phosphorous acid, and a bipolar graphite plate with flow channels for fuel and oxidant. Their advantages include: are tolerant of carbon dioxide (up to 30%), have stable electrolyte characteristics with low volatility even at operating temperatures as high as 200°C, and operate at low temperature, but at higher temperatures than other low temperature fuel cells. Thus, they produce higher grade waste heat that can potentially be used in cogeneration applications. PAFC's major disadvantages are: can tolerate only about 2% carbon monoxide and only about 50 ppm of total sulfur compounds, are using a corrosive liquid electrolyte at moderate temperatures, resulting in material corrosion problems, have a liquid electrolyte, needing liquid handling problems (the electrolyte slowly evaporates over time), allowing product water to enter and dilute the electrolyte, are large and heavy, cannot auto-reform hydrocarbon fuels, and have to be warmed up before they are operated or be continuously maintained at their operating temperature. In PAFCs, hydrogen reacts with oxygen, and the cell overall reaction is the one of given in Equation (2.28) (the fuel cell basic reaction). Thus, the fuel cell produces water that accumulates at the cathode. This product water must be continually removed to facilitate further reaction. For 5 kW or similar power applications, passive cooling (heat sinks and fins) may be sufficient to cool the system. The high temperatures also have the advantage of transporting the product water as steam instead of as liquid. PAFCs have about a third of the performance of modern polymer electrolyte membrane fuel cells (PEMFCs), in terms of power per membrane area in W·cm^{-2}, because PEMFCs have stronger acids in their electrolytes and because the thinner polymer membranes have much lower ohmic losses. Finally, PAFCs must be kept above 45°C even when not in use because below this temperature the acid solidifies and expands, risking damage to electrodes or silicon carbide matrix. These two reasons of low power density and finicky temperature conditions explain why, after a few bus demonstrations, PAFCs have been relegated to stationary applications.

Solid oxide and molten carbonate fuel cells are higher temperature cell types, suitable for applications above 600°C, so more exotic materials can and must be used for the electrolyte. With an SOFC, the electrolyte is a ceramic oxide and conduction takes place by oxygen ion (O^{2-}) hopping through the electrolyte. On the other hand, the MCFCs use molten ionic alkali mixtures s electrolytes, immobilized in porous lithium aluminate matrixes, while the conducting species are carbonate ions. Lithiated nickel oxide is the current choice for the cathode, and nickel, cobalt or cooper is used for anode. Both of these fuel cells find their utility in stationary power applications, where efficiency gains can be realized by using the exhaust stream and its high grade waste heat to drive a gas turbine bottoming cycle or provide cogenerated heat. Their high operation temperatures offer the possibility of internal reforming, where natural gas and steam are introduced directly into the fuel cell and steam-reforming and water–gas cleanup occur automatically. SOFCs can even use carbon monoxide as fuel without reforming. The high temperatures mean that noble metal catalysts are not needed, but also bring with them their own materials problems. Neither of these fuel cells has the high power density needed for vehicle power. Molten carbonate fuel cells operate at about 1200°F or 650°C. Each cell can produce up to between 0.7 and 1.0 V. The MCFC advantages are: support spontaneous internal reforming of light hydro carbon fuels, generate high-grade waste heat, have fast reaction kinetics (react quickly), have high efficiency and do not need noble metal catalysts. Their disadvantages include: require the development of suitable materials that are resistant to corrosion, are dimensionally stable, have high endurance and lend themselves to fabrication, have a high intolerance to sulfur, the anode in particular cannot tolerate more than 1–5 ppm of sulfur compounds (primarily H_2S and COS) in the fuel gas without suffering a significant performance loss, have a liquid electrolyte, which introduces liquid handling problems and require a considerable warmup period. Corrosion is a particular problem and can cause nickel oxide from the cathode to dissolve into the electrolyte, loss of electrolyte, deterioration of separator plates, and dehydration or flooding of the electrodes. All of these corrosion effects result in a decline in performance, limit cell life and can culminate in cell failure. Use of a platinum catalyst overcomes some of these problems, but eliminates an important cost-saving advantage. Dimensional instability can cause electrode deformation that alters the active surface area and may cause loss of contact and high resistances between components. Molten carbonate fuel cells can operate using pure hydrogen or light hydrocarbon fuels. When a hydrocarbon, such as methane, is introduced to the anode in the presence of water, it absorbs heat and undergoes a steam reforming reaction:

$$CH_4 + H_2O \Rightarrow 3H_2 + CO \qquad (2.35)$$

When using other light hydrocarbon fuels, the number of hydrogen and carbon monoxide molecules may change but in principle the same products result. The overall cell reactions at the anode and cathode are:

$$2H_2 + O_2 \Rightarrow 2H_2O$$

$$CO + \frac{1}{2}O_2 \Rightarrow CO_2 \qquad (2.36)$$

Notice that the first reaction is the hydrogen reaction and occurs regardless of fuel, while the second one is the carbon monoxide reaction and occurs only when using a hydrocarbon fuel, so the fuel cell produces water, regardless of fuel, and carbon dioxide if using a hydrocarbon fuel. Both product water and carbon dioxide must be continually removed from the cathode to facilitate further reaction. MFCFs offer higher fuel-to-electricity efficiencies than lower temperature fuel cells, approaching 60%, and greater fuel flexibility, while the high operating temperatures are making them candidates for cogeneration applications, in which the overall thermal efficiencies can approach 85%. The MCFCs main research efforts are focused on the lifetime and endurance increases, and the reduction of the long-term performance degradation. The main determining factors are electrolyte loss, cathode dissolution, electrode creepage and sintering, separator-plate corrosion, and catalyst poisoning for internal reforming.

SOFCs, an emerging major high temperature alternative, resulting in fast electrochemical kinetics and no need for noble metal catalyst are often referred as the third fuel-cell generation. SOFCs are all-solid-state systems, including the electrolyte, operating at 1000°C for adequate ionic and electronic conductivity of various cell components. The fuel may be gaseous hydrogen, H_2/CO mixture, or hydrocarbons because of high temperature operation makes in situ reforming of hydrocarbons with water vapor possible. Solid oxide fuel cells use an electrolyte that conducts oxide (O^{2-}) ions from the cathode to the anode. This is the opposite of most types of fuel cells, which conduct hydrogen ions from the anode to the cathode. The electrolyte is composed of a solid oxide, usually zirconia (stabilized with other rare earth element oxides like yttrium), and takes the form of a ceramic. Common configurations include tubular and flat (planar) designs. The designs differ in the extent of dissipative losses within cells, in the manner of sealing between the fuel and oxidant channels, and in the manner that cell-to-cell electrical connections are made in a stack of cells. Metals such as nickel and cobalt can be used as electrode materials. Each cell can produce between 0.8 and 1.0 V. The advantages of solid oxide fuel cells are that they: support spontaneous internal reforming of hydrocarbon fuels, since oxide ions, rather than hydrogen ions travel through the electrolyte, the fuel cells can in principle be used to oxidize any gaseous fuel, operate equally well using wet or dry fuels, generate high-grade waste heat, have fast reaction kinetics, have very high efficiency, can operate at higher current densities than molten carbonate fuel cells, have a solid electrolyte, avoiding problems associated with handling liquids, can be fabricated in a variety of self-supporting shapes and configurations, do not need noble metal catalysts. Their disadvantages are: require the development of suitable materials that have the required conductivity, remain solid at high temperatures, are chemically compatible with other cell components, are dimensionally stable, have high endurance and lend themselves to fabrication, few materials can operate at high temperatures and re-main solid over long periods of time. Furthermore, the selected materials must be dense to prevent mixing of the fuel and oxidant gases, and must have closely matched thermal expansion characteristics to avoid delamination and cracking during thermal cycles, have a moderate intolerance to sulfur, still fabrication processes are not fully set, and not yet a mature technology. Solid oxide fuel cells are more tolerant to sulfur com-pounds than are molten carbonate fuel cells, but overall levels must still be limited to 50 ppm. This increased sulfur tolerance makes these fuel cells attractive for heavy fuels. Excess sulfur in the fuel decreases performance. Combining the anode and cathode reactions, the overall cell reactions are the one of the Equation (2.36), with the same operation specifications.

A direct methanol fuel cell is an exception to the rule that fuel cells are categorized by their electrolytes; while in this case, it is the fuel that defines the fuel cell. Dilute methanol is flowed through the anode as the fuel and broken down to protons and electrons and water. Methanol is chosen because it is one of the few widely available fuels that is electro-active enough to use in a fuel cell, ethanol can also be used but with poorer efficacy, due to its poorer electrical activity. Because a this type of fuel cell run directly on liquid fuel would offer dramatic advantages in overall system density since neither low-density hydrogen nor bulky reformers would be needed. Methanol can be made relatively easily from gasoline or biomass, and although it only has a fifth the energy density of hydrogen by weight, as a liquid it offers more than four times the energy per volume when compared to hydrogen at 250 atmospheres. In a direct methanol PEM fuel cell, the cells are supplied with a liquid mixture of methanol and water at the anode, and air at the cathode. At 266°F or 130°C, a noble catalyst immediately decomposes the methanol according to the reaction:

$$CH_3OH + H_2O \Rightarrow 6H^+ + CO_2 + 6e^- \qquad (2.37a)$$

Oxygen, from the air, ionizes and reacts with the hydrogen to form water:

$$\frac{3}{2}O_2 + 6e^- + 6H^+ \Rightarrow H_2O \qquad (2.37b)$$

Combining the anode and cathode reactions, the overall cell reaction are resulting in pure water and carbon dioxide. The technology is still in its infancy, but holds great promise for the future. One major problem is that the oxidation of methanol produces intermediate hydrocarbon species which poison the electrode. The other major problem is that the exchange current for methanol is much lower than that for hydrogen, about six orders of magnitude, meaning that the oxidation of methanol at the anode becomes as slow as the oxygen electrode reaction, and large over-potentials are required for high power output. Also, there is high crossover of the methanol through the electrolyte, meaning that the fuel molecules diffuse directly through the electrolyte to the oxygen electrode. As much as 30% of the methanol can be lost this way, severely compromising power. The cell voltage is about 1.185 V.

The polymer electrolyte membrane fuel cells, known also as solid polymer (electrolyte) fuel cells or proton exchange membrane fuel cells (PEMFCs), are the most elegant type in terms of design and operation modes. Since they have the ability to deliver such high power densities at this temperature they can be made smaller which reduces overall weight, cost to produce and specific volume, while the use of an immobilized electrolyte membrane leads the simplification in the production process that in turn reduces corrosion, this provides for longer stack life. This immobilized proton membrane is really just a solid-state cation transfer medium. Proton exchange membrane (PEM) (or "solid polymer") fuel cells are currently the most promising type of fuel cell for automotive use and have been used in the majority of prototypes built to date. The PEM is named for the solid-state exchange membrane that separates its electrodes. PEM fuel cells use an electrolyte that conducts hydrogen ions (H^+) from the anode to the cathode, and the electrolyte is composed of a solid polymer film that consists of a form of acidified Teflon. In fact this membrane is just a hydrated solid that promotes the conduction of protons. Although many different types of membranes are used, by far that most common is Na on. Other types of membranes being researched are: polymer-zeolite nano-composite PEM, sulfonated poly-phosphazene membranes and phosphoric acid-doped poly-bisbenzoxazole high temperature ion-conducting membrane. Although the Na on membrane is so commonly used it is considered an industry standard, and all new membranes are compared to it. The Na on layer is essentially a carbon chain, which has a Fluorine atom layer attached to it. This is considered Teflon. A branch is formed off of this basic chain (also made of carbon atoms surrounded by Fluorine), the frequency of these side chains are reflected in the types of Na on. The side chains increase the hydrophilic effect of the material. This property of the Na on allows for up to 50% increase in dry weight. PEM fuel cells operate at lower temperatures, typically in the range of 70°C–90°C (160°F–195°F) and a pressure of 15–30 psig. Each cell can produce up to about 1.10 V. The membrane allows for the transfer of protons and thus permits the general fuel cell process, which is similar to the PAFC. Hydrogen at the anode separates the electron and proton, freeing them to travel throughout the fuel cell. The electron flows externally, while the proton travels though the conductive membrane to the cathode. The major advantages of PEM fuel cells are: they are tolerant of carbon dioxide (PEM fuel cells can use un-scrubbed air as oxidant and reformate as fuel), operate at lower temperatures, which are simplifying materials issues, and provide for quick start-up and increase safety, use a solid, dry electrolyte, which eliminates liquid handling, electrolyte migration and electrolyte replenishment problems, use a non-corrosive electrolyte (pure water operation minimizes corrosion problems and improves safety), have high voltage, current and power density, operate at low pressure which increases safety, have good tolerance to differential reactant gas pressures, are compact, robust and rugged, have relatively simple mechanical design, and use stable materials of construction. Their major disadvantages include: can tolerate only about 50 ppm carbon monoxide, can tolerate only a few ppm of total sulfur compounds, need reactant gas humidification, use an expensive platinum catalyst and use an expensive membrane that is difficult to work with. Notice that the humidification is energy intensive and increases the complexity of the system. The use of water to humidify the gases limits the operating temperature of the fuel cell to less than water's boiling point and therefore decreases the potential for co-generation applications. In PEM fuel cells hydrogen with oxygen are reacting, the anode, cathode and the overall cell reactions are the ones described in Equations (2.28) through (2.30). Thus, this type of fuel cell produces water that accumulates at the cathode. This product water must be continually removed to facilitate further reaction. Notice that PEM fuel cells can also run using methanol fuel directly, rather than hydrogen. Although the energy released during this reaction is less than when using pure hydrogen, it results in a much simpler fuel storage system and circumvents the need to produce hydrogen.

Biological fuel cells are one type of genuine fuel cell that does hold promise in the very long term and can represent breakthrough scientific and commercial advances in fuel cell technologies. These would normally use an organic fuel, such as methanol or ethanol. However, the distinctive "biological" aspect is that enzymes, rather than conventional "chemical" catalysts such as platinum, promote the electrode reactions. Such cells replicate nature in the way that energy is derived from organic fuels. However, this type of cell is not yet anywhere near commercial application, and is not yet suitable for detailed consideration in an application-oriented book such as this. However, the biological fuel cell should be distinguished from biological methods for generating hydrogen, which is then used in an ordinary fuel cell, as discussed in the next chapter section.

The applications of fuel cells vary depending of the type of fuel cell to be used. Ones of the characteristic essentials of a fuel cell are very simple, with few if any moving parts. This can lead to highly reliable and long-lasting systems. Fuel cells are very quiet energy conversion equipment, even those with extensive extra fuel processing equipment. This is very important in both portable power applications and for building and local power generation in combined heat and power schemes. Since fuel cells are capable of producing power anywhere in the 1 W to 10 MW range they can be applied to almost any application that requires power. On the smaller scale they can be used in personal electronics, such as cell phones, personal computers, laptops, tablets and any other type of personal electronic equipment. In the 1–100 kW range a fuel cell can be used to power vehicles, both domestic and military, public transportation is also a target area for fuel cell application, along with any building and industrial energy system applications. And finally, in the 1–10 MW or even higher range fuel cells can be used to convert energy for distributed power applications (microgrids, grid power quality improvements, supply security, load shedding). Since fuel cells can be used anywhere in the power spectrum their development will have an immediate impact in their prospective power range. One of the major applications for the fuel cell in the future will likely be that of domestic and public transportation. The fuel cells are well adapted to this application because use of fuel cells are reduce the design complexity of a vehicle or a building energy system or improve the overall system efficiency through cogeneration capabilities. In the low scale range the fuel cell has a great advantage over batteries in that they do not need to be recharged, only fueled, and they have much higher power densities than current commercialized batteries. Since they can provide more power per area the cell can be smaller while applying the same power, thus saving space considerably. In a large-scale setting the fuel cell can be used to assist in increasing the efficiency of the current turbine power plant. By using the hot exhaust from the fuel cell and transferring it to a turbine-power cycle the overall practical efficiency of the system can reach up to 80%.

One of the distinct features of fuel cells, regardless the type, is the *scalability*. The unique characteristic of fuel cells is that their high efficiency and other attributes are nearly unaffected by the size of the plant. That means fuel cells are scalable to all sizes with, more or less, the same high efficiency, low emissions and costs. In addition, modular installations of fuel cells can help them to match load and increase their reliability. Most fuel cells, especially high operating temperature ones, can be used in hybrid systems to produce further electricity and/or in cogeneration systems to produce heating and/or cooling as well as electricity. Moreover, at least some types of fuel cells have demonstrated the following characteristics: fast response to load changes, unattended operation, good off-design load operation, reliability and high availability. All these attributes and characteristics make fuel cells ideal candidates for some major power applications. However, before this commercialization can be realized, some significant improvements are required. The most important barrier is cost. In order to reduce the cost of fuel cells, new construction methods and materials must be developed. Mass production and the economy of scale can reduce cost significantly, but some mass markets have to be in place to support it. Also for each application, suitable durability, endurance, reliability, longer operation periods, specific power and power density need to be achieved, especially for high temperature fuel cells. This includes transient operation and operation in extreme ambient conditions. Other obstacles to overcome can be enumerated as follows: lack of familiarity of markets with fuel cell technology (especially the power generation industry) and the lack of hydrogen production, storage and distribution infrastructure for hydrogen-fueled fuel cells. Fuel cell technology is highly multidisciplinary and the development requires engagement and embracement of most of the engineering fields, from electrochemistry to manufacturing and from thermodynamic to material science and control, and is experiencing a tremendous growth.

2.4.2.1 PEM Fuel Cell Stack Construction and Design Considerations

Individual fuel cells have a maximum output voltage on the order of 1 V. For most practical fuel cell applications, unit cells must be combined in a modular fashion into a cell stack to achieve the voltage and power output level required for the application. Generally, the stacking involves connecting multiple unit cells in series via electrically conductive interconnects, while different stacking arrangements have been developed. Different designs of fuel cell stacks use fuel cells of varying dimensions and in varying quantities. Though a wide range of fuel cell geometries has been considered, most fuel cells under development now are either planar (rectangular or circular) or tubular (either single- or double-ended and cylindrical or flattened). The most common fuel cell stack design is the so-called planar-bipolar arrangement (Figure 2.8 depicts a PEMFC stack). Individual unit cells are electrically connected with interconnects. In this design, the fuel cell's flat plate electrolyte and electrodes are located parallel to each other and the individual cells are connected in series (bipolar plates).

The planar fuel cells enjoy high power density and their fabrication is simpler. But their structural integrity and sealing at high operating temperatures are serious challenges. In order to distribute air and fuel evenly within each cell, flow channels can be integrated into the interconnections. Also, there are two more functions for the interconnections: (a) to separate fuel and oxidant of two adjacent unit cells and (b) to provide electrical connection between adjacent unit cells. In terms of the flow direction of the reductant and oxidant, there are three common configurations: the co-flow, in which the two streams are parallel and in the same direction, the counter-flow, in which the two streams are parallel and in opposite directions, and cross-flow, in which the two streams, are perpendicular to

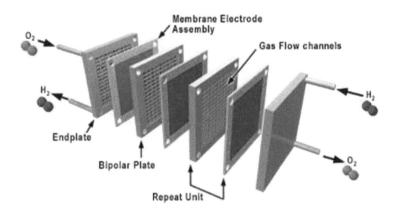

FIGURE 2.8
Planar fuel cell stack diagram.

each other. Also, manifolds are required to distribute fresh air and fuel between cells and collect depleted streams from cells. These manifolds can be either external or integrated in the fuel cell stack design. Because of the configuration of a flat plate cell, the interconnect becomes a separator plate with two functions: (1) to provide an electrical series connection between adjacent cells, specifically for flat plate cells and (2) to provide a gas barrier that separates the fuel and oxidant of adjacent cells. In many planar-bipolar designs, the interconnections also include channels that distribute the gas flow over the cells. The planar-bipolar design is electrically simple and leads to short electronic current paths (which helps to minimize cell resistance). Physically, each fuel cell consists of a membrane electrode assembly (MEA), which consists of the anode, cathode, electrolyte and catalyst, sandwiched between two flow field plates made of graphite. The MEA is the heart of the fuel cell. The MEA consists of a solid polymer electrolyte membrane sandwiched between two porous carbon electrodes. A platinum catalyst is integrated between the membrane and the electrodes. The electrode assemblies often include integral seals where they contact adjacent components. The MEA plates channel the fuel and air to opposite sides of the MEA. The MEA and flow field plates are presented in greater detail in the next section. Coolant is used to regulate the fuel cell reaction temperature. To facilitate this, cooling plates are placed between each fuel cell. These cooling plates channel the coolant past the fuel cells to absorb or supply heat as required. Seals between the graphite plates ensure that the oxidant, fuel and coolant streams never mix within the fuel cells. Electrical endplates are placed at either end of the series of flow field plates. These endplates are connected to the terminals from which the output power is extracted, and typically include the fluid and gas interface connections. The entire sequence of plates is held together by a series of tie rods or other mechanical means. The electrodes provide the interface between the reactant gases and electrolyte. As such they must allow wet gas permeation, provide a reaction surface where they contact the electrolyte, be conductive to the free electrons that flow from anode to cathode, and be constructed of compatible materials. A catalyst, often an expensive fuel cell component, being a noble metal or rare earth is added to the surface of each electrode where it contacts the electrolyte in order to increase the rate at which the chemical reaction occurs. A catalyst promotes a chemical reaction by providing ready reaction sites but is not consumed in the process. Humidification of the reactant gases is an important aspect of PEM fuel cell operation. Without adequate humidification,

ion conduction cannot occur and fuel cell damage can result. The amount of water that a gas can absorb is highly dependent on the humidification temperature, particularly at low pressure. Hotter gases can hold more water than colder gases. Since the goal of humidification is to saturate the reactant gases with as much water as possible, the gases must be humidified at or near the fuel cell operating temperature.

In solid state electrolyte fuel cells, in order to provide mechanical strength for a planar cell, each cell should be self-supported or supported by an external support. In a self-supported cell, either the electrodes or electrolyte should be thick enough to support other components. The three types of self-supported fuel cells are electrolyte supported, anode supported, and cathode supported. The problem with electrolyte supported cells is the high ionic resistance in the electrolyte due to its greater thickness, requiring a high operating temperature of the system. In electrode supported cells, due to their thinner electrolyte and lower ionic resistance, the operating temperature can be reduced significantly. Alternatively, cells can be externally supported by the interconnections, which results in stronger structure and thinner cell components, but cell support requirements may impose some limitations in flow channel design. Recently, anode supported fuel cells have been receiving the most attention due to their better thermal and electrical conductivity, mechanical strength, and minimal chemical interaction with the electrolyte. In tubular fuel cell design, a whole cell is in the form of a tube, the cathode being the inner layer, and the electrolyte and anode are deposited on top of the cathode. This configuration is mainly used in solid oxide fuel cells. This design solved the sealing problem and provides excellent mechanical integrity for the stack. In this design, the electrodes and electrolyte are in the form of different layers on a tube. The latest development in this design is flat-tube design, which is less expensive and has higher power density. This design is similar to tubular design only with flattened tubes which incorporates ribs inside the cell. The ribs reduce the current path, like a bridge for electrons, decreasing the cell internal resistance. They can also help to reduce the thickness of cathode and its over-potential. Another innovative design is the integrated-planar SOFC (IP-SOFC), which is a combination of the two aforementioned designs. This type of SOFC enjoys the lower fabrication cost of the planar SOFC and thermal-mechanical strength of the tubular SOFC. However, the tubular SOFCs suffer both low specific power density and volumetric power density due to longer current path and higher electrical resistance. Also, their fabrication cost is high, and there is not much potential to reduce this cost. Last but not least, due to the relatively high thickness of the electrolyte, their operating temperature is high. These problems are partially solved in planar SOFCs.

Practical fuel cell design focuses on achieving a high power output per area of membrane, scaling the active membrane area to a useful size, and making the overall stack suitably compact for its intended use. Critical areas of concern are seals, flow field pattern tolerances and cell alignment. As with any commercial product, the resulting design must be reliable, easy to be implemented in the manufacture process, economically viable and have a significant operating life time. The fuel cell structure and design is primarily determined by its type and often application. On some fuel cell stacks, humidifiers are integrated into the stack itself. On other fuel cell stacks, humidifiers are separate, external components. Internal humidifiers draw water directly from the stack cool-ant stream and results in a simple, well integrated system with excellent temperature matching characteristics. External humidifiers are most commonly of either a membrane or a contact design. External humidifiers draw water either from the stack cool-ant stream or from a separate humidification water circuit. Each fuel cell stack or system include power conditioning and control units. On all but the smallest fuel cells the air and fuel will need to be circulated through them stack using pumps or blowers. Often compressors will be used, which will sometimes be accompanied by the use of intercoolers, as in internal combustion engines. Practical fuel cell systems require additional equipment to regulate the gas and fluid streams, provide lubrication, operate auxiliary equipment, manage the electrical output and control the process. The direct current (DC) output of a fuel cell stack will rarely be suitable for direct connection to an electrical load, and so some kind of power conditioning is nearly always needed.

2.4.3 Actual Fuel Cell Operation

The efficiency of fuel cells is often touted as one of the primary benefits of the technology. Although this is true in principle, it is important to distinguish between fuel cell stack efficiency and fuel cell system efficiency. Fuel cell stack efficiency is usually taken to be the mean the actual efficiency of the cell electrochemical reaction, as discussed earlier in this chapter. Fuel cell system efficiency relates to the overall performance of a fuel cell system. A fuel cell stack can only operate if provided with pressurized air and hydrogen and flushed with coolant. Auxiliary fuel cell system equipment are reducing the cell output power and the system overall efficiency. In an ideal world, the theoretical optimum fuel cell voltage of 1.229 V would be realized at all operating currents. In reality fuel cells achieve their highest output voltage at open circuit (no load) conditions and the voltage drops off with increasing current draw. This is known as polarization and is represented by the cell polarization curve, the cell voltage vs. current diagram. The polarization curve characterizes the cell voltage as a function of current. The current, in turn, depends on the size of the electrical load placed across the fuel cell. In essence the polarization curve shows the electrochemical efficiency of the fuel cell at any operating current since the efficiency is the ratio of the actual cell voltage divided by the theoretical maximum of 1.229 V. The characteristics of the cell voltage-current density graphs result from four major irreversibilities, leading to the departure of the ideal cell behavior and output voltage. Losses that are taking place in an actual fuel cell, each having a different effect on the theoretical voltage of the fuel cell, are listed below. *Activation losses*, these losses are caused by the slowness of the reaction taking place on the surface of the electrodes. A proportion of the voltage generated is lost in driving the chemical reaction that transfers the electrons. *Ohmic losses* are referring to the voltage drop due to the resistance to the flow of electrons through the material of the electrodes. This loss varies linearly with cell current density. *Concentration losses* are the ones that result from the change in concentration of the reactants at the surface of the electrodes as the fuel is used. *Fuel crossover losses* are losses that result from the waste of fuel passing through the electrolyte and electron conduction through the electrolyte. This loss is typically small, but can be more important in low temperature fuel cells. The mass-transport losses that occur due to mass transport/concentration problems are those directly related to the fuel pressure issues. If the hydrogen is being used at a very vigorous rate at the anode then the partial pressure of the hydrogen drops, thus slowing the reaction rate. This is also the same case that occurs at the cathode with oxygen. The net effects are the cell voltage reduction. Ohmic losses, additional losses due to the electrical resistance of the electrodes, and the resistance to the flow of ions in the electrolyte, are the simplest to understand and to model are also affecting the overall fuel cell, stack or system efficiency. The activation polarization efficacy is customarily expressed by a semi-empirical equation, called Tafel equation:

$$\eta_{act} = \frac{RT}{\alpha nF} \ln\left(\frac{i}{i_0}\right)$$

(2.38)

Here i is the rate of mass transport at the electrode surface, α is the electron transfer coefficient of the reaction at the electrode, and i_0 is the exchange current density. If all the losses that we have looked at, activation, fuel crossover, mass transport and ohmic losses, are combined, then the actual operational graph of a fuel cell can be produced.

The thermal efficiency of an energy conversion device or system is defined as the amount of useful energy produced relative to the change in stored chemical energy (commonly referred to as thermal energy) that is released when a fuel is reacting with an oxidant. In the previous chapter section we defined the theoretical efficiency of a fuel cell. The efficiency of an actual fuel cell can be expressed in terms of the ratio of the operating cell voltage to the ideal cell voltage. The actual cell voltage is less than the ideal cell voltage because of the losses associated with cell polarization and the Joule loss. The thermal efficiency of the fuel cell can then be written in terms of the actual cell voltage and can be expressed as:

$$\eta_{th} = \frac{\text{Useful Output Energy}}{\Delta H} = \frac{0.83 \times V_{cell}I}{E_{ideal}I} = \frac{0.83 \times V_{cell}}{E_{ideal}} \tag{2.39}$$

As mentioned earlier, the ideal voltage of a fuel cell operating reversibly with pure hydrogen and oxygen in standard conditions is 1.23 V. Thus, the thermal efficiency of an actual fuel cell operating at a voltage of V_{cell}, based on the higher heating value of hydrogen, is given by:

$$\eta_{th} = \frac{0.83 \times V_{cell}}{1.23} = 0.675 \cdot V_{cell} \tag{2.40}$$

A fuel cell can be operated at different current densities; the corresponding cell voltage then determines the fuel cell efficiency. Decreasing the current density increases the cell voltage, thereby increasing the fuel cell efficiency. In fact, as the current density is decreased, the active cell area must be increased to obtain the desired amount of power. Fuel cell operation is influenced by various thermodynamic and electrochemical variables, such as temperature, pressure, gas concentration, reactant utilization, current density, etc., which directly influence the cell potential and voltage losses. Changing the fuel cell operating parameters can have either a beneficial or a detrimental impact on fuel cell performance and on the performance of other system components. Changes in operating conditions may lower the cost of the cell, but increase the cost of the peripheral components. Generally, a compromise in the operating parameters is made to meet the required application.

2.5 Hydrogen Energy and Economy

Hydrogen is advocated for quite a long time as an environmentally friendly and a powerful energy storage medium and fuel. Among the most important advantages of using the hydrogen as energy storage medium are: is the lightest element and a very stable molecular compound, on volumetric basis as compressed gas can store several times more energy than in compressed air, it reacts easily with oxygen to generate energy and to form water, which environmentally harmless, can be easily used in fuel cells, and has a long industrial application history. However, the hydrogen has quite a few disadvantages as energy storage medium, such as is flammable and explosive, requiring special containers and transportation, is highly diffusive, due to its specific energy content, and being the lightest element high-pressure and large containers must be used for significant mass storage. Liquefaction could simplify storage but no less than 30% of charged energy would thereby be consumed. Hydrogen is flammable within a wide range of concentrations in air (4%–75% of the volume) whereas, for example, the range for methane is 5%–15%, and it also has very low ignition energy (about 0.07 times that of the methane). Gaseous hydrogen (H_2) can be manufactured with the process of electrolysis, consisting of an electric current applied to water separates it into components O_2 and H_2. The oxygen has no inherent energy value, but the higher heat value (HHV) of the resulting hydrogen can contain up to 90% of the applied electric energy, depending on the technology. This hydrogen can then be stored and later combusted to provide heat or work, or to power a fuel cell. Compression to a storage pressure of 350 bar, the value usually assumed for automotive technologies, consumes up to 12% of the hydrogen's HHV if performed adiabatically, although the loss approaches a lower limit of 5% as the compression approaches an isothermal ideal. Alternatively, the hydrogen can be stored in liquid form, a process that costs about 40% of HHV, using current technology, and that at best would consume about 25%. Liquid storage is not possible for automotive applications, because mandatory boil-off from the storage container cannot be safely released in closed spaces (i.e., garages). Hydrogen can also be bonded into metal hydrides using an absorption process. The energy penalty of storage may be lower for this process, which requires pressurization to only 30 bars. However, the density of the metal hydride can be between 20 and 100 times the density of the hydrogen stored. Carbon nanotubes have also received attention as a potential hydrogen storage medium. A major issue of the hydrogen use as energy medium is the hydrogen embrittlement (grooving), consisting of hydrogen diffusion through metal matrices, leading to small cracks compromising the hydrogen storage container quality. However, despite of its disadvantages, its higher storage capacity, abundance and relative easy way to be produced through electrolysis from water make the hydrogen a strong candidate for energy storage medium. For these reasons, some scientists are suggesting that the hydrogen widespread use may transform our economy into the *hydrogen economy*.

In the context of the hydrogen economy, hydrogen will be an energy carrier, rather than a primary energy source. It may be generated through electrolysis or other industrial processes, using the energy harnessed by wind or solar power conversion systems, or chemical methods and used as fuel with almost no harmful environmental impacts. However, the establishment of the hydrogen economy is requiring that the current hydrogen storage and transportation are solved and suitable materials for storage are readily available. A hydrogen economy may lead to widespread of the use of renewable energy based power generation, avoiding the use of expensive and environmentally harmful fossil fuels. Weather a hydrogen economy is evolving in the near or far future is strongly dependent on the technological advances in hydrogen-based storage and transportation. Proponents of the hydrogen economy are making the cause based on the fact that hydrogen is the cleanest end-user energy source, especially in transportation applications and one of the most abundant elements in the nature. Almost every country can become energy independent in this scenario. Critics of this full transition to the hydrogen economy are arguing that the transition cost is prohibitive, and a transition in intermediate steps may be more economically viable, with the transition focusing on more locally than regionally, entire country or globally. A future and extensive hydrogen infrastructure to be established in the same ways as the energy distribution networks were established in the early twenty-first century. Since hydrogen is currently produced in industry for a variety of reasons, and occasionally as a by-product of other processes, there is a relevant body of information on how to obtain hydrogen. One such process is the steam reforming of natural gas. This method of hydrogen formation is most efficiently used with light hydrocarbons such as methane and naphtha.

TAs mentioned in previous paragraph one of the most used and least expensive way to obtain hydrogen is to extract it from the natural gas, primarily methane. The most common method for the conversion of methane to hydrogen involves the use of the steam reforming, followed by the water–gas shift reaction. It now provides some 95% of all the hydrogen produced in the United States. The first step in this procedure is the purification, the elimination of impurities, such as sulfur, from the methane-rich natural gas. The methane is then reacted with steam at a relatively high temperature, using nickel oxide as a catalyst. This process is called steam reforming, and can be formulate as following:

$$CH_4 + H_2O \Rightarrow CO + 3H_2 \tag{2.41}$$

This phase can be followed by a second step in which air is added to convert any residual methane that did not react during the steam reforming and to produce additional hydrogen.

$$2CH_4 + O_2 \Rightarrow 2CO + 4H_2 \tag{2.42}$$

The phase is often followed by the water–gas shift reaction at a somewhat lower temperature that produces more hydrogen from the CO and steam, as:

$$CO + H_2 \Rightarrow CO_2 + H_2 \tag{2.43}$$

As discussed earlier in this chapter, the driving force for any reaction is the standard Gibbs free energy change, occurring as the result. This is the difference between the sum of the standard Gibbs free energies of formation of the products and the sum of the standard Gibbs free energies of formation of the reactants. A number of countries, especially the United States and China, have large amounts of coal that they can use as an energy source. It is possible through to involve the steam in a reaction with solid carbon instead of with methane to produce syngas, which is a mixture of CO and hydrogen. The resultant CO can then be reacted with steam in the water–gas shift reaction, just as is done in the case of steam reforming of methane. The power plants that generate hydrogen from coal by this two-step process have overall efficiencies of about 35%. But if they capture the effluent CO_2 from the water—gas reaction, the efficiencies can rise to above 40%. The thermal behavior of these various reactions can also be determined from data on the standard enthalpies of the species in these reactions. If the hydrogen is to be used in a low temperature fuel cell, the gas mixture resulting from the water–gas reaction also generally undergoes a further step, called methanation, in which the remaining CO is converted back into methane, which is recycled. This is necessary because CO poisons the platinum catalysts that are typically used in such fuel cells. It should be noted that all of these gas phase reactions produce products that consist of mixtures of gases. The separation of hydrogen from the other gas components must also be done, and there is a need for the development of better selective membranes for this purpose.

The second major method for the hydrogen production is the electrolysis of water by imposing a voltage between two electrodes within it that exceeds its thermodynamic stability range. The result is the evolution of hydrogen gas at the negative electrode, and oxygen gas at the positive electrode. Both of these gases have a significant commercial value. It is easy to design systems to collect them separately.

Relatively pure hydrogen can be produced by the electrolysis of water, and this method appears to be attractive in the long run. There is an enormous amount of water on the earth, with the potential to supply an almost limitless supply of hydrogen. About 4% of the hydrogen currently used in the world is produced by electrolysis. The problem with this apparently attractive scenario is that the electrolytic decomposition of water to produce hydrogen is currently quite expensive. Thus, it is only an attractive large-scale option where the cost of electricity is relatively low. An advantage of water electrolysis to produce hydrogen is that instead of requiring large central facilities, there can be distributed generation using smaller units. They can be located at places near where the hydrogen will be used in order to reduce transportation costs. Although the electrolytic production of hydrogen is significantly more expensive than obtaining it from natural gas, it has the advantage that the resulting gas can be of significantly greater purity. This can be especially important when the hydrogen is used in low temperature fuel cells with polymeric solid electrolyte membranes. Even minor amounts of impurity species, such as CO, can cause problems by absorbing on the surfaces of the platinum catalysts that are typically used to assist the conversion of H_2 molecules to H^+ ions and electrons at the negative electrode. This requires additional treatment, and results in higher costs. Large commercial electrolyzers now produce hydrogen at about 30 bar pressure and a temperature of 80°C, and have energy efficiencies of 80%–90%. The major source of loss is connected with the processes that take place at the positive electrode, where oxygen is evolved. An additional method that can be used to produce hydrogen from water is to thermally decompose it by heating to a very high temperature. It is also possible to produce hydrogen by chemically decomposing water. Many species form oxides when in contact with water. In general, however, the oxide that forms produces a protective surface layer that prevents further reaction with the water. We have to notice that there also are several developing processes for effectively using the biomass for the production of hydrogen. The two major processes are anaerobic digesters and pyrolysis gasifier, the former being useful in the kW range and the later one being useful at the MW range.

Despite the fact that the predominant methods used today are the catalytic steam reforming of natural gas and partial oxidation of heavy oils, electrolytic and thermal decomposition of water are more suitable for the use of hydrogen for energy storage. In this case hydrogen will be produced from water using the energy generated by either large hydro plants or base-load nuclear or coal power plants. It is even possible to use renewable photovoltaic, solar or wind-generated energy for hydrogen production. After transmission to the consumer, hydrogen is used as a primary fuel for peak energy generation or simply as an alternative to other fuels. Combustion as a fuel would result in the recombination of hydrogen and oxygen to form water, thus completing the cycle. The options exist for the storage of hydrogen: compressed gas, chemical compounds, liquid hydrogen and metallic hydrides. Since the cost of liquefaction is considerable, the most attractive concept for the bulk storage of hydrogen produced from substantial non-oil-based primary energy sources is compressed gaseous hydrogen in underground caverns, where it can be stored in a similar way to natural gas. Alternative systems for using hydrogen include its chemical combination with other elements to form compounds which are more amenable to storage. Liquid hydrogen has a mass energy density three times greater than oil, while its use is attractive for heavy surface transport and aircraft, allowing improvement in payloads and extending range. Because of its low density, liquid hydrogen may be less attractive than other materials for its energy storage density on a volumetric basis. Liquid ammonia and methane appear to be the most efficient materials for storing hydrogen on both a weight and volume basis. To store liquid hydrogen and other cryogenic fluids, as well as gas under pressure, reliable low-mass storage vessels will be necessary.

2.6 Summary

Large-scale introduction of renewables is changing the requirements on the electrical grid. The grid needs to handle electricity production at new locations and the variation in time of electricity generation is changing as well. The intermittent behavior of the major renewables, i.e., solar and wind energy systems, requires either complementary power sources that can balance power supply, a shift in energy demand or deployment of electrical energy storage. There is a wide range of storage technologies and the choice of technology is dependent on which problem to solve, and in many cases, what resources are locally available. Off-grid systems are systems for an isolated area like an island or a village or even a house or a single device. These are mainly used where connection to the national, or local, grids is too expensive. The interest in off-grid systems has increased due to the decrease in cost of small-scale renewables, like solar panels and small wind turbines. To balance demand and supply in an off-grid system a combination of several energy sources is beneficial but in most cases energy storage systems such as battery banks, fuel cell stacks or supercapacitors are required. Batteries and fuel cells are electrochemical energy storage and energy conversion devices, having a broad range of applications. Fuel cells are an interesting alternative for conventional power generation technologies because of their high efficiency and low environmental effects. Primary cells, secondary cells and fuel cells are the major sources of electrical energy. In the case of secondary cells, the cell reaction producing electrical energy (discharge process) can be reversed by applying an external source of current (the charging process). In the case of primary cells, the discharge process is irreversible and cannot be reversed. In fuel cells, the chemical energy associated with the oxidation of fuels (reducing agents) is directly converted into electrical energy more efficiently than in other conventional processes. Fuel cells convert the energy of combustion directly to electrical energy. A fuel cell is a galvanic cell in which the chemical energy associated with the oxidation of reducing agents (fuels) is directly converted into electrical energy. Scalability, quiet operation, easier carbon capture, and possible water production are some of their attributes. Based on the type of electrolyte, fuel cells can be categorized into six groups: polymer electrolyte fuel cell, alkaline fuel cell, phosphoric acid fuel cell, molten carbonate fuel cell, direct methanol fuel cell, and solid oxide fuel cell (SOFC). Each type of fuel cell is suitable for certain application. Stationary and distributed power generation, transportation, and portable applications are the main candidates for fuel cell utilization. Among these fuel cells, due to their high operating temperature, solid oxide fuel cell are especially suited for stationary electricity generation, while low temperature types are better candidates for transportation and residential applications. Today battery and fuel cell technologies are mature, either there are challenges that must be overcome and have several applications from power system, distributed generation to the transportation systems and portable electronics. Battery research and development are focusing on the reduction of inactive to active components to improve energy density, increasing conversion efficiency and rechargability, maximizing the system performances under more stringent operation conditions, enhancing safety and reducing their environmental effects. Fuel cell systems are offering opportunities for powering electric and hybrid vehicles, as replacement or in tandem with combustion engines, and for use in utilities or for large portable applications. For long time hydrogen was advocated as energy storage medium and fuel in the twenty-first-century energy portfolio.

Questions and Problems

1. Explain the differences between shallow- and deep-cycle batteries and list for each type main characteristics and applications.

2. List the major advantages and disadvantages of fuel cells.

3. List major fuel cell types and the most suitable applications for each type.

4. What are the distinguishing features between primary galvanic cells, secondary cells, and fuel cells?

5. What is the significance of depth-of-discharge (DoD), and how this parameter is affecting the battery life cycle?

6. What gases did we use to feed the fuel cell to produce electricity? Can you name any by-products that may have been produced by fuel cells?

7. The lead-acid battery (a secondary type cell) can be recharged and reused for a few years after which it cannot be recharged and reused. Can you think of reasons why it cannot be used any further?

8. Describe the battery memory effect.

9. What is the difference between charge efficiency and energy efficiency?

10. The standard free energies of combustion of CH_4 and CH_3OH are −818 and −706.9 kJ/mol, respectively. Calculate the EMFs of these electrochemical cells. The standard molar enthalpies for the combustion reactions are −890.4 and −7640 kJ/mol, respectively. Calculate the values of the efficiencies for CH_4 and CH_3OH.

11. If the capacity of a battery is 120 Ah and is charged at a rate C/6, how long will take to be fully charged?

12. A battery stack, used as an electrical energy storage system, is designed for 20-MW peak power supply for duration of 4 hours. This energy storage system uses 600 Ah batteries operating at 420 V DC. Estimate the stack minimum number of batteries and the current in each during peak operation.

13. For a lead-acid battery having a nominal Puekert capacity of 60 Ah, assuming 1.2 Pukert coefficient plot the capacity for different discharge rates and for different currents.

14. The voltage of a 12.5 V lead-acid battery is measured as 10.5 V when it delivers a current of 50 A to an external (load) resistance. What are the load and the battery internal resistances? What is the battery voltage and current through a load of 7.5 Ω?

15. A car battery with an open-circuit voltage of 12.8 V is rated at 280 Ah. The internal resistance of the battery is 0.25 Ω. Estimate the maximum duration of current flow and its value through an external resistance of 1.85 Ω.

16. A battery has an internal resistance of 0.02 Ω per cell needs to deliver a current of 25 A at 150 V to a load. If the cell electrochemical voltage is 1.8 V, determine the number of cells in series.

17. A battery in an industrial facility lasts 7.5 years when it is repeatedly discharged to 50% DoD. Determine its life if the repeated DoD is increased to 66.7%.

18. If there are two options in procuring 120 Ah battery cells for a battery system in an industrial facility, requiring frequent discharging and charging. One cell has an average cell voltage of 1.25 V during discharging phase and 1.40 V during charging phase, and a C/D ratio of 1.12, while the second one has an average voltage of 1.35 V during discharge, 1.5 V during charge, and a C/D ratio of 1.16. The cells have the same cost per Ah rating, select the most appropriate cell and justify your answer.

19. A battery lasts 2500 charge-discharge cycles of 100% DoD. Determine the battery approximate cycle life at 60%, 40%, and 20% DoD.

20. Under standard (normal) operation conditions a fuel cell, hydrogen-oxygen type generates a 3.5 A current and a 0.85 V. What are the fuel cell internal resistance and its energy efficiency?

21. Propane (C_3H_8) can be used in a fuel cell and the combustion products are CO_2 and water. Write a balanced chemical reaction for the combustion process. The standard free energy of combustion is -2108 kJ/mol of C_3H_8. Since five oxygen molecules (oxidation state zero) are reduced to the oxidation state of -2, 20 electrons are involved in the process ($O_2 + 4H^+ + 4e^- \rightarrow 2H_2O$). What is the emf of the fuel cell?

22. A 2.45-MW hydrogen-oxygen fuel cell stack, designed to produce electrical energy for a small electric network (microgrid), has an efficiency of 74%. Calculate the flow rates (kmol/s) of hydrogen and oxygen, and the fuel cell stack losses.

23. A hydrogen-oxygen has a liquid waver as product when it operates at 0.825 V. What is the electrical energy output in kJ/kg·mol of hydrogen, and the cell efficiency?

24. A hydrogen-oxygen fuel cell stack produces 60 kW of DC power at an efficiency of 63%, with water vapor as product. What is the hydrogen mass flow rate, in g/s, and the cell voltage?

25. A fuel cell system, designed to power a sensor monitoring node, must deliver 12 V at 1.2 kW for 10 days. The manufacturer cell specifications are: open circuit voltage 1.18 V, internal cell resistance 0.25 Ω, and there is a linear relationship between load voltage and current. How many cells must be connected in series?

References and Further Readings

1. F. Bueche, *Introduction to Physics for Scientists and Engineers*, McGraw-Hill, New York, 1975.
2. C.A. Vincent and B. Scrosati, *Modern Batteries* (2nd ed.), Arnold, London, UK, 1997.
3. G. Boyle, *Renewable Energy—Power for a Sustainable Future*, Oxford University Press, Oxford, UK, 2012.
4. J.O. Besenhard (ed.), *Handbook of Battery Materials*, Wiley-VCH, Weinheim, Germany, 1999.
5. D. Linden and T.B. Reddy, *Handbook of Batteries* (3rd ed.), McGraw-Hill, New York, 2002.
6. J. Larminie and A. Dicks, *Fuel Cell Systems Explained* (2nd ed.), Wiley, Chichester, UK, 2003.
7. M. Winter and R.J. Brodd, What are batteries, fuel cells, and supercapacitors? *Chemical Review*, 2004, Vol. 104, pp. 4245–4269.
8. R. O'Hayre, S-W. Cha, W. Colella, and F. B. Prinz, *Fuel Cell Fundamentals*, Wiley, 2006.
9. V. Quaschning, *Understanding Renewable Energy Systems*, Earthscan, London, UK, 2006.

10. F.A. Farret and M.G. Simões, *Integration of Alternative Sources of Energy*, Wiley-Interscience, New York, 2006.
11. R.A. Ristinen and J.J. Kraushaar, *Energy and Environment*, Wiley, Hoboken, NJ, 2006.
12. S.C. Singhal and K. Kendall, *High Temperature Solid Oxide Fuel Cell, Fundamental, Design and Applications*, Elsevier, Amsterdam, the Netherlands, 2006.
13. J. Andrews and N. Jelley, *Energy Science, Principles, Technology and Impacts*, Oxford University Press, Oxford, UK, 2007.
14. A. Ter-Gazarian, *Energy Storage for Power Systems*, The IET Press, London, UK, 2008.
15. A. Vieira da Rosa, *Fundamentals of Renewable Energy Processes* (2nd ed.), Academic Press, Amsterdam, the Netherlands, 2009.
16. B.K. Hodge, *Alternative Energy Systems and Applications*, Wiley, Hoboken, NJ, 2010.
17. B. Everett and G. Boyle, *Energy Systems and Sustainability: Power for a Sustainable Future* (2nd ed.), Oxford University Press, Oxford, UK, 2012.
18. F.S. Barnes and J.G. Levine (eds.), *Large Energy Storage Systems Handbook*, CRC Press, Boca Raton, FL, 2011.
19. R.A. Dunlap, *Sustainable Energy*, Cengage Learning, Stamford, CT, 2015.
20. V. Nelson and K. Starcher, *Introduction to Renewable Energy* (*Energy and the Environment*), CRC Press, Boca Raton, FL, 2015.
21. M. Martín (ed.), *Alternative Energy Sources and Technologies*, Springer, Cham, Switzerland, 2016.
22. R.A. Huggins, *Energy Storage Fundamentals, Materials and Applications* (2nd ed.), Springer, Cham, Switzerland, 2016.
23. R. Belu, *Industrial Power Systems with Distributed and Embedded Generation*, The IET Press, Stevenage, UK, 2018.
24. W. Vieltich, A. Lamm, and H. A. Gasteiger, *Handbook of Fuel Cells*, Vol. 1 & 2, Wiley, NY, 2003.

3

Biomass, Biofuels, Waste-to-Energy Recovery

3.1 Introduction, Bioenergy Concepts and Issues

Bio-renewable energy resources, often referred to as biomass, are organic materials of recent biological origin that can be used for energy generation. The definition is deliberately broad with the intent of only excluding fossil fuels from the organic material wide varieties, arising from the biotic environment. Bio-renewable energy resources are classified as either dedicated energy crops or waste materials or. In generic terms, the biomass is the term used for all organic material originating from plants (including algae), trees and crops and is essentially the collection and storage of the solar energy through photosynthesis. Biomass energy is the conversion of the biomass energy into useful forms of energy such as heat, electricity and liquid fuels. Bioenergy biomass comes either directly from the land, as dedicated energy crops, or from residues generated in the processing of crops for food or other products such as pulp and paper, and wood industry, construction and demolition wood or other materials, pallets used in transportation, or the clean fraction of municipal solid wastes or industry wastes. Not all biomass is directly used to produce energy, being rather converted into intermediate energy carriers, the so-called biofuels, such as charcoal (high energy density solid fuel), ethanol (liquid fuel), or producer-gas (biomass gasification). Wastes are materials, usually from industrial, agricultural or human processes, discarded because they have no apparent value or represent a nuisance or a pollutant to the environment. Dedicated energy crops are plant species grown specifically for the bio-based energy products, i.e., for purposes other than food and/or feed. Biomass includes all materials of organic origin (e.g., all natural living or growing materials, their residues and resulting waste of their processing or uses). Therefore, all plants and animals, their residues and wastes or materials that are resulting from their transformation and processes (e.g., paper to cellulose), food industry organic wastes, or wastes of households and industrial production are qualifying as biomass. Biomass appears in various forms, which are simultaneously produced in organisms. Cellulose, a polysaccharide, consisting of pure glucose chains, which have been connected by hydrogen in the crystal bonds, is the most frequent found organic substance. Of all renewable energy sources the largest contribution, in the short and medium range, is expected to come from biomass. Fuels derived from energy crops are not only potentially renewable, but are also similar in origin to the fossil fuels (in fact biomass of millions of years ago) to provide direct substitution. They are storable, transportable, available and affordable and can be converted into a wide variety of energy carriers using existing and novel conversion technologies, having the potential to be a significant energy contributor.

Biomass energy is by far the largest renewable energy source, representing about 10% of the world total primary energy supply and over 10% of the global renewable energy supply.

However, biomass energy represents only about 1% of the total fuel used for electricity production. Biomass for energy is the main contributor to renewable energy around the world, with almost 15% of total energy consumption in deriving from biomass. Biomass is in fact a term that covers a broad range of often very different products, although all are of organic origin. Many of these products can be used as a source of energy, either for electricity or heat production, or as a feedstock for biofuels production. Among all the renewable energy sources, biomass is unique as it effectively stores the solar energy. It is the only renewable carbon source that can be converted into convenient solid, liquid and gaseous fuels through specific conversion processes. Biomass generates about the same amount of carbon dioxide as do fossil fuels (when burned), but from a chemical balance point of view, every time a new plant grows, carbon dioxide is removed from the atmosphere. The net carbon dioxide emission is close to zero as long as plants continue to be replenished for biomass energy purposes. If the biomass is converted through gasification or pyrolysis, the net balance results in removal of carbon dioxide. Biomass is often not considered a new energy source, given the role that it has played, and continues to play in energy mix, in most developing countries, where the biomass still accounts for about one-third of primary energy use while in the poorest ones up to 90% of all energy is supplied by biomass. Over two billion people cook by direct biomass combustion, an inefficient use of biomass fuels, largely from natural forests, which can further contribute to deforestation and environmental degradation. Although global records of biomass heating versus power uses are not available, in Europe two thirds of biomass is used for heating. For the transportation sector, production of fuel ethanol for vehicles reached over 40 billion liters in 2006, with significant annually increases since in the United States, Brazil, France, Germany and Spain. Since 2006, the United States became the leading fuel ethanol producer, with over 20 billion liters production, ahead of longstanding leader Brazil, with new production plants is coming in place every year. However, U.S. ethanol production is not keeping up with demand, and the ethanol imports increased during that period. Since 2007, most gasoline sold in the United States is blended with some percentage of ethanol. Biodiesel production also increased significantly, both in the United States and worldwide. Half of world biodiesel production continued to be in Germany. Significant production increases also took place in Italy, Brazil and the United States. In Europe, supported by new policies, biodiesel gained broader acceptance and market. Last but not least, one of the major biomass advantages is the opportunity to make decentralized facilities which can help local communities and use raw materials that are not used very efficiently in many places and areas.

The potential of bioenergy crops is huge, since over 3000 plants out the tens of thousands existing species are used as food, energy and other feedstock sources, with about 300 cultivated plants, and more than 60 plant species of critical importance. It is a vital task to increase the plant varieties that are used in bioenergy, through identifying, adapting, breeding, testing or employing biotechnology or genetic engineering to enlarge the planet flora potential. The introduction of new crops can lead to improvements in the biological and environmental condition of soils, water, vegetation and landscapes, while increasing the biodiversity. They can be converted trough first-generation biofuel technologies (e.g., combustion, and ethanol from sugar or starch crops, biodiesel from oil crops) or through the second-generation biofuel technologies (e.g., synthesis of lignocellulosic crops and their residues to liquid or gaseous fuels). A large potential and opportunity is expected from microalgae fuel production, often characterized as the third-generation biofuel technology. Theoretically, the microalgae can produce over ten times more biomass per unit area than land crops, can be grown in saline, brackish or waste waters or in bioreactors and can be converted to oil, ethanol or hydrogen. The overall bioenergy contribution to the world energy demands increased from about 467 EJ per year in 2005 to over 600 EJ by 2017. Most biomass-based electricity production occurs in developed countries, however, several developing countries, India, Brazil, other Latin American/Caribbean and African countries are generating large electricity shares from bagasse combustion from sugar alcohol production. Denmark, Finland, Sweden and the Baltic countries provide substantial shares (up to 50%) of district heating fuel from biomass. In many developing countries, small-scale power and heat production from agricultural waste is quite common. Biomass pellets have become more common, with about 6–7 million tons consumed in Europe, about half for residential heating and half for power generation, often in small combined heat and power generation (CHP) systems.

Forest and agricultural biomasses are divided into two groups according to their origin: biomass from traditional forest or agricultural exploitations, and biomass from energy crops. Dedicated energy crops are high-fiber crops grown specifically for their cellulose and hemicellulose productivity. Harvesting may occur on an annual basis (e.g., switchgrass), on a 5–7 year cycle for fast-growing trees (e.g., hybrid poplar) or on longer time periods for regular trees. Lignocellulosic crops are divided into herbaceous energy crops and short-rotation woody crops (SRWC). Herbaceous crop species have little or no woody tissue. The above-ground growth of these plants usually lives for only a single growing season. Herbaceous crops include annuals and perennials plant species. Annuals die at the growing season end, being replanted in the spring. Perennials die each year in temperate climates but re-establish themselves each spring from rootstocks. Both annual and perennial herbaceous energy crops are harvested on an annual basis, if not more frequently, with yields averaging 5.5–11 Mg/ha·yr, with maximum yields up to 25 Mg/ha·yr in temperate regions. As with trees, yields are higher in tropical and subtropical regions. Herbaceous crops more closely resemble hardwoods in their chemical properties than softwoods. Their low lignin content makes them relatively easy to be delignified, which improves accessibility of the carbohydrate in the lignocellulose. The hemicellulose contains mostly xylem, which is highly susceptible to acid hydrolysis compared to the cellulose. Agricultural residues are susceptible to microbial degradation, destroying their processing potential in a matter of days if exposed. Herbaceous crops have high silica content compared to woody crops, which can present problems during processing. SRWCs are fast growing woody biomass, displaying rapid juvenile growth, suitable for use in dedicated feedstock supply systems, wide site adaptability, good pest and disease resistance. Woody crops grown on a sustainable basis are harvested on a rotation of 3–10 years. Woody crops include hardwoods, trees that are classified as angiosperms (flowering plants, such as willow, oak and poplar) and softwoods, trees that are classified as gymnosperms, most of trees known as evergreens (e.g., pine, spruce and cedar). Hardwoods can re-sprout from stumps, which is reducing their production costs compared to softwoods. Advantages of hardwoods in industry processing include: higher density for many species, easy delignification and accessibility of wood carbohydrates, the presence of hemicellulose high in xylem, easily removable, low ash content, particularly silica, compared to softwoods and herbaceous crops, and high acetyl content compared to most softwoods and herbaceous crops, an advantage in the process of the acetic acid recovery. Softwoods are fast growing but their carbohydrates are not as accessible for chemical processing as the hardwood carbohydrates. Softwoods have considerable value as construction lumber and pulp, are usually available as waste material, logging and manufacturing residues. Logging residues, consisting of a high proportion of branches and tops, contain considerable high-density compression wood, not easily delignified. Logging residues are more suitable as boiler fuel or for other thermochemical treatments than as feedstocks for chemical or enzymatic processing.

Waste materials that are qualifying as bio-renewable resources include municipal solid waste, agricultural residues, agricultural processing by-products, and manure. Municipal solid waste (MSW) is whatever is thrown out in the garbage, including materials that do not qualifying as bio-renewable resources, e.g., glass, metal or plastics. Agricultural residues are simply the crop parts discarded after harvest such as corn stover (husks and stalks), rice hulls, wheat straw and bagasse (fibrous materials of the sugarcane milling). Food processing wastes are the effluents from a variety of industries ranging from breakfast cereal manufacturers to alcohol breweries, in the form of dry solids and watery liquids. Concentration of the animals into large livestock facilities has led to treat the animal wastes in similar ways to human wastes. Waste materials share common traits other than the difficulty of characterizing them because of their variable and complex composition. Thus, waste biomass presents special problems to engineers who are tasked with converting such quite unpredictable feedstocks into reliable source for power, high-quality fuels or chemicals. The major virtue of waste materials is their very low cost. By definition, waste materials have little economic value and often can be acquired for little more than the cost of transportation from origin site to a processing plant. Increasing costs for solid waste disposal and sewer discharges and restrictions on landfilling certain wastes allow some wastes to be acquired at negative cost; a bio-renewable resource processing plant is paid by a company to dispose its wastes. For this reason, the most economically attractive opportunities in bio-renewable resources are the waste feedstocks.

3.1.1 Biomass as Fuel and Solar Energy Storage

From thermodynamic point of view any plant can be considered a thermal engine, converting and storing the solar energy into chemical energy as complex organic molecules. Photosynthesis is a process, taking place in plants, through which the solar energy is used to produce glucose, basically converting water and carbon dioxide into high-energy carbohydrates (sugars and starches) and oxygen as a by-product. The site of photosynthesis is in the chloroplast, an organelle found in the leaves of green plants. Photosynthesis, which occurs in chloroplasts, uses solar energy to combine carbon dioxide and water into energy-rich organic molecules (glucose or fructose) and releases oxygen into the environment. The main functions of chloroplasts are to produce sugar or glucose ($C_6H_{12}O_6$), storing in this way the solar energy. Chloroplasts contain the pigment chlorophyll that absorbs most of the visible spectrum radiation, reflecting only green and yellow light wavelengths. The conversion efficiency of such process is very low, in the range of 0.1%–1.5%, incomparable lower than the common efficiencies of the made-man systems or devices. Biomass represents and natural medium to store solar energy. Plants and algae can trap the solar energy through photosynthesis and store this energy into the carbohydrate molecule chemical bonds. It must be noted that fossil fuels, such as coal and petroleum, are also biomass products. During photosynthesis, plants are using solar energy, CO_2 for the atmosphere, minerals and water from the ground to produce primarily carbohydrates (Equation 3.1) and oxygen, and by further biosynthesis, a large number of less oxygenated compounds including lignin, triglycerides, terpenes, proteins, etc. Plants are using their leaves' chlorophyll as catalysts, and large amount of solar energy to convert carbon dioxide (CO_2) and water into glucose and fructose during the photosynthesis processes:

$$6CO_2 + 6H_2O \xrightarrow[\text{Chlorophyll}]{\text{Sun light}} C_6H_{12}O_6 + 6O_2 \qquad (3.1)$$

Example 3.1: Estimate the solar energy and glucose formation efficiency, through the photosynthesis energy balance equation, by using the data for the reactants, CO_2 and water and the reaction products, glucose ($C_6H_{12}O_6$) and oxygen of Equation (3.1), given in Table 3.1.

Solution: The energy balance related to this reaction (Equation 3.1) in the stationary states of the open system is expressed:

$$6 \cdot \Delta H_{CO_2} + 6 \cdot \Delta H_2O - \Delta H_{Glucose} - 6 \cdot \Delta H_{O_2} - E_{SR} = 0$$

Here, ΔH_{XYZ} indicates the molar formation enthalpy of the chemical compound (XYZ) and E_{SR} is the highly exergetic solar radiation absorbed by the system (plat leaves, here) during the process of producing carbohydrates.

$$E_{SR} = \Delta H_{Glucose} + 6 \cdot \Delta H_{O_2} - 6 \cdot \Delta H_{CO_2} - 6 \cdot \Delta H_2O = -1264.0 - 0 + 6 \times 394.1 + 6 \times 286.0$$

$$= 2816.6 \text{ kJ/mol}$$

The glucose formation efficiency is calculated as:

$$\eta_{Glucose} = \frac{\Delta H_{Glucose}}{E_{SR}} = \frac{1264}{2816.6} = 0.4487 \text{ or } 44.9\%$$

These values are higher than the same values estimated from the laboratory data.

The detailed chemical process to convert CO_2 and water to glucose or fructose, by using solar energy is rather complicated, but the final products are glucose/fructose and oxygen. On average, the capture efficiency of incident solar radiation in biomass is 1% or less, but it can be as high as 15%, depending on the type of plant. The carbon (e.g., CO_2) and mineral (K, N, P) cycles are closed after decomposition of biomass or waste products, if disposed on land or after processing, consumption, degradation, or combustion. Consequently, the life cycle of biomass as renewable feedstock has almost a neutral effect on CO_2 emission. Based on this fact, biomass is considered an intrinsically safe and clean material, with unlimited availability and high potential to be used as a renewable resource for the production of energy and alternative fuels, new materials in technical applications, and organic materials and chemicals. At present, forestry and agricultural residues and municipal waste are the main feedstocks for the generation of electricity and heat from biomass. In addition, a very small share of sugar, grain and vegetable oil crops are used as feedstock for the production of liquid biofuels. The photosynthesis products have in average higher energy of about 16 MJ/kg, in the case of pure carbohydrates, the energy amount per unit of mass are depending on the oxidation degree of the carbon, zero when fully oxidized as for CO_2, about 16 MJ/kg as carbohydrates, and 55 MJ·kg when fully reduced as for methane (CH_4).

TABLE 3.1

Thermodynamic Data of the Photosynthesis Reactants

Compound	Molar Mass (g/mol)	ΔH (kJ/mol)
CO_2	44	−394.1
H_2O	18	−286.0
$C_6H_{12}O_6$	180	−1264.0
$C_6H_{10}O_5$	160	−789.0
O_2	32	0
H_2	2	0

Even the all biomass of the biosphere, all living matter on Earth living represents only a tiny fraction of total Earth mass at human scale; it stores an enormous amount of energy and has a critical role in maintain the atmosphere structure and composition. This reaction has several intermediate steps, where the chlorophyll plays significant roles, is highly endo-thermic (absorbing energy) with a free energy of formed glucose/fructose of 2816 kJ/mol. From the glucose and fructose, the plants are forming very complex organic molecules, carbohydrates with high chemical energy content through this general chemical reaction:

$$x CO_2 + x H_2O \Rightarrow \text{Carbohydrate Molecules} + x O_2 \qquad (3.2)$$

The plants use glucose and fructose to produce a vast number of chemical substances needed for growth and life processes. Such reactions and processes have very efficiencies; the plants can be planted and grow with low capital investments and technologies, while by producing carbohydrates and oxygen and by removing CO_2 from the atmosphere can have a critical role in the atmosphere composition. Biomass combustion is in theory neu-tral to the atmosphere, because it returns the same number of CO_2 molecules as removed through the biomass production processes. Biomass is a renewable energy source because the plants and crops grow at shorter time scales compared to the human timescales. However, from the applicability point of view, the biomass reproducibility time periods of various biomass forms are very important, ranging from several months for grains and grass, to 2–5 years for fast growing timber and to tens of years for timber and forests, with longer times in case of the ones of northern temperate regions than the ones in the south-ern temperate areas. In the photosynthesis physical process, the leaf pigment molecules absorb light photons and the excited electrons are transferred to the adjacent molecules, rather than de-excited and emitting a photon, so the electron energy can be used in com-plex chemical reactions to produce carbohydrates and oxygen. The maximum efficiency of this process is about 33%, and only about 50% of the incoming solar radiation can be absorbed, while leaf losses are about 25% (reflection and transmission), so the actual effi-ciency in the laboratory conditions is about 10%. However, the actual efficiency is much lower, as only 33% of solar radiation is absorbed during the growing plant period, from which about 60% is converted into biomass (the rest, 40% is used by plants for respiration), and multiplying these factors together leads to about 0.5% efficiency.

Example 3.2: At mid-latitude regions, during summer the average solar energy density is about 1000 kWh/m², estimate the biomass yield per hectare.

Solution: The amount of the solar energy per hectare is:

$$\text{Solar energy per hectare} = 1000 \ \frac{kWh}{m^2} \times 10{,}000 \ m^2 = 10 \times 10^6 \ kWh \text{ or } 36 \times 10^{12} \ J/ha$$

Then by first estimating the yield in carbohydrate biomass, the average biomass yield per hectare is:

$$\text{Solar energy per hectare} = 0.05 \ (.5\% \ \text{efficiency}) \times 36 \times 10^{12} J/ha = 180 \times 10^9 \ J/ha = 180 \ GJ/ha$$

The biomass yield is approximately calculated as:

$$\text{Biomass yield} = \frac{180 \ GJ/ha}{16 \ MJ/kg} \approx 11{,}250 \ kg/ha \text{ or } 11.25 \ t/ha$$

3.2 Biomass Potentials and Uses

The organic matters that are making up plants are known the generic term of biomass. Fuels from biomasses encompass several forms, ranging from wood (solid), to biogas and alcohol (liquid) biofuels. Biomass can be used to produce heat, electricity, transportation fuels, or chemicals needed in various industry. Industry is by far the biggest biomass user, with over 51% biomass, the transportation being the second biomass user, with about 24% biomass, used to produce ethanol and biodiesel, electricity utilities are using about 12% of biomass for power generation, producing about 1% of the world electricity, while the residential sector uses 13% of the biomass supply. The energy content of dry biomass ranges from 15,904.9 kJ/ kg for straws to 17,579.1 kJ/kg for wood. Domestic biomass resources include agricultural, forestry and industry wastes, as well as the municipal solid wastes, providing the opportunity for local, regional and national energy self-sufficiency across the globe. The biomass-derived energy does not have the negative environment impacts of the non-renewable energy sources. Biomass contains energy that is produced when plants are processed into other materials such as paper and animal wastes. Woody plants consist of 20%–30% hemicelluloses. They are also polysaccharides that consist not only of pure glucose chains but other sugars as well. The wood pulp lignin constitutes about 30% of the woody plants. Compared with cellulose and lignin the remaining biomass forms play a smaller role in bioenergy. Starch, sugar as well as fats, proteins and dyes constitute only a small fraction of the biomass production. Biomass is one of the most important renewable energy resources. It can be used diversely as an energy carrier and can be converted to a variety of energy forms, such as heat, steam, electricity, hydrogen, ethanol, methanol and methane. Conversion product selection depends upon a number of factors including direct heat or steam needs, conversion efficiencies, used equipment and environmental impact of conversion process, waste stream and product use. Compared to other fossil fuels methane produces few atmospheric pollutants and generates less carbon dioxide per unit energy because methane is comparatively a clean fuel. The trend is toward its increased use for appliances, vehicles, industrial applications and power generation. Ethanol is also becoming quite a popular biomass derived fuel.

Evaluation of biomass resources as potential feedstocks requires information about plant composition, heating value, bulk density and production yields. Compositional information is reported in terms of organic components, proximate analysis or ultimate analysis. However, this information is scattered and often is showing large variations. Analysis in terms of organic components reports the types and amounts of plant chemicals, e.g., proteins, oils, sugars, starches and lignocellulose (fiber). The composition varies widely among plant parts. For example, corn grain is mostly starch with a relatively small amount of fiber while corn stover, that part of the crop left on the field, is mostly fiber (~84%) with very little starch content. The information is useful in designing biochemical processes that convert plant components into commodity chemicals and fuels. Often the fiber, a polymeric composite of cellulose, hemicellulose and lignin, is reported in terms of these three constituents, useful in processes in which the carbohydrate fractions are breaking down into sugars. Proximate analysis is important in developing thermochemical biomass conversion processes. Proximate analysis reports the yields of various products obtained upon heating the material under controlled conditions; these products include moisture, volatile matter, fixed carbon and ash content. Because biomass moisture content is highly variable, being easily determined by gravimetric methods (weighing, heating at 100°C, and reweighing), the biomass proximate analysis is usually reported on a dry basis. Volatile matter is that biomass fraction that decomposes is released as gases by heating a sample at moderate temperatures, about 400°C, in a non-oxidizing environment. Knowledge of volatile matter is important in designing biomass burners and gasifiers. The remaining fraction is a mixture of solid (fixed) carbon and mineral matters (ash), distinguished by further sample heating in the presence of oxygen, converting the carbon into carbon dioxide only the ash remains. Ultimate analysis is simply the major elemental biomass composition: carbon, hydrogen, oxygen, nitrogen, sulfurs, chlorines, the moisture and ash contents, or about how dry is the biomass. This information is important in the mass balance estimates on the biomass conversion processes. Compared to fossil fuels, biomass is characterized by relatively high oxygen content, which is reducing the heating value and represents a challenge in converting these compounds into hydrocarbon substitutes. Usually, the one carbon mole-based generic molecular formula is employed for performing process mass balances. Heating value is the net enthalpy released in the reaction a fuel with oxygen under isothermal conditions (equal starting and ending temperatures). If water vapor formed during reaction condenses at the process end, the latent enthalpy of condensation contributes to the higher heating value (HHV). Otherwise, the water latent enthalpy does not contribute and the lower heating value (LHV) prevails. Methane is produced from biomass by thermal or biological gasification. Application of thermal processes is limited to feedstock with either low water content or those having the potential to be mechanically dewatered inexpensively. Feedstocks containing 15% of total solids require all of the feed energy for water removal. Thermal processes for methane production are only economic at large scales and generate a mixture of gaseous products, composed primarily of methane and carbon dioxide with some traces that must be upgraded to the methane. The major limitation of biological gasification is that conversion is usually incomplete, often leaving as much as 50% of the organic matter unconverted.

3.2.1 Biomass Potential and Applications

Heat and electricity production are dominating the bioenergy uses, while the main growth bioenergy markets are European Union, North America and Southeast Asia, especially in power generation from biomass wastes, residues and biofuels. Two key industrial sectors for the biomass combustion and gasification applications for power generation are paper and pulp and sugarcane industries. Biomass power generation by advanced combustion technology and co-firing schemes is a growing market worldwide. Mature, efficient and reliable technology is available to convert biomass into power, while several biomass combustion schemes are rapidly improving due to the biomass availability and modern conversion technology. Competitive performances compared to fossil fuels are possible where lower cost residues are available, such as for co-firing, where the investment costs are minimal. Specific national policies, carbon taxes or renewable energy support are accelerating the development. Gasification technology (integrated with combined cycles) offers even better methods for power generation from biomass in the medium term and can make power generation from energy crops competitive in many areas of the world once this technology has been proven on a commercial scale. Gasification, in particular larger scale circulating fluidized bed concepts, also offers excellent possibilities for co-firing schemes.

Biomass, a renewable heat source can be used for small-, medium- and large-scale heat and power generation applications. Pellets, chips and various agricultural and forestry by-products can be used for bio-heat production. Pellets are offering higher energy density and standard fuel use methods, such as automatic heating systems, offering convenience for the end-users. The construction of new plants to produce pellets and the installation of millions of burners, boilers or stoves and appropriate logistics to serve the consumers may result in significant pellet market growths. Modern stoves and boilers operated with chips, wood pellets and wood logs are optimized with respect to the efficiency and emissions. However, there still are significant works to be done, in particular, the improvements in fuel handling methods, automatic control and maintenance requirements. Rural areas present a significant market potential for the applications of such systems. There is a growing interest in the district heating plants which currently are run mainly by energy companies and sometimes by cooperatives for small-scale systems. The systems applied so far generally use forestry and wood-processing residues but the application of the agro-residues is an important aspect. Direct biomass combustion is an established conversion technology for heat production at commercial scales. Hot combustion gases are produced when the solid biomass is burned under controlled conditions, and are often used directly for product drying, but more commonly the hot gases are used through heat exchangers to produce hot air, hot water or steam. The most often implemented combustor uses a grate to support a bed of fuel and to mix a controlled amount of combustion air, improving the overall system operation and performances. Sophisticated designs are permitting the overall combustion process to be divided into stages, drying, ignition and combustion of volatile constituents, and burn-out of char, with separate control of conditions for each phase. Grates are proven and reliable, can tolerate a wide range in fuel quality (moisture content and particle size), and are controllable and efficient. The goal of reducing emissions is one of the driving forces of the current developments. Significant improvement in efficiencies can be achieved by installing systems that generate both useful power and heat (cogeneration plants have a typical overall annual efficiency of 80%–90%). CHP is generally the most profitable choice for power production with biomass if heat, as hot water or as process steam, is needed. The increased efficiencies reduce fuel input and overall pollutant emissions compared to separate systems for power and heat, and also realize improved economics for power generation where expensive natural gas and other fuels are displaced. The technology for medium-scale CHP from 400 kW to 4 MW is now commercially available in the form of the Organic Ranking Cycle (ORC) systems or steam turbine systems. The first commercially available units for small scale CHP (1–10 kW) are on the market, and a breakthrough for the gasification of biomass in the size between 100 and 500 kW is in the final stage of commercialization.

The biomass-based power generation has increased in recent years mainly due to the implementation of favorable political frameworks in many countries. In EU, for example the biomass-based electricity generation (solid biomass, biogas and biodegradable fraction of municipal solid waste) is increasing at 5% rates, with similar yearly projected growth rates through 2025. However, most biomass power plants in operation are characterized by low boiler and plant efficiencies, being still expensive to be built. The main challenges therefore are to develop more efficient and lower-cost systems. Advanced biomass-based power generation units require fuel upgrading, combustion and cycle improvements, and better flue-gas treatment. Future technologies have to provide better environmental protection at lower cost through advanced biomass preparation, combustion and conversion processes with post-combustion cleanup. Such systems include fluidized bed combustion, biomass-integrated gasification, biomass externally fired gas turbines, generating steam, then using the steam to power a turbine for generating electricity. Even though the steam production by biomass combustion is quite efficient, the steam conversion to electricity is less efficient. Where the electricity generation is to be maximized, the steam turbine must exhaust into a vacuum condenser, with overall improved conversion efficiencies in medium and large power plants. Low-temperature heat (up to 50°C), available from the condenser is non-usable in many applications, being wasted into the atmosphere or local waters. Notice that the average conversion efficiency of U.S. steam plants is about 18%. Since 1979, in the United States has been installed over 7000 MWe capacity of wood-fired power plants. Where there is a need for both heat and electricity, the biomass processing, such as wood kilning, sugar or palm oil processing, power plants can be set to provide high-temperature steam, by taking some steam directly from the boiler, by extracting partially expanded steam from special designed turbines, or by arranging for the steam turbines to produce exhaust steam at the required temperature. All options are reducing the generated electricity, though the overall efficiency is increased, above 50%. Bioelectricity, particularly in the CHP form, is in the mainstream of current technological trends.

Aside from the large number of small individual heating systems, more than 1000 MW of bioelectricity units are currently used in EU, through conventional or advanced combustion methods. Biomass, in the form of industrial, agricultural and municipal solid waste, is used to generate electricity with conventional steam turbines. The United States has an installed biomass electricity generation capacity of about 8000 MW. Such units are usually up to 20 MW, and relatively capital intensive and energy inefficient, but are providing a cost-competitive power where a low-cost biomass is available or through CHP applications. However, such technology is not predicted to expand considerably in the future because of limited low-cost biomass supplies. Less capital-intensive and more efficient technologies are needed to make usable higher cost biomass sources, such as biomass from dedicated energy crops to become economic competitive. Higher efficiency and lower capital costs can be realized by efficient gas turbines. Present efforts are focused on biomass integrated gasifier/gas turbines, while the steam technology is still considered a good option. Biomass steam technology-based electricity and CHP systems are competitive with fossil fuel produced electricity where biomass residues are available at low or no cost, but not if the biomass fuels have to be purchased at market prices. In this case, other reasons for the biomass utilization are the electricity price structures, policy, etc. Steam technology is an acceptable alternative for biomass to electricity plants where good price structures exist. However, the environment and other benefits of using biomass are not maximized and price support is needed, which in many instances may not be an option. The biomass electricity generation cost-effectiveness is improved if conversion efficiencies increase and capital costs decrease. Increasing conversion efficiencies helps to maximize environmental benefits, associated environmental tax credits by reducing the fossil fuels dependency. However, because the steam technology is in essence fully developed, there is a limited scope for finding improvements. New conversion technologies are therefore important, e.g., gasification and pyrolysis. Biomass can provide a diverse energy source, potentially improving energy security through the oil and natural gas substitution. The use of domestic bioenergy resources can contribute to the energy mix diversification. Biomass imports, from distributed international sources, can also contribute to energy diversification, especially if lignocellulosic resources and bioenergy products derived from them are considered. The international bioenergy market is expected to have wider ranges of net suppliers, therefore the import of bioenergy products is not or less affected by the geopolitical concerns as oil and natural gas imports are. Biomass used as an energy source is of interest due to the following envisaged benefits and main advantages:

1. Biomass is a renewable, sustainable and relatively environmentally friendly source of energy.
2. A huge array of diverse materials, frequently stereochemically defined, are available from the biomass giving the user many new structural features to exploit.

3. Increased use of biomass would extend the lifetime of diminishing crude oil supplies.

4. Biomass fuels have negligible sulfur content and, therefore, do not contribute to sulfur dioxide emissions that cause acid rain.

5. The biomass combustion produces less ash than the coal combustion, and the produced ash can be used as a soil additive on farms, etc.

6. The combustion of agricultural, forestry and municipal solid wastes for energy production is an effective waste use, reducing significantly the waste disposal problems, particularly in municipal areas.

7. Biomass is a domestic resource, not subject to world price fluctuations or the supply uncertainties as of imported fuels.

8. Biomass provides a clean, renewable energy source, improving the environment and energy security.

9. Biomass use is a way to reduce the carbon dioxide emissions.

However, there are a few major disadvantages of the use and applications of biomass, such as:

1. Biomass production is periodic with uncertain yields, basically one or two annual harvests, usually in summer or fall, depending significantly on the weather conditions, crops being available for only up to 4 month interval, in which all the materials must be harvested, pre-processed and stored to reduce losses.

2. Biomass is a diffuse energy source with very low conversion efficiencies, 1.5% or even lower with an annually power yields of about 1 W/m^2, requiring large land areas for commercially developments.

3. Further processing steps are needed, such as grinding, fermentation or combustion in order to make usable the biomass chemical energy.

4. Biomass is in strong competition with food production, and socioeconomic reasons may prevent the use of large agricultural land for biomass production.

5. Biomass, depending on the type and origin, has a highly variable energy density.

The major biomass types are agricultural, forest and marine biomass, energy crops, biomass from animal wastes, industrial and municipal wastes. Agricultural biomass is the residue from the field crops (e.g., stalks, branches, leaves, straw, or pruning wastes), processing by-products of the agricultural products (e.g., residues from cotton ginning, olive pits, fruit pits). The biomass from animal waste includes mainly waste from intensive livestock operations, poultry, pig or cattle farms or slaughterhouses, and it is a rich fuel source. Forest energy biomass consists of firewood, thinning and logging residues, materials cleared for forest fire protection, and by-products from wood industries. The municipal waste consists of solid wastes, such as garbage wastes, swage liquids, effluents from institutional activities, city and commercial wastes. There are several ways these wastes are recycled and materials are recovered for gas, liquid and solid fuels. Energy crops refer to the selected species of trees and shrubs for fuel production, harvestable in shorter time intervals. The energy crops depend on land and water availability and the plant management. The fuelwood is used directly into stoves or boilers or is processed into methanol, ethanol and producer gas. There are several species suitable for energy plantation, such as *Acacia, Dalbergia sissoo, Eucalyptus, Prosopis juliflora, Leucaena leucocephala*, etc. Floating water plants (e.g., water hyacinths) are pest plants in many rivers, lakes and ponds in tropical and semitropical regions. Their growth rates are very high, with net productivity of up to 25 tons of dry product per acre per year. A large potential and opportunity is expected from the microalgae bioenergy production, often characterized as the third bioenergy generation. The microalgae can produce over ten times more biomass than land crops per unit area, grows in saline, brackish or waste waters or in bioreactors and can be converted to oil, ethanol or hydrogen.

3.2.2 Biomass Conversion and Utilization Methods

Biomass can be converted into several secondary energy carriers (electricity, gaseous, liquid and solid fuels or heat) using various conversion methods or is used directly for heat and electricity generation. The raw materials are converted into usable energy forms by direct combustion, by bio- or thermochemical processes. Combustion results in heat or electricity, while biomass gasification results in gaseous fuels, used for various products: heat, electricity, synthetic natural gas (SNG), transport fuels and chemicals. If heat or electricity is required, combustion and gasification are the competing processes. If heat is the only product required, the direct combustion is the choice. Small-scale heat producing plants suffer from poor economics, especially if high emission standards must be met. Large-scale biomass gasification for SNG production and the subsequent distribution to users where the SNG is burned to produce heat is too an attractive choice. If electricity is the desired product, biomass combustion and gasification are the choices. The ways that biomass is converted for electricity or for biofuels are derived and adapted from the techniques used for the combustion of the fossil fuels or the methods used for chemical and physical transformation of the fossil fuels to other forms of fuels (e.g., coal gasification or liquefaction). The biomass conversion methods to fuels and electricity are distinguished in thermal, chemical and biochemical conversion, as shown in Figure 3.1. Bioenergy is generated through the conversion of the solid biomass, gaseous biomass, or biofuels (liquid biomass). During chemical reactions, energy is either released to the environment (exothermic reactions) or absorbed from the environment (endothermic reactions). During chemical reactions, bonds are

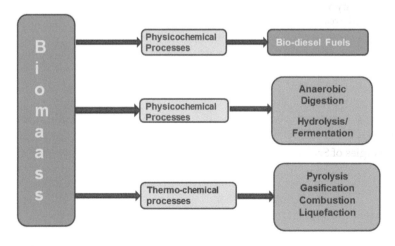

FIGURE 3.1
Bioenergy conversion flow diagram.

broken into the reactants and new bonds are made in the products. Bond-breaking is an endothermic process and bond-making is an exothermic process. The average bond dissociation energies of some chemicals are shown in Table 3.2. To generate energy, all biomass types are undergoing thermochemical processes, following a similar chemical equation:

$$Biomaterial + O_2 \Rightarrow CO_2 + Water + Heat \tag{3.3}$$

For any chemical reaction, the overall energy change, the reaction enthalpy (ΔH), is the difference of the energy absorbed in bond-breaking and the energy released in bond-making, being expressed by:

$$\Delta H = \sum BE(\text{bonds broken}) - \sum BE(\text{bonds formed}) \tag{3.4}$$

Here, BE is the bond energy of a reactant or a product. Combustion energetics can be estimated from the bond energies for all the common classifications of fossil fuels or biofuels. The amount of energy released is dependent on the oxidation state of the carbons in the hydrocarbon which is related to the hydrogen-to-carbon ratio. More hydrogen per carbon (H/C) exists, the lower the oxidation state and the more energy will be released during the oxidation reaction. Thus the greater is the H/C ratio, the more energy release on that combustion process. Combustion is a reaction of material with oxygen. In the case of the combustion of fossil fuels or biofuels, the combustion reaction is simply a burning process. In the combustion reaction, the species reacting with the oxygen are oxidized (oxygen is very electronegative element). Fossil fuels and biofuels are composed primarily of hydrocarbons (molecules containing carbon hydrogen bonds). In these molecules carbon is in a much reduced state. During the combustion reaction, the hydrocarbon molecules are converted to carbon dioxide and water, while thermal energy is released.

Example 3.3: Determine the thermal energy released during the methane combustion.

Solution: Applying the relationship of Equation (3.4) for the methane combustion, with the bond energy values of Table 3.2 the heat released during the methane combustion is determined from the balance between the input energy need to break the bonds (endothermic process) and the energy released on forming the bonds (exothermic process). First the methane combustion reaction is given by:

$$CH_4(g) + 2 \cdot O_2(g) \Rightarrow CO_2(g) + 2 \cdot H_2O(g)$$

TABLE 3.2

Bond Energies of Selected Chemical Components

Bond	Bond Energy (kJ/mol)	Bond	Bond Energy (kJ/mol)
H–H	432	C = O	799
O=O	494	C–C	347
O–H	460	C=C	611
C–H	410	C=C (aromatic)	519
C–O	360	N=O	623

$$\Delta H = \left[4\left(C\text{--}H \text{ at } 410 \text{ kJ/mol} \right) + 2\left(O\text{=}O \text{ at } 494 \text{ kJ/mol} \right) \right]$$

$$- \left[2\left(C\text{=}O \text{ at } 799 \text{ kJ/mol} \right) + 4\left(O\text{--}H \text{ at } 460 \text{ kJ/mol} \right) \right]$$

$$= -810 \text{ kJ/mol}$$

Every mole of methane (16 g) releases 810 kJ of thermal energy (heat) on the complete combustion (burning) in excess of oxygen.

3.2.3 Biomass Processes and Upgrading Procedures

There are several pretreatment techniques, from the well-established mechanical techniques, consisting of simply chopping, chipping or milling the raw feedstock into ready to use materials for the subsequent conversion, to less well-established thermomechanical or thermochemical upgrading techniques that are increasing the biomass energy density. Pelletization, torrefaction and pyrolysis technologies are such examples. Pellets are small wood-based cylinders 6–12 mm in diameter and 10–30 mm in length, produced by compressing wood sawdust. The high pressure of the press causes the temperature of the wood to greatly increases, causing the wood lignin content to form glue that is binding the pellet together as it cools. Pelletizing is an efficient energy densification technique as pellets typically have a bulk density of 650 kg/m^3, about 3.3 times higher than industrial softwood chips. Moreover, due to their very low water content, pellets also have a high net calorific value (or lower heating value) of about 17 MJ/kg, that is about 17% higher than wood chips. This property alone is making the material pelletization economically viable, reducing the transport and storage costs. Since pellets are mostly produced from sawdust, a sawmilling co-product, the quantity of the produced pellets depends on the volume of timber consumed in the wood industry. Biomass briquettes are fabricated in a similar way as pellets and have a typical dimension of 3–10 cm. Unlike pellets, which can be used for automaticallycharged stoves and boilers, briquettes require manual charging, making them a less user-friendly fuel. Torrefaction is a thermal process, involving slowly biomass heating at 200°C–300°C in the absence of oxygen. This degrades the biomass into a completely dry coal-like product that has lost the original biomass fibrous structure, and hence significantly improving grindability, net calorific value, from 19 to 23 MJ/kg and energy density. Torrefaction is a highly densification efficient means, with torrefied products retaining about 92% of the original feedstock energy. In addition, torrefaction transforms hygroscopic feedstocks into a hydrophobic material. This represents a significant advantage over traditional dried biomass such as pellets, since torrefied feedstock can be transported over long distances and stored outside without absorbing any moisture, hence without reductions of its calorific value. Although torrefaction is an old technique, it is not fully commercially available as a means of pre-treating method for biomass-to-energy production chains. While the torrefied biomass can be produced from a wide variety of biomass while yielding similar product properties, this upgrading technique is usually applied to wood feedstocks.

  The process of biomass harnessing energy can be compared to the heat generation from burning coal. During combustion process, the biomaterial and oxygen are combined in a high temperature environment to produce CO_2, water vapor and thermal energy. The generated heat depends on factors, such as climate, conversion process and biomaterial species, in average about 20 MJ/kg of thermal energy of fuel substrate are generated. Moisture in biomaterial lowers the heat content, fuels burn better when dry, and for combustion enhancement, the biomass water content must not exceed 20%. Processing biomass by grinding or drying is making it more suitable for combustion. Thermochemical processes involve the pyrolysis, liquefaction, gasification and supercritical fluid extraction methods. Thermochemical biomass conversion is one of the most common and convenient biomass energy conversion methods. This includes combustion, gasification, liquefaction and carbonization. In all these processes, pyrolysis plays a key role in the reaction kinetics and in reactor design and in determining product distribution, composition and properties. The thermochemical process products are divided into gases, vapors and tar components and carbon-rich solid residues. Biomass thermal conversion has received special attention since it leads to useful products and simultaneously contributes to solving pollution problems from biomass accumulation. Biochemical processes are essentially microbial digestion and fermentation. Biomass gasification is basically the solid biomass conversion into a gas mixture called producer gas through partial combustion in a gasifier. A biochemical process converts biomass to ethanol and methanol. Transportation fuels from biomass are at present mainly derived from sugar- or starch-containing crops (e.g., sugar-cane or maize). From lignocellulosic crops, advanced technologies are the conversion via gasification to methanol and hydrogen, the conversion to ethanol using hydrolysis and fermentation, and finally the conversion to long-chain hydrocarbon fuels. The process of converting biological material into energy begins with harvesting and processing, followed by thermochemical procedures where heat energy and chemical catalysts convert biological material into intermediate compounds. There are three common thermochemical processes: (1) combustion, requiring sufficient oxygen for oxidation, (2) gasification, requiring insufficient oxygen, in order to prevent complete oxidation, and (3) pyrolysis, which occurs in the oxygen absence, being the most common ways for the biomass utilization.

Direct combustion is used for cooking, space heating, providing heat to industrial processes and steam generation for electricity production, and being the simplest and earliest method of converting the chemically stored energy into heat. In direct combustion, the hydrogen and carbon in a fuel combined with oxygen, a process which releases heat. It is the most common thermochemical method of converting biomass to energy. Biomass burns in three successive but overlapping stages, first, the contained water evaporates and then there is distillation and probably combustion of volatiles and finally the high temperature reaction of fixed carbon with oxygen. Only the last two steps release heat and provide energy. Besides the traditional uses (cocking, space heating, etc.), direct combustion is industrially used for the generation of steam and electricity. However, direct combustion of biomass often requires adapted and special equipment to satisfy the constraints imposed by the properties of biomass feedstock materials, higher ash fuels, often low heating value of the biomass, due to high moisture and ash contents. Variation in feedstock composition and size distribution also creates difficulties in controlling combustion. Finally, biomass causing high corrosion or erosion is requiring special materials of construction. Every day, municipal, commercial and light industrial wastes are converted to energy in hundreds of biomass burning facilities throughout the world by direct combustion. More than 90% of the world solid wastes are processed into energy by biomass burning and refuse-fired energy systems. The efficiency of direct biomass combustion is usually low, often between 20% and 30% or often even much lower. *Co-firing* process consists of the combustion of a small percentage, up to 10% of locally produced biomass and coal, producing steam at high temperatures in the boiler and the power plant efficiency not degrades. Combustion is the conversion of a fuel into chemical compounds, the so-called *combustion products* by combination with an *oxidizer*. It is an *exothermic* chemical reaction (i.e., a reaction that releases energy as it occurs), and may be represented symbolically by:

$$\text{Fuel} + \text{Oxidizer} \Rightarrow \text{Combustion Products} + \text{Energy} \tag{3.5}$$

In Equation (3.4), the fuel and the oxidizer are reactants, i.e., the substances presented before the reaction takes place. Combustion process, consisting of the biomass and the oxygen reaction, resulting in carbon dioxide (CO_2), water and thermal energy (heat) can be represented by a relationship as:

$$C_6H_{10}O_5 + 6O_2 \Rightarrow 6CO_2 + 5H_2O + \text{Heat} \tag{3.6}$$

Example 3.4: Calculate the heat released by the complete combustion of $C_6H_{10}O_5$ (biomass), represented by Equation (3.6).

Solution: The energy balance related to the combustion chemical reaction of Equation (3.6), with the thermodynamic data of Table 1.1 is:

$$\text{Heat} = -\Delta H_{C_6H_{10}O_5} - 6 \times \Delta H_{O_2} + 6 \times \Delta H_{CO_2} + 5 \times \Delta H_2O = -789.0 - 0 + 6 \times 394.1 + 5 \times 286.0 =$$

$$= 3006.6 \text{ kJ/mol or } 3 \text{ MJ/mol}$$

Most of the forestry biomass and agricultural waste derived energy is coming from the wood combustion. There is a constant drive to improve the combustion efficiency to 30% or higher and to reduce pollutant emissions. The major developments are the large CHP systems. Direct combustion for heat or driving a steam cycles are mature technologies, new developments are toward better overall thermodynamic efficiencies of the steam cycle and biomass powder firing in ceramic gas turbines. The produced heat amount depends on the humidity of the biomass, the level of excess air required and whether or not complete combustion is accomplished. However, the modern combustion technology is well advanced, permitting widespread industrial applications. Two boiler types are commonly in use today: boiler with fixed or traveling grates, and boilers with fluidized-beds. The former type is very common, ranging from the household boiler to large-scale 50 MW industrial furnaces, and can accommodate heterogeneous combustible materials in terms of composition, humidity and granularity. On the other hand, the load following is rather difficult. In a fluidized-bed, the combustible particles, together with the granular bed material, are carried by a constant flow of gas in upward direction. The fuel is constantly injected into this bed. The bed itself constitutes the major heat capacity of the system and therewith stabilizes the process. In this way, effective heat and mass transfer are being taken care of. Such a system can combust a wide range of materials including fuels of non-biological origin. From an investment point of view, this fluidized-bed technology becomes attractive at plant sizes larger than 10 MW. Their main advantage is the possibility to use mixtures of various types of biomass and/or to co-fire them with other fuels. Nevertheless, compared to grate-fired boilers, their operation in partial load is problematic.

Biomass co-firing (or co-combustion) involves supplementing existing fossil-based (mostly pulverized coal) power plants with biomass feedstock. There are three types of biomass co-firing: direct co-firing, the biomass is combusted directly in the existing coal furnace, indirect co-firing, the biomass undergoes a pre-gasification conversion and the resulting syngas is combusted in the coal furnace, and parallel co-firing where the biomass is combusted in a separate boiler, with utilization of the steam produced within the main coal power station steam circuits. Over the past decade, direct co-firing has been successfully demonstrated with many technology options and with a wide range of biomass feedstocks (wood, herbaceous biomass, crop residues and energy crops). Direct co-firing is achieved in two ways:

1. The raw solid biomass is pre-mixed, in granular, pelletized or dust form, with the coal in the coal handling system or the ORC engine is similar to steam engine but works with low boiling temperature organic oil as a process fluid instead of steam.
2. The milled biomass to a size of 5 mm and is directly injected into the pulverized coal firing system.

Direct co-firing of a range of liquid biomass materials (e.g., vegetable oil, tallow) in existing plants is also practiced on a commercial basis, albeit at much smaller scale than for the solid materials. In most cases, the biomass co-firing ratio is limited to 5%–10% on a heat input basis, and this is controlled by the availability of biomass and in some cases, by site-specific plant constraints. In one or two cases, co-firing ratios of up to 25% have been achieved. Direct co-firing in large-scale modern coal plants is today the most cost effective use of biomass for power generation. This technology only requires minor investment to adapt handling and feeding equipment without noticeably affecting boiler efficiency, provided the biomass is not too wet and has been pre-milled to a suitable size. Furthermore, electric efficiencies for the biomass-portion range from 35% to 45%, which is generally higher than the efficiency of biomass dedicated plants. In spite of the significant progress achieved in co-firing over the last decade, biomass properties pose several challenges to coal plants that may affect their operation and lifetime. Most of the potential issues faced by co-firing are associated with the biomass ashes which are very different from coal ashes. Problems arise mainly at increased co-firing ratios and with biomass materials with high ash contents. The technical risks are mainly associated with the increased ash deposition on surfaces in the boiler and in SCR catalysts (thus reducing the efficiency of the system), and with the impact of flue gas on gas cleaning equipment. The contamination of ashes by alkaline metals is relatively well understood and, in Europe, this has largely been recognized in performance standards for the utilization of ashes in the manufacture of building products. Indirect and parallel co-firing options are designed to avoid biomass-related contamination issues, but have proven much more expensive than the direct co-firing approach as additional infrastructure is needed. Parallel co-firing units are mostly used in pulp and paper industrial power plants. The indirect option faces issues regarding the cooling and cleaning of the syngas.

Gasification techniques consist of biomass heating to convert in combustible gas, volatile components, and ash. The gasification technology varies based on the gasification agent or the reactor, but it is often more demanding because of feedstock specifications. Municipal solid waste and agricultural residues are the common gasification feedstocks. Under controlled conditions, characterized by low oxygen supply and high temperatures, biomass materials can be converted into a gaseous fuel known as *producer gas*, consisting of carbon monoxide, hydrogen, carbon dioxide, methane and nitrogen. Producer gas is the mixture of combustible and noncombustible gases. The quantity of gases constituents of producer gas depends upon the type of fuel and operating condition. The heating value of producer gas varies from 4.5 to 6 MJ/m^3 depending upon the quantity of its constituents. In energy terms, the conversion efficiency of the gasification process is in the range of 60%–70%. The gasification reaction is an endothermic one, and the heat is supplied to the system by external sources, occurring in two endothermic steps. Biomass is first heated to over 700°C, which vaporizes the volatiles, e.g., hydrogen, CO, CO_2, and other hydrocarbon gases. The by-products are charcoal and ash. In the second step, the charcoal is gasified through reactions with oxygen, steam and hydrogen at higher temperatures. The gasification products include syngas, bio-charcoal and tar. The specific amount of each depends on the feedstock, oxidizing agent and the process conditions. Syngas, consisting of CO, CH_4 and other hydrocarbons is used for heating, electricity generation and as CHP fuel or for the production of ethanol, diesel and chemicals. Gasification process has high conversion efficiencies and is well suited for large power plants to achieve full potential. The process required heat can be brought to the reaction zone through the reactor wall, by the bed material itself or through a hot process gas stream. Due to the fact that no air (oxygen) is taken up into the process, a product gas with middle or high calorific value is produced (10–18 MJ/m^3). Such high calorific value gas is attractive since volume streams are reduced making down-stream processing like gas cleaning, compression or any catalytic process relatively simple and therefore cheaper. In this respect a wealth of highly advanced processes are being developed or are in the demonstration phases at a sizeable scale. The other way to generate the heat necessary for the gasification reaction is to partially biomass combustion giving the most direct supply of heat to the gasification process itself. The overall gasification reaction, producing carbon monoxide, hydrogen, thermal energy and some CO_2 and H_2O in the fuel gas stream is:

$$2 \cdot C_6H_{10}O_5 + O_2 \Rightarrow 12 \cdot CO + 10 \cdot H_2 + 3.70 \text{ MJ/kg} \qquad (3.7)$$

Pyrolysis (carbonization) is the process of the thermal degradation of carbonaceous material in the absence of air or oxygen, being on major process of thermochemical conversion, broadly defined as a system which the carbonaceous materials are thermally decomposed. Pyrolysis is usually carried out at near atmospheric pressures and temperatures of up to about 1100°C. Several factors, such as the final temperature, the type of biomass, the heating rate, amount of oxygen present and the equipment design all affect the yield and the product make up. Generally charcoal yields are 30%–40% of the dry biomass feed. Gaseous product yields based on the weight of the dry mass fed, ranging from 5% to 20% depending on temperature. The main constituents of the gas are usually carbon monoxide, carbon dioxide, hydrogen and methane with higher proportions of hydrogen at elevated temperatures. Temperatures in the range of 350°C–800°C are the most used in the pyrolysis processes. Gas, liquid and solid products (char or coke) are produced in the pyrolysis reactions, but the amounts of each are determined by controlling the reaction temperatures and residence time. Either a solid carbon material or a liquid, produced by pyrolysis process can be of interest in combustion with existing systems for large-scale electricity production. The solid pyrolysis product, char, has properties similar to coal, therefore is easily accommodated as a renewable or a CO_2 emission free energy carrier, and can be mixed with coal. Same applies to the liquid pyrolysis product, but in this case separate injection technology is needed in coal fired boilers. If biomass fuel is to be imported, to meet national goals for renewable energy or to achieve emission reduction, the pyrolysis can be a step to be taken at the location of the biomass production, so only highly concentrated energy carriers are transported. The output of the desired product can be maximized by careful control of the reaction conditions. Further, the pyrolysis is applied to reduce the size of waste streams, like electronic scrap, plastics, etc. In some processes, even high calorific value gas is generated, precious metals are recovered and environmentally hazardous wastes are removed. Technologies combining pyrolysis and gasification are developed to overcome the drawbacks of both technologies separately by combining their advantages. Like in the case of advanced gasification, these processes need demonstration at realistic scales. Methane, ethane and methanol, produced through pyrolysis, are used in gas turbines, space and water heating and in transportation. Overall efficiency of 35% by weight can be achieved by maximizing the output of the solid product through the implementation of longer reaction times and low temperatures (350°C). Heat for the process in traditional kilns is produced by burning the gas and liquid by-products. Higher temperatures and shorter residence times are used to produce pyrolysis oils. Optimum conditions are approximately 500°C with reaction times up to 2 seconds. The short reaction time is achieved by rapid quenching of the fuel, which prevents further chemical reactions taking place. Preventing additional chemical reactions also allows higher molecular weight molecules to survive. The fast heating and cooling requirements of raw materials create process control problems, which are most often dealt with by fine milling the feedstock. Pyrolysis is usually carried out at near atmospheric pressures and temperatures up to about 1100°C. Several factors such as the final temperature, the type of biomass, the rate of heating, the proportion of oxygen present, and the equipment design all affect the yield and the product make up. Generally charcoal yields are 30%–40% of the dry biomass feed. Gaseous product yields based on the weight of the dry mass fed range from 5% to 20% depending on temperature.

Hydrolysis and *fermentation* processes are producing liquid fuels, such as methanol and ethanol, which are commonly used in combustion engines, gas turbines, steam generation as a gasoline and diesel additives. Acid hydrolysis is a well-established process and nearing commercialization. Enzymatic hydrolysis is at the later stages of research and development, and is starting to be demonstrated at larger scale. Separation can be performed by distillation. However, this is very energy intensive, other novel and less energy intensive options are being explored. The process whereby hydrolysis enzyme production, cellulose hydrolysis, hexose fermentation and pentose fermentation all take place in different steps (i.e., in different bioreactors) is called *separate hydrolysis and fermentation* (SHF). Processes exist which combine these steps are in development to make the overall process potentially quicker and cheaper.

Liquefaction is another technique that uses high temperatures to convert biomass. It can be both a direct and indirect process of thermo chemical conversion. The former is usually catalytic, the feed being first converted into a gaseous intermediate from which liquid fuels are then synthesized. In one of the direct liquefaction processes, the biomass slurry containing the catalyst is fed to a high pressure (but not more than 280 kg/cm²) and medium temperature (340°C) reactor where a liquefaction reaction takes place in a reducing gas atmosphere, with carbon monoxide and/or hydrogen also being fed to the reactor. A series of complex reactions occur simultaneously, such as pyrolysis, gasification and volatilization. Hydrogasification by pyrolysis in the presence of the hydrogen produces mainly methane and water. The gas, containing the combustibles, is removed and can be used as a heat source to pre-dry the wood feedstock. The liquid-solid mixture is then separated by distillation. The bottoms containing residual solids can be recycled together with a portion of the oil which is not removed as product. After the oil product is separated, vacuum distillation allows the desired product. The product composition depends on the biomass feed. In the direct liquefaction process, there are two stages. In the first, the biomass is gasified to an intermediate species which in the second is converted to methanol, gasoline or polymerized liquid hydrocarbons. The intermediate mixture could be synthesis gas (carbon monoxide and hydrogen), light olefins (ethylene and propylene), etc. For example, ethanol synthesis has been in use for several decades.

Anaerobic decomposition occurs naturally in the materials at the municipal and agricultural waste sites, in the sewage treatment facilities and in animal waste treatment plants. The decomposition process produces methane, carbon monoxide and dioxide in various proportions. The resulting combustion gas has a low heating value, but can be mixed with other fuels and can be used in gas turbines or steam power units.

A conventional reactor is mixed, fed once or more per day, heated to a temperature of 35°C and operated at a retention time of 20–30 days and loading rate of 1.7 kg VS $m^3 \cdot d^{-1}$. Under these conditions about 60% reduction in organic matter is achieved corresponding to a methane yield of 0.24 m^3 per kg VS added. The biogas composition is typically 60% methane and 40% carbon dioxide with traces of hydrogen sulfide and water vapor. The conventional design is being replaced by more innovative designs influenced primarily by feed suspended solids content. The objectives of these designs are to increase solids and microorganism retention, decrease reactor size and reduce process energy requirements. Improved designs have increased possible loading rates 20-fold, reduced residence times and improved process stability. In the subsequent chapter section the anaerobic digestion is discussed.

Combustion, pyrolysis, and gasification have many common similarities but differ in their end uses and product ratios. When choosing a suitable mechanism for energy production, in the selection must be considered the desired final products, such as gas, biochar, or only heat, and their end uses, such as electricity generation, heat, or transportation fuel.

3.2.4 Renewable Methane from Biomass

Biomass gasification involves thermal conversion to simple chemical building blocks that can be transformed into fuels, products, power, and hydrogen. Resource potential estimates for terrestrial biomass give 22 EJ while for feedstocks like grass, wood, and seaweed it is 7 EJ. The potential for marine biomass is huge at greater than 100 EJ per year. Biogas contains 50%–70% methane and 30%–50% carbon dioxide as well as small a amounts of other gases with calorific value of about 21–24 MJ/m^3. Bio-methane can be used as vehicular fuel. As bio-methane-genesis decomposes organic matter with production of useful energy products, anaerobic digestion of organic matter is receiving increased attention. Solid and agricultural wastes release undesired methane into the atmosphere due to anaerobic digestion in landfills, lagoons or stockpiles. Treatment and recovery of this gas in reactors would reduce this source of atmospheric methane. An attractive option for treatment of the organic fraction of these wastes is to separately treat organic fraction by composting and applying the stabilized residues in land as a soil amendment. The residues can reduce the water needs and can prevent the erosion. As bio-methane-genesis decomposes organic matter with production of useful energy products, anaerobic digestion of organic matter is receiving increased attention. Solid and agricultural wastes release undesired methane into the atmosphere due to anaerobic digestion in landfills, lagoons or stockpiles. Treatment and recovery of this gas in reactors would reduce this source of atmospheric methane. An attractive option for treatment of the organic fraction of these wastes is to separately treat organic fraction by composting and applying the stabilized residues in land as a soil amendment. The residues would reduce water needs and prevent erosion. Nevertheless, fermentable household waste is comprised of materials rich in sugars, minerals, and proteins that could be used for other processes as substrates or raw materials. On the other hand, less studies and research were conducted about the effectiveness of anaerobic digestion for methane (CH_4) production and anaerobic fermentation for hydrogen (H_2) production, for example from dried and shredded food waste, and this are requiring additional research efforts in order to improve the commercialization of the methane and hydrogen biomass and waste production.

3.3 Biofuels

Biofuel is fuel produced directly or indirectly from biomass such as fuelwood, charcoal, bioethanol, biodiesel, biogas (methane) or bio-hydrogen. However, most people associate biofuel with liquid biofuels (bioethanol, biodiesel and straight vegetable oil). In this chapter the term *biofuels* refers to liquid biofuels used for transportation or industrial heat generation. Biomass seems to be the only renewable alternative for liquid transportation fuel. Biofuels are fuels obtained from renewable feedstock and available in liquid as well as gaseous form. First-generation biofuels were produced from fermentation of grains, cereals, sugar, starch, crops such as sugar cane, sugar beet, wheat, corn and vegetable oil based biofuels such as pure vegetable oils and Biodiesel produced from oil seed crops such as rapeseed, soybean, palm, tallow oils, waste cooking oil and animal fats. The second-generation biofuels are fuels produced from lignocellulosic materials such as plant stalks, leaves, wood, tall grasses, crop residues, ethanol produced by using enzymes, synthetic diesel via gasification process, liquid fuels from agricultural waste and switch grass. The major difference between first- and second-generation biofuels is in the resources used in production and methods of production. Biofuels are an important contribution to our energy supply, and have historically made up the majority of the energy requirements and demands, prior to era of the fossil fuels. Fuels from biomass are directly related to the ethanol, biodiesel, biomass power, and industrial process energy supply. Biofuels derive from both biomass and biogas sources and includes biodiesel, methanol, butanol, and ethanol, with the latter two as the most common sources. Although fermentation via lignocellulosic material can produce bioethanol, most biofuels originate or are converted from once-living organisms through agricultural processes or anaerobic digestion. The energy content of biofuel varies by fuel source but produces around 20 MJ of energy per liter for ethanol and 34 MJ per liter for biodiesel, values that change depending on the plant species and their specific energies. Biofuels are widely used as transportation fuels, but also for heat and electricity production. There are basically two major classes of biofuels: ethanol and biodiesel. Bio-methanol is a fuelwood-based methyl alcohol obtained by the destructive distillation of wood or by gas production through gasification. Bio-methanol is a substitute product for synthetic methanol and is extensively used in the chemical industry (methanol is generally manufactured from natural gas and, to a lesser extent, from coal). The diagram of the conversion process of biomass-to-biofuel is shown in Figure 3.2.

Ethanol is an alcohol fuel (ethyl alcohol) made by fermenting the sugars and starches, that are found in plants and then distilling the fermenting products. Any organic material containing cellulose, starch, or sugar can be made into ethanol. More than 90% of the ethanol produced in the United States is from corn. So the ethanol is obtained through sugar fermentation (sugarcane), by starch hydrolysis or cellulose degradation followed by sugar fermentation, and finally by subsequent distillation. Obtaining alcohol from vegetable raw materials has a long tradition in agriculture. The sugar fermentation derived

FIGURE 3.2
Flow diagram of the biomass-to-biofuel conversion.

from agricultural crops using yeast to produce alcohol, followed by distillation, is a well-established commercial technology. Alcohol is also efficiently produced from starch crops, such as wheat, maize, potato and cassava, first obtaining glucose by starch hydrolysis, and then fermenting it to produce alcohol. The goal of directed use of cellulose-containing biomass from agriculturally utilized species for producing alcohol has not yet been practiced on a large-scale level. Production has been confined to the use of wood, residues and waste materials. Biodiesel is a fuel made by reacting chemically the alcohol with vegetable oils, animal fats, or recycled greases. Most biodiesel today is made from soybean oil, often blended with petroleum diesel in ratios of 2% (B2), 5% (B5), or 20% (B20), but is also used as pure biodiesel (B100). Biodiesel fuels are compatible with and can be used in unmodified diesel engines with the existing fueling infrastructure. Biodiesel exceeds diesel in cetane index (diesel performance rating), resulting in superior ignition, has higher flashpoint and is more versatile where safety is an issue. Horsepower, acceleration, and torque are comparable to the oil diesel. Biodiesel has the highest Btu content of any alternative fuel, though it is slightly less than that of diesel, which may have a small impact on vehicle range and fuel economy.

Legislative mandates are enacted or proposed regarding pollutant emission reductions from transportation fuels, while there are several ways to achieve it, the biofuels are a major one. This raises questions about the possible implications of substituting biofuels for oil fossil fuels. In particular, biofuels are often touted as a method of saving our planet from a human produced climate changes. However, there are a few concerns about biofuels. The water required to produce biofuels is often debated with advocates arguing that ethanol production requires less water than gasoline, while opponents note that a water shortage throughout the world is already having a detrimental effect on agricultural productivity. One of the main biofuel criticisms is that the production of biodiesels may be energetically inefficient. In other words, the biofuel production requires more energy than can be obtained burning them. There are claims that producing ethanol from corn or switch-grass, and biodiesel from soybeans or sunflower requires higher energy inputs from fossil fuels than is gained from these biofuels. Such low efficiencies in the biofuel production are arising as a consequence of the very low photosynthesis efficiency, the slow growth of the crops, and the energy-intensive extraction processes. However, the lower estimates are often considered misleading because are neglecting the by-product energy values, and considering them the biofuels are more economically viable (especially when government subsidies help lower the direct cost to consumers, at the expense of indirect costs in the form of taxes). Because about 15% of the world population is undernourished it might be not wise to remove agricultural land from needed food production. In order to continue increasing biofuel production may inevitable put on the land needed for growing food or destroy forests which would only further contribute to climate changes by releasing carbon sequestered in woodland. These are ethical constraints on the biofuels' production and raises concerns about potential food shortages and food price inflation. A viable alternative to using land-based crops for producing biofuels is to use algae, or microalgae (photosynthetic microorganisms), as the major source of the biofuel. The ability to grow on non-agricultural land, using untreated water (potentially even treating it in the process), and consuming industrial CO_2 in the process, can help to develop renewable bioenergy in a sustainable and responsible way. Microalgae have attracted attention because under conditions of nitrogen deprivation, they produce large amounts of lipids (oil) with dry weights reported to be as high as 60%, opening the possibility of using algae for biodiesel. The fact that they yield significantly more biomass when compared to terrestrial plants has only added to the recent hype. However, under conditions of high stress, the nitrogen deficiency required to produce high lipid contents, the yields are drastically reduced. Optimizing the growth rate and oil production for different algal species is an important area of research.

Biogas can also be used as fuel in internal combustion engines. The CNG technology that is currently available in India can be used in both ways as biogas and an automotive fuel. Wood gas is the third alternative representing standardized fuel made from biomass. This technology does not lend itself well to being used in domestic cook-stoves, but larger stoves, used in bakeries, cafeterias or restaurants can be based on it. However, wood gas is currently used as fuel in internal combustion engines for generating electricity. Many such units are today in operation. Biogas based electricity generation should be seriously considered by planners and administrators as a means of supplying electricity to remote villages. The electricity demand of a village is usually not very high, and supply of this electricity from a central generating facility is very expensive because of the system capital expense. There are also losses and theft of electricity when it is transmitted over longer distances. The village level generators are operated by the villagers themselves. They can generate electricity as and when they want and can use it for whatever purpose they want. The issues and problems of alcohol for fuels are similar to those of biodiesel. Currently, alcohol for fuels is made from molasses, a free by-product of the sugar industry. As the cost of sugarcane, its harvest, transport, and processing are borne by sugar, the present cost of alcohol is low. But if crops like sugarcane, sugar beet or sweet sorghum are grown exclusively for alcohol production, the above mentioned costs would have to be borne by alcohol, which then would not be so cheap.

3.3.1 Solid, Liquid and Gaseous Biofuels

Biofuel is a type of fuel whose energy is derived from biological carbon fixation. Biofuel, in a most common understanding, is a liquid energy fuel that can be produced from biomass conversion or carbon fixation through photosynthesis. Biofuels include fuels derived from biomass conversion, as well as solid biomass, liquid fuels and various biogases. Although fossil fuels have their origin in an ancient carbon fixation, they are not considered biofuels by the generally accepted definition because they contain carbon that has been *out of the carbon cycle* for a very long time. Biofuel is considered practically carbon neutral, as the biomass absorbs roughly the same amount of carbon dioxide during growth, as when burnt. Biofuels are fuels produced directly or indirectly from biomass such as fuelwood, charcoal, bioethanol, biodiesel, biogas or bio-hydrogen. Ethanol is the most widely used biofuel in the United States today. It is usually available at almost all local gas stations. Bioethanol is an alcohol made by fermentation, from carbohydrates produced in sugar or starch crops such as corn or sugarcane. Cellulosic biomass, derived from non-food sources such as trees and grasses, is also being developed as a feedstock for ethanol production. Ethanol can be used as a transport fuel in its pure form, but it is usually used as a gasoline additive to increase octane and improve vehicle emissions. Current plant design does not provide for converting the lignin plant materials to fuel components by fermentation. The second main biofuel is biodiesel, being one of the most important biofuels, made from vegetable oils and animal fats. Biodiesel can be used as a fuel for vehicles in its pure form, but it is usually used as a diesel additive to reduce levels of particulates, carbon monoxide, and hydrocarbons from diesel-powered vehicles. Biodiesel is produced from oils or fats using transesterification and is the most common biofuel in Europe. First-generation biodiesel has traditionally been produced from oil seeds. The oil is extracted from the seed using mechanical operations, providing heat, and typically using a solvent. There are several procedures, but most of them are making use of any of these three options. Therefore, once the oil is extracted, it is separated from the solvent using flash distillation. The high difference in the boiling point between the oil and the solvent simplifies the process. As in the case of the bio-ethanol, seeds and grain are food, which pushed the industry to substitute the oil source to nonedible seeds such as cooking oil. In this case, the impurities must be taken into account before processing the oil. The biodiesel production from oil is based on the transesterification reaction of the oil with alcohols. Basically the oil consists of three chains of hydrocarbons whose viscosity is difficult to process. Therefore, the idea is to break it down into three chains, which not only is reducing the viscosity of the mixture, but also in this way the properties of the product match those of the crude-based diesel. The transesterification is an equilibrium reaction between the oil and alcohols. For economic reasons, methanol has been used for a long time. From the technical point of view, it provides high yield to biodiesel, fatty acid methyl ester, and quick reaction times.

There are four basic groups of plant species rich in lignin and cellulose that are suitable for conversion into solid biofuels (as bales, briquettes, pellets, chips, powder, etc.): annual plant species (e.g., cereals, pseudo-cereals, hemp, kenaf, maize, rapeseed, mustard, sunflower, and whole plant reed canary grass), perennial species harvested annually (e.g., reeds family), fast-growing tree varieties (e.g., poplar, aspen or willow) with a perennial harvest rhythm (short cutting cycle), short rotation coppice, and tree species with a long rotation cycle. The raw materials are often used directly after mechanical treatment and compaction, or are converted to other types of biofuels. The lignocellulose plant species offer the greatest potential within the worldwide biomass feedstock array. The major important basic processes are for conversion of lignocellulose biomass into fuels suitable for electricity production: biomass direct combustion to produce high grade heat, advanced gasification to produce fuel gas of medium heating value, and flash pyrolysis to produce bio-oil, with the possibility of upgrading to give hydrocarbons similar to those in mineral crude oils. The production of methanol or hydrogen from woody biomass feedstock (such as lignocellulose biomass woodchips from fast growing trees) via processes that begin with thermochemical gasification can provide considerably more useful energy per hectare than the production of ethanol from starch or sugar crops and vegetable oils like rapeseed methyl ester (RME, fuel derived from rapeseed oil). However, biomass-derived methanol and hydrogen are more expensive than conventional hydrocarbon fuels, unless oil prices rise to a level that is far higher than expected prices in coming decades. The most promising thermochemical conversion technology of lignocellulose raw materials currently available is the production of pyrolytic oil or *bio-oil* (or bio-crude oil), currently produced by flash or fast pyrolysis processes at up to 80% weight yield. It has a heating value of about half that of conventional fossil fuels; however, it can be stored, transported and used everywhere, where the conventional liquid fuels are used, such as boilers, kilns and turbines. It can be easily upgraded through hydro-treating or zeolite cracking into hydrocarbons for combustion applications such as gas turbines or further refined into gasoline and diesel as transportation fuels. In addition, there is significant unexploited potential of bio-oils for the extraction and recovery of specialized chemicals. The bio-oil technologies are evolving rapidly with improving process performance, larger yields and better quality products. Catalytic upgrading and extraction are also showing considerable potential for transportation of fuels and chemicals; however, such technologies are still in earlier development and research stages. The utilization of these products is of major importance for the industrial applications of these technologies, with investigations underway in several laboratories and companies. The economic viability of these processes is promising in the medium term, and their integration into conventional energy systems presents no major problems.

Charcoal is manufactured by traditional slow pyrolysis processes or as a by-product from flash pyrolysis. It can be used industrially as a solid fuel/reductant, for liquid slurry fuels, or for the activated charcoal manufacture. Biofuels have an average energy content that is approximately equal to that of brown coal (20 MJ/kg). There are no expected environmental impacts or any disadvantage created by mixing some alcohol percentage with diesel or gasoline fuel, and no legal constraints at moment. There are several methods for processing wood residues to make them cleaner, easier to use or to transport, the most common being the charcoal production. It is worth mentioning that the conversion of wood fuel to charcoal does not increase the fuel energy content. Charcoal is often produced in rural areas and transported for use in urban areas. The wood is heated in the absence of sufficient oxygen, so a full combustion is not occurring, allowing pyrolysis to take place, driving off the volatile gases and leaving the charcoal. The removal of the moisture means that the charcoal achieves higher specific energy content than wood. Other biomass residues such as millet stems or corncobs can also be converted to charcoal. Charcoal is produced in a kiln or pit. A typical traditional earth kiln comprises the fuel to be carbonized, which is stacked in a pile and covered with a layer of leaves and earth. Once the combustion process is underway the kiln is sealed, the charcoal is removed when the process is complete and kiln cooled. A simple improvement to the traditional kilns is adding a chimney and air ducts, allowing for a sophisticated gas and heat circulation. There is a little capital investment and a significant increase in yield is achieved.

Briquetting, a common used method for solid biomass and waste, is the compression of loose biomass materials. Many waste products, such as wood residues and sawdust from the timber industry, municipal waste, bagasse from sugar cane processing, or charcoal dust, are briquetted to increase compactness and transportability. Briquetting is often a large-scale commercial activity and often the raw material will be carbonized during the process to produce a usable gas and also a more user-friendly briquette. Pellets are more highly compressed loose biomass material than briquettes (ca. 650 kg/m^3, diameter 6 mm, length 30–40 mm, ash contents 1% and water contents less than 10%). Pellets are an important and rapidly growing biofuel for the production of heat and electricity. Pellets are developing quickly in the coming years, to the benefit of the environment and local economies. Increased utilization of wood, agricultural residues, straw and industrial by-products in the Nordic EU countries, all over Europe and in North America uses are indicating that pellets have proven to be a realistic alternative to fossil fuels. The substitution of fossil fuels with pellets, for heating of buildings and co-generation of electricity, can play an important role in fulfilling the emission reduction and the commitments for a secure supply of energy. The potential for expansion of the pellet industry is significant. The Swedish Pellet Producers Association has estimated a market in Europe of 4–5 million tons per year within the next 5 years. Pellets will make an important contribution to the world strategy for renewable energy sources**.**

3.3.2 Biomass to Ethanol Conversion

Ethyl alcohol or ethanol (C_2H_5OH), sometimes also called grain alcohol, is the alcohol that can be used as fuel in (modified) internal combustion engines, having higher heating value as fuel quoted typically at 29.847 MJ/kg (108, 500 BTU/gal) about 2/3 of that of gasoline. The ethanol production begins with the sugar fermentation, which is extracted from sugarcane, sugar beets or sorghum. A variety of carbohydrates (starches, hemicellulose and cellulose) can also serve as feedstock if the carbohydrates are broken into sugars that can undergo fermentation. Simple sugars are carbohydrates, while the glucose and fructose are monosaccharides and sucrose is a disaccharide of the two combined with a bond. Glucose and fructose have the same molecular formula ($C_6H_{12}O_6$) but glucose has a six-member ring and fructose has a five-member ring structure. Fructose is known as the fruit sugar as its make source in the diet is fruits and vegetables. Honey is also a good source. Glucose is known as grape sugar, blood sugar or corn sugar as these are its rich sources. Sucrose, having a formula similar to those of simple sugars is carbohydrate types. Glucose and fructose are monosaccharides and sucrose is a disaccharide of the two combined with a bond. Glucose and fructose have the same molecular formula ($C_6H_{12}O_6$) but glucose has a six-member ring and fructose has a five-member ring structure. Fructose is known as the fruit sugar as its make source in the diet is fruits and vegetables. Honey is also a good source of fructose. Sucrose, having molecular formula ($C_{12}H_{22}O_{11}$), is what is known as sugar or table sugar, typically extracted as cane or beet sugar. If sucrose is treated with acid or heat, it hydrolyzes to form glucose and fructose. This mixture of sucrose, glucose and fructose is also called invert sugar. Ethanol is used to improve the gasoline octane ratings and emission characteristics, usually in the 10% blend format, the E10 (10% ethanol mixed with 90% gasoline). Anaerobic fermentation processes may be used not only to produce gases but also to produce liquid fuels from biological raw materials. The ability of yeast and bacteria to ferment sugar-containing materials to form alcohol is well-known from beer, wine and liquor manufacture. If the initial material is sugarcane (glucose), or in the case of sucrose the fermentation reactions are expressed then as:

$$C_6H_{12}O_6 + H_2O \Rightarrow 2C_2H_5OH + 2CO_2 + \text{Heat} \tag{3.8}$$

And, respectively,

$$C_{12}H_{22}O_{11} + H_2O \Rightarrow 4C_2H_5OH + 4CO_2 + \text{Heat} \tag{3.9}$$

However, the released heat during these reactions is very small with almost all the sugar energy is stored into alcohol. The energy content of ethanol, heat combustion is 30.5 MJ/kg (higher enough to substitute gasoline), compared with that of glucose 15 MJ/kg and its octane rating is 89–100. With alternative fermentation bacteria, the sugar may be converted into butanol, having formula $C_2H_5(CH_2)_2OH$. In almost of sugar-containing plant material, the glucose molecules exist in polymerized form such as starch or cellulose, of the general structure ($C_6H_{10}O_5$)n. Starch or hemicellulose is degraded to glucose by hydrolysis, as shown in Figure 3.3, while the lignocellulose resists degradation owing to its lignin content. Lignin glues the cellulosic material together to keep its structure rigid, whether it is crystalline or amorphous. Wood has high lignin content (about 25%), and straw also has considerable amounts of lignin (13%), while potato or beet-starch contains very little lignin. Some of the lignin seals may be broken by pretreatment, ranging from mechanical crushing to the introduction of swelling agents causing rupture. The hydrolysis process, in earlier times, hydrolysis was always achieved by adding an acid to the cellulosic material. Acid recycling is incomplete; with low acid concentration the lignocelluloses is not degraded, and with high acid concentration the sugar already formed from hemicellulose is destroyed.

The outcome of the glucose fermentation process is a water-ethanol mixture. When the alcohol fraction exceeds about 10%, the fermentation process slows down and finally halts.

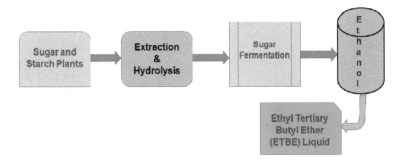

FIGURE 3.3
Ethanol production process.

Therefore, an essential step in obtaining fuel alcohol is to separate the ethanol from the water. Usually, this is done by distillation, a step that may make the overall energy balance of the ethanol production negative. The sum of all agricultural energy inputs (fertilizer, vehicles and machinery), process inputs (cutting, crushing, pretreatment, enzyme recycling, heating for different process steps from hydrolysis to distillation) and transportation energy is around 1.5 times the overall energy outputs (alcohol and fertilizer if it is utilized). However, if the inputs are domestic fuels, combustion of agricultural residues, and if the alcohol produced is used to displace imported oil products, the balance might be quite acceptable from a national economic point of view. If, further, the process lignin-containing materials are recovered and used for process for distillation, then such energy should be counted not only as input but also as output, making the total input and output energy roughly balance. Furthermore, more sophisticated process design, with cascading heat usage and parallel distillation columns operating with a time displacement such that heat can be reused from column to column, could reduce the overall energy inputs to 55%–65% of the outputs. Radically improved energy balances can emerge if the distillation is replaced by a less energy-intensive separation method.

Example 3.5: Hydrolysis and fermentation of fructose, known as the fruit sugar, produces ethanol and carbon dioxide, through the chemical reaction of Equation (3.8). Determine the amount of ethanol produced from 200 kg of fructose.

Solution: The molecular masses, expressed in kg/kg·mol of carbon, hydrogen, oxygen and hydroxyl, OH are as:

$$C = 12.01 \frac{kg}{kg \times mol}, H = 1.008 \frac{kg}{kg \times mol}, O = 16.0 \frac{kg}{kg \times mol}, OH = 17.008 \frac{kg}{kg \times mol}$$

The molecular mass of fructose, water, carbon dioxide and ethanol are then equal to:

$$Fructose\left(C_6H_{12}O_6\right) = 12 \times 12.01 + 12 \times 1.008 + 6 \times 16.0 = 252.216 \frac{kg}{kg \cdot mol}$$

$$Water\left(H_2O\right) = 18.016 \frac{kg}{kg \cdot mol}$$

$$CO_2 = 44.01 \ \frac{kg}{kg \cdot mol}$$

and

$$Ethanol(C_2H_5OH) = 6 \times 12.08 + 6 \times 1.008 + 16.0 = 46.068 \ \frac{kg}{kg \cdot mol}$$

From 200 kg of fructose the amount of the ethanol produced is:

$$Ethanol \ Amount = \frac{2 \times 46.068}{252.216} \times 200 = 73.061 \ kg$$

The ethanol density is 789 kg/m³, then the volume of the ethanol produced is:

$$Produced \ Ethanol \ (Volume) = (73.061 / 789) \times 1000 = 92.6 l$$

The ethanol fuel can be stored and used in the transportation sector much the same way as gasoline. It can be mixed with gasoline or can fully replace gasoline in spark ignition engines with high compression ratios. The knock resistance and high octane number of ethanol make this possible, and with preheating of the alcohol (using combustion heat that is recycled), the conversion efficiency can be improved. Several countries presently use alcohol and gasoline blends with up to 10% ethanol. This does not require any engine modification. Changing the gasoline Otto engines is inconvenient in a transition period, but if alcohol/ethanol distribution networks are implemented and existing gas stations modified, then the car engines could be optimized for alcohol fuels without regard to present requirements. A possible alternative to spark ignition engines is compression ignition engines, where autoignition of the fuel under high compression (a ratio of 25) replaces spark or glow plug ignition. With additives or chemical transformation into acetyl, alcohol fuels could be used in this way. Ethanol does not blend with diesel oil, so mixtures do require the use of special emulsifiers. However, diesel oil can be mixed with other biofuels without problems, e.g., the plant oils (rapeseed oil, etc.) presently in use in Germany. The leading feedstock for the biomass ethanol production are grains, the limited grain availability for ethanol production is the major factor in the gasoline replacement by the ethanol. For example, if the entire U.S. corn production is used for ethanol, only 10% of the gasoline is replaced. However, the use of other biomass feedstock, such as cellulosic feedstock may be an option. The conversion of the cellulosic biomass to ethanol consist of three primary polymers (cellulose, hemicellulose and lignin) that are making up the cell walls of a plant, having typically, depending on the plant species 30%–50% for cellulose, 20%–25% for hemicellulose and 10%–25% for lignin. These three polymers are setting the strength of mature cell walls and are also the reasons that cellulosic biomass is difficult to breakdown in order to extract the sugars, through hydrolysis process of the cellulosic biomass. Ethanol combustion follows the general combustion processes as any fuel, regardless the type. The combustion is the process of rapidly fuel burning, a rapid chemical reaction with oxygen (or with air) that produces heat and light. Combustion is an example of oxidation reaction, a reaction with oxygen. The products of combustion reactions usually include carbon oxide, water (liquid or vapor), carbon oxide (CO) and often some traces. Combustion can be described as complete, in excess of oxygen, and the fuel (hydrocarbon) is the limiting reagent or incomplete, when insufficient oxygen is presented, the limiting reagent in this case.

Example 3.6: If 1.0 kg of sugar is completely fermented to ethanol, what is the theoretical yield of ethyl alcohol and its heat content? The ethanol density is 0.79 g/mL (grams or milliliters)

Solution: From Equation (3.6), there 2 mol of ethanol to each mol of glucose (1:2), so convert the mass of glucose to moles by using the compound's molar mass:

$$1000 \text{ g} \times \frac{1 \text{ mol glucose}}{188.156 \text{ g}} = 5.552 \text{ moles of glucose}$$

$$5.552 \text{ moles og glucose} \rightarrow 11.104 \text{ moles of ethanol}$$

To convert the ethanol moles to *grams*, use the ethanol *molar mass*, so:

$$\text{Ethanol Mass (kg)} = \frac{11.104 \times 46.07 \text{ g}}{1000} = 0.51156128 \text{ kg or} \approx 0.512 \text{ kg of ethanol}$$

Assuming a 30.5 MJ/kg energy density, then the heat of the above ethanol yield is:

$$\text{Heat} = 0.512 \times 30.5 = 15.602 \text{ MJ}$$

Ethanol is ethyl alcohol obtained by sugar fermentation or by starch hydrolysis or cellulose degradation followed by sugar fermentation, and by subsequent distillation. Obtaining alcohol from vegetable raw materials has a long tradition in agriculture. The fermentation of sugar derived from agricultural crops using yeast, followed by distillation to produce alcohol is a well-established technology. Alcohol can also be produced efficiently from starch crops, wheat, maize, potato or cassava. Glucose produced by the starch hydrolysis can also be fermented to produce alcohol. The goal of directed use of cellulose-containing biomass from agriculturally utilized species for producing alcohol has not yet been practiced on a large-scale engineering scale. Production has been confined to the use of wood, residues and waste materials. The challenge today is to assemble these technologies into commercial applications. Several wood wastes and residues are available, while high-yielding woody crops provide lignocellulose at lower cost than agriculture crops. Approaches for the improvement of acetone butanol-ethanol fermentation are focusing on the development of hyper-amyl lytic and hyper cellulolytic clostridial strains with improved potential for biomass-to-butanol conversion. These new strains are producing about 60% more butanol than the parental wild strains, suggesting significant economic advantages for producing butanol. Bioethanol can be used as a fuel in different ways: directly in modified engines, blended with gasoline, or after transformation into ethyl-tertio-butyl ethanol. In the first case it does not need to be dehydrated, whereas blends need anhydrous ethanol, requiring neutral alcohol with a very low level of impurities. Such products ether resulting from a catalytic reaction between bioethanol and isobutylene, or are processed from methanol. Bioethanol can also be the alcohol used for the biodiesel production, thus giving ethyl esters instead of methyl esters, while the other constituent is vegetable oil from rape, sunflower, soybeans or other sources. Notice that, the oil companies and car manufacturers do often not like bioethanol blends, primarily because of their lack of water tolerance and volatility and the need to put a label on the pumps to inform the consumer. Bioethanol can be used in an undiluted form or mixed with motor gasoline. Gasohol is the term used in the United States to describe a maize-based mixture of gasoline (90%) and ethyl alcohol (10%), not be confused with gasoil, an oil product used to fuel diesel engines.

Vegetable oils and fats, in contrast to the simple carbohydrate building blocks, glucose and fructose, exhibit a number of modifications to the molecule structure, being also more energy valuable. From the point of view of plant cultivation of energy crops, the oil yields per hectare is rather important than quality aspects. The use of vegetable oils as fuels is not recent the oils are used as a material for burning and for lighting for very long time. The inventor Rudolph Diesel used peanut oil to fuel one of his engines at the Paris Exhibition of 1900. Biodiesel is increasingly used as a transportation fuel, and there are more than 280 plant species with considerable oil contents in their seeds, fruits, tubers and/or roots. The oil is extracted by compressing the seeds (sunflower, rapeseed) mechanically with or without preheating. Pressing with preheating allows for up to 95% of the oil removal, whereas without preheating the amount is considerably less. Extraction by using solvents is more efficient, removes nearly 99% of the oil, but it has higher energy consumption. Raw vegetable oils are refined prior to their use in engines or transesterification. In order to meet the requirements of diesel engines, vegetable oils can be modified into vegetable oil esters (transesterification). The transesterification procedure includes the production of methyl esters, RME (rape methyl ester) or SME (sunflower methyl ester), and glycerol (glycerin) by processing the plant oils (triglycerides), alcohol (methanol) and catalysts (aqueous sodium hydroxide or potassium hydroxide). Cracking is another option for modifying the triglyceride molecule of the oil. However, the cracking products are very irregular and more suitable for gasoline substitution; the process has to be conducted on a large scale; costs are considerable; and conversion losses are also significant. These negative aspects together with the much lower efficiency of gasoline engines make the cracking process of minor interest. The so-called Veba process is another triglyceride converting procedure. During the refining of mineral oil to form the conventional fuels, gasoline, diesel, propane or butane, up to 20% rapeseed oil is added to the vacuum distillation phase. The molecules are cracked and the mixture is treated with hydrogen, while the generated fuel molecules are not different from conventional fuel molecules. Advantages of this process are that no glycerin is produced as a by-product, and the produced fuel is similar to the standardized fuels and the same distribution system and handling is used. The disadvantages are: higher hydrogen consumption and the biodegradability reductions. Pure plant oils, particularly when refined and des-limed, can be used in pre-chamber (indirect injected) and swirl-chamber diesel engines as pure plant oil or in a mixture with diesel. Pure plant oil cannot be used in direct injection diesel engines, used in standard tractors and trucks, because engine coking occurs after some operation time. The addition of a small proportion of plant oils to diesel fuel is possible for all engine types, but tends to increase the engine deposits. The transesterification of vegetable oils permits the utilization of these oils in existing engines, either as a 100% substitute or in blends with mineral diesel oil. Vegetable oil esters are well suited for mixture with or the replacement of the diesel fuels, being effective in eliminating injection problems in direct injection diesel engines.

3.3.3 Biodiesel or Vegetable Oil

Biodiesel, often called a flexible fuel, is miscible with conventional diesel, being used in the existing engines. Biodiesel is defined as mono-alkyl (ethyl/methyl) ester of long chain fatty acids obtained from transesterification of triglyceride with alcohol, usually methanol and ethanol giving alkyl ester and glycerol. Biodiesel is a clean-burning fuel produced from vegetable oils, animal fats, or grease. Biodiesel is derived from renewable lipid sources like vegetable oils or animal fats. The conversion of oil to methyl or ethyl ester can be carried out by means of different processes, including transesterification, pyrolysis and emulsification. The process for ester formation from any oils or animal fats is transesterification, and esterification if it is based on fatty acids. Biodiesel is similar to the conventional diesel fuel in its characteristics, is biodegradable, nontoxic, with lower emissions than fossil diesel. Its chemical structure is that of fatty acid alkyl esters. In addition, it gives cleaner burning and has less sulfur content, and thus reducing emissions. Because of its origin from renewable resources, it is more likely that it competes with petroleum products in the future. In order to use biodiesel as a fuel is mixed with diesel fuel to create a biodiesel-blended fuel. However, *biodiesel* term is referring to the pure fuel before blending. Biodiesel can be used directly as pure biodiesel (B100) or in blends with diesel, in ratios of 2% (B2), 5% (B5), or 20% (B20). Biodiesel fuels are compatible with unmodified diesel engines with the existing fueling distribution infrastructure. Biodiesel can be obtained from various products, such as rape (rape methyl ester), soy (soy methyl ester), fat from deep-friars and animal leftovers, as the so-called the second type generation biodiesel. Biodiesel exceeds diesel in cetane number (50), diesel fuel performance (resulting in superior ignition), similar viscosity, higher flashpoint, making it more versatile where safety is concerned. Horsepower, acceleration and torque are comparable to diesel. Biodiesel has the highest Btu content of any alternative fuel, though lower than that of diesel, which may have a small impact on vehicle range and fuel economy. Triglycerides conversion into methyl or ethyl esters through transesterification process reduces the molecular weight to one-third that of the triglyceride, viscosity by factor of eight and marginally increases the volatility. Biodiesel esters contain about 10% oxygen, which helps the engine combustion. Different types of oil esters are used as biodiesel. Problems in fuel systems can occur due to the biodiesel dissolving effect on certain plastics and rubbers. The biodiesel properties are grouped by several criteria, the most important are those influencing the engine processes, such as ignition, starting, formation and burning of the fuel-air mixture, exhaust gas formation and quality, the heating value, cold weather properties (cloud point, pour point and cold filter plugging point), transport and depositing (oxidative and hydrolytic stability, flash point, induction period, microbial contamination, filterability limit temperature, wear of engine parts, lubricity, cleaning effect, viscosity and compatibility with materials used to manufacture the fuel system). However, one of the most serious obstacles to use biodiesel as an alternative fuel is the production complicated and costly purification processes.

Oil from a large variety species, recycled cooking, processing oils and animal fats oil are used in biodiesel production. All these forms are usually processed before are used in engines, to avoid maintenance issues or the engine lifetime reductions. Vegetable oils also have a high viscosity, containing fatty acids, phospholipids and other impurities that hinder their direct use as fuel. In order to avoid the engine modifications that are required for use, and substantially improve their characteristics as fuel, they are transformed into methyl or ethyl esters, which have similar properties to diesel. Oil extraction is the first phase in any biodiesel production process, separated in two production scales: centralized production in large industrial facilities, and small-scale processing units, in decentralized cold processing type on rural area facilities. Unprocessed vegetable oils do not meet biodiesel specifications and are also not a legal motor fuel. There are three basic biodiesel production processes: base-catalyzed, direct acid-catalyzed transesterification, and oil direct conversion to fatty acids and then to biodiesel. The base-catalyzed transesterification is the most common process, because it occurs at low temperature and pressure, having about 98% efficiency, minimal side reactions and direct biodiesel conversion, without any special reactor materials. Due to the fact that biodiesel is made for a variety of materials there are variations in properties and characteristics. The esterification is the chemical modification of vegetable oils into oil esters, suitable for use in engines. Vegetable oils are produced from oil crops using pre-pressing and extraction techniques, with the by-product the *cake* (protein), which is a valuable feedstuff for animal feeding. Esterification is needed to adapt the oil properties to the requirements of diesel engines. This process eliminates glycerides in the presence of an alcohol and a catalyst (usually aqueous sodium hydroxide or potassium hydroxide). Methyl esters are formed if methanol is used while ethyl esters are formed if ethanol is used. The most common vegetable oil ester for biofuel is RME (rape methyl ester). The generic and simplified biodiesel chemical reaction is:

$$\text{Plant Oil (Triglyceride)} + CH_3OH \text{ (Methanol)} \Rightarrow \text{Methyl Ester} + \text{Glycerol} \qquad (3.10)$$

Biodiesel quality depends on factors, such as its chemical and physical characteristics, being influenced by the feedstock quality, the fatty acid composition of the parent vegetable oil or animal fat, production process, process materials, the post-production parameters and by the handling and storage. Given the fact that the diesel engines are designed to use diesel fuel, the biodiesel physicochemical properties should be similar to those of conventional diesel. The criterion of biodiesel quality is the inclusion of its physical and chemical properties into the requirements of the adequate standard. Quality standards for biodiesel are updated, due to the evolution of compression ignition engines, emission standards, or the reevaluation of the feedstock eligibility for use in biodiesel production. The standards for the biodiesel quality are based on various factors, including characteristics of the existing diesel fuel standards, the types of diesel engines most common in the region, the emissions regulations, the development stage and the climatic properties of the region or country where it is produced or used, and the purpose and motivation for the biodiesel use. Cetane number is the dimensionless indicator that is characterizing the fuel ignition quality for compression ignition engines (CIE). Since in the CIE burning of the fuel-air mixture is initiated by compression of the fuel, the cetane number is a primary indicator of fuel quality as it describes the ease of its self-ignition. The cetane number ranges from 15 to 100, limits that are given by the two reference fuels used in the cetane number experimental determination: a linear-chain hydrocarbon, hexadecane ($C_{16}H_{34}$), very sensitive to ignition, having a cetane number of 100, and a strongly branched-chain hydrocarbon, having the same chemical formula $C_{16}H_{34}$, with high resistance to ignition, having a cetane number of 15.

The elemental composition of the biodiesel consists of carbon (C), hydrogen (H) and oxygen (O), the C/H ratio and the chemical formula of diesel and biodiesel depends on feedstock and production chemical procedures. The elemental composition of biodiesel varies slightly depending on the feedstock it is produced from. The most significant difference between biodiesel and diesel fuel composition is their oxygen content, which is from 10% to 13%, the biodiesel being also in essence free of sulfur. Unlike petroleum fuels, composed of hundreds of hydrocarbons (pure substances), biodiesel is composed solely of some fatty acid ethyl and methyl esters; their number depends on the feedstock used to manufacture biodiesel and is between 6 and 17. The fatty acid methyl and ethyl esters in the composition of biodiesel are made up of carbon, hydrogen and oxygen atoms that form linear chain molecules with single and double carbon-carbon bonds. The molecules with double bonds are unsaturated. The heat of combustion (heating value) at constant volume of a fuel containing only the elements carbon, hydrogen, oxygen, nitrogen and sulfur is the quantity of heat liberated when a unit quantity of the fuel is burned in oxygen in an enclosure of constant volume, the products of combustion being gaseous carbon dioxide, nitrogen, sulfur dioxide and water, with the initial temperature of the fuel and the oxygen and the final temperature of the products at 25°C. The unit quantity can be mol, kilogram or normal square meter. Thus the units of measurement of the heating value are kJ/kmol, kJ/kg. The volumetric heat of combustion, i.e., the heat of combustion per unit volume of fuel, can be calculated by multiplying the mass heat of combustion by the density of the fuel (mass per unit volume). The volumetric heat of combustion, rather than the mass heat of combustion is important to volume-dosed fueling systems, such as diesel engines.

Fuel density (ρ), measured in a vacuum, is strongly influenced by temperature, the quality standards state the determination of density at 15°C. Fuel density affects fuel performance, as some of the engine properties, such as cetane number, heating value and viscosity are strongly connected to density. The fuel density also affects the combustion quality. Diesel engine fueling systems (pump and the injectors) meter the fuel by volume, the density affects the fuel mass, reaching the combustion chamber, thus the energy content of the fuel dose, altering the fuel/air ratio and the engine power. Knowing the density is necessary in the biodiesel manufacturing, storage, transportation and distribution process, being an important parameter in the design of these processes. The density of esters depends on the molar mass, the free fatty acid and water content and the temperature. The biodiesel density is usually higher than that of diesel fuel, being dependent on fatty acid composition and purity. Biodiesel being made up of a few methyl or ethyl esters, having similar densities, the biodiesel density varies between tight limits. Contamination of the biodiesel significantly affects its density; therefore density is an indicator of contamination. The viscosity of liquid fuels is their property to resist the relative movement of their composing layers (viscosity is the reverse of fluidity). Viscosity is one of the important biodiesel properties, influencing the engine starting, the spray quality, the size of the particles, the penetration of the injected jet, the quality of the fuel-air mixture combustion, and the fuel lubricity, as some the fuel system elements are only lubricated by the fuel (pumps and injectors). Biodiesel quality is affected by oxidation during storage (air contact) and hydrolytic degradation (water contact). The two processes are characterized by the biodiesel oxidative stability and hydrolytic stability. Biodiesel oxidation can occur during storage or within the vehicle fuel system. The biodiesel stability refers to two issues: long-term storage stability (aging) and stability at elevated temperatures or pressures as the fuel is recirculated through an engine fuel system. For biodiesel, the storage stability, referring to the fuel ability to resist chemical changes during long-term storage is very important. These changes consist of oxidation due to contact with oxygen from the air. Biodiesel composition affects its stability in contact with air. Unsaturated fatty acids, especially the polyunsaturated ones, have higher oxidation tendency. The chemical equation for the combustion of biodiesel is as follows:

$$2 \cdot C_{19}H_{34}O_2 + 53 \cdot O_2 \Rightarrow 38 \cdot CO_2 + 34 \cdot H_2O \tag{3.11}$$

To find the enthalpy of combustion for this reaction, characterizing one of the biodiesel complete combustion the previous combustion energy equation is used:

$$\Delta H_0^{\text{Combustion}} = \sum n \times \Delta H_0^{\text{formation}} - \sum n \times \Delta H_0^{\text{reactants}} = 10{,}761 \text{ kJ/mol}$$

This is the calculated theoretical enthalpy of the complete combustion for the biodiesel.

3.3.4 Methanol Production

Methanol (methyl alcohol, wood alcohol, or wood spirits, or abbreviated, MeOH) is the simplest alcohol, a light, volatile, colorless, flammable liquid with a distinctive odor. At room temperature it is a polar liquid, miscible with water, petrol and organic compounds. It burns with an almost invisible flame and is biodegradable. Without proper conditions, methanol attracts water while stored. Methanol is a safe fuel, with toxicity similar to that of gasoline, being also fast biodegradable. It can be used in the engines with minor modifications. For example, low-percentage methanol-gasoline blends (up to 3%) are used in conventional spark-ignition engines with no technical changes. Methanol is an attractive compound, for both direct fuel for the fuel cell and automotive industry. However, it is still mainly used as a feedstock for chemical synthesis for the production of various chemicals. Methanol is produced from fossil fuels, coal, oil and natural gas or by biomass conversion. More than 75% of the methanol produced is based on a natural gas. Common for all resources is the intermediate syngas production, synthesized into methanol. The use of alcohol fuels in heavy duty applications is investigated by several engine manufacturers. Like ethanol more fuel, about twice the amount is needed, but the compression ratio can be increased. Methanol is toxic, being aggressive toward some materials. Emissions are of the same magnitude as from gasoline engines, though NOx emission is slightly lower. On the other hand formaldehyde emissions could cause problems and unburned fuel is toxic because of the methanol. Methanol can also be used in fuel cells. But currently it is a problem to get high efficiencies, because of minor leakages through the electrolyte. The power output is not as high as from hydrogen fuel cells, the process bring slower. Studies on bio-methanol produced from biomass reports production efficiencies about 54%. In nature MeOH is produced via anaerobic metabolism by many bacteria. It is also formed as a by-product during the ethanol fermentation process. MeOH also occurs naturally in many plants. Biomass is converted to MeOH via thermochemical and biotechnological pathways as shown in the diagrams of Figure 3.4.

The methanol production involves three steps: (1) synthesis gas preparation (reforming); (2) ethanol synthesis; and (3) methanol purification. There are two most common procedures of the biomass methanol production. First one, the thermochemical conversion procedures to the MeOH production are basically the same as for fossil feedstock, such as coal or natural gas. The biomass is gasified and the resulting synthesis gas, a mixture of CO, H_2 and CO_2, is adapted to the quality requirements of the MeOH synthesis. The formation of MeOH, an exothermic reaction, is enhanced by high pressures and low temperatures. For reasons such as the process simplification, the investment cost and the production energy consumption reduction, several alternatives used for the methanol production from biomass are in the development, in the commercial implementation phase and/or under research. Among these new procedures are: direct methane oxidation, liquid-phase methane oxidation, or conversion through mono-halogenated methane. Notice that the methanol production is subjected to a thermodynamic equilibrium that is limiting the methanol produced per reactor pass, leading to an unconverted gas high fraction. In the production schemes, the synthesis gas preparation and methanol synthesis are examined carefully, to ensure a feasible gas composition in order to maximize the methanol conversion. The second method, the biochemical path, is via methane formation through anaerobic digestion. This process is well developed due to the extensive biogas production from municipal waste or landfill sites. The biogas must be cleaned to obtain a gas with high methane content, and then MeOH is produced from the methane. Recently a biochemical path, by using methanol-trophic bacteria has been investigated for methanol

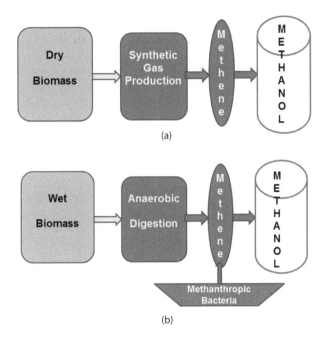

(a)

(b)

FIGURE 3.4
Methane production diagram: (a) from dry biomass and (b) from wet biomass.

production. In the first production method, during the methane synthesis the following chemical reactions are taking place:

$$CO + 2H_2 \Rightarrow CH_3OH$$

$$CO_2 + 3H_2 \Rightarrow CH_3OH + H_2O \qquad (3.12)$$

$$CO_2 + H_2 \Rightarrow CO + H_2O$$

In the alternative methane production method, the direct oxidation reaction is:

$$2 \cdot CH_4 + O_2 \Rightarrow 2 \cdot CH_3OH \qquad (3.13)$$

The conversion efficiency of the synthesis gas to methanol step is about 85%, so for example an overall wood to methanol energy efficiency in the range of 40%–45% is common. Improved catalytic gasification techniques raise the overall conversion efficiency to some 55%. The currently achieved efficiency is about 50%, but not all life-cycle estimates of energy inputs have been included or performed. The octane number of methanol is similar to that of ethanol, but the heat of combustion is less, up to 18 MJ·kg⁻¹. However, the engine efficiency of methanol is higher than that of gasoline, by at least 20% for today car engines, an *effective energy content* of 22.5 MJ·kg⁻¹ being quoted. Methanol is mixed with gasoline in standard internal combustion engines, or is used in specially designed Otto or diesel engines, such as a spark ignition engine run on vaporized methanol, with the vaporization energy being recovered from the coolant flow. Uses are similar to ethanol, but with differences in the assessment of environmental impacts, from production to use (e.g., toxicity of fumes at filling stations). The gasification can be made in closed environments, where all emissions, ash and slurry are collected. Cleaning processes in the methanol formation steps recover most catalysts in reusable forms, but other impurities have to be disposed-off with the gasification products. Precise schemes for waste disposal have not been formulated, but it seems unlikely that all nutrients could be recycled to agriculture or forestry as in the case of ethanol fermentation. However, the ammonia production by a process similar to the one yielding methanol is an alternative use of the synthesis gas. More research and fundamental studies aiming to better understand the way in which methanol production relies on lignin degradation are ongoing in research centers and companies. The combustion of the methanol is expressed by the following exothermic chemical reaction:

$$2 \cdot CH_3OH(l) + 3 \cdot O_2(g) \Rightarrow 2 \cdot CO_2(g) + 4 \cdot H_2O(g) + \text{Heat} \qquad (3.14)$$

Here, g stands for gas, and l for liquid. The heat of methanol combustion is just the heat divided for molar coefficient in the chemical reaction, with the sign as specified by the Hess Law. Equation (3.14) is the balanced equation for the complete combustion of the liquid methanol in the oxygen, yielding to the carbon dioxide and water, similar to any fuel, for example as previous discussed for the ethanol combustion. Heat of combustion is known as the combustion enthalpy.

> **Example 3.7:** Find the heat amount released during the reactions of the complete methanol combustion to produce carbon dioxide, liquid water and heat.
>
> **Solution:** The balanced reaction equation of the methanol complete combustion (in excess of oxygen) with carbon dioxide and liquid water product is the same as Equation (3.14):
>
> $$2 \cdot CH_3OH(l) + 3 \cdot O_2(g) \Rightarrow 2 \cdot CO_2(g) + 4 \cdot H_2O(l) + \text{Heat}$$
>
> The enthalpy of combustion, for the complete combustion of methanol to produce carbon dioxide and liquid water is estimated by:
>
> $$\Delta H_0^{\text{Combustion}} = \text{Reactants} - \text{Products} = (-210.0) - \left[-393.5 - 2 \cdot (-285.8) \right] = 755.1 \text{ kJ/mol}$$
>
> Notice in the above estimate, the liquid water enthalpy was used, not the one for vapor (gaseous water) is 241.5 kJ/mol, not 285.8 kJ/mol, an often mistake in energy estimates.

3.3.5 Biogas Production

Biogas consists of similar proportions of methane and carbon dioxide and is produced by the anaerobic fermentation of wet organic feedstock in a process called bio-methanization. Bio-methanization process has certain significance for the disposal of organic residues and waste products in the processing of agricultural products and in animal husbandry. Here, environmentally relevant aspects are of primary concern. Biogas derives from the breakdown of biomass under anaerobic conditions. Biogas sources include agricultural waste in the natural environment, municipal solid waste, landfill, or sewage. Fermentation, another type of anaerobic digestion, can also generate biogas. Biogas contains mostly methane (55%–90%) but can include carbon dioxide and hydrogen sulfide depending on its source. This flammable mixture may be used as a fuel, such as ethanol from sugar canes; it can also be purified to a natural gas equivalent (98% methane). Each cubic meter of methane contains approximately 50 MJ of energy, or 4–7 kWh of heat energy per cubic meter of biogas. When combusted, the gas or fuel are releasing this energy for transportation, heating, or power generation. However, its extraction from waste such as landfills and its use for electricity generation reduce direct atmospheric emissions. The following relationship represents the chemical reaction for the methane combustion:

$$CH_4 + 2O_2 \Rightarrow CO_2 + 2H_2O + \text{Heat} \tag{3.15}$$

Example 3.8: Estimated the heat released during the complete combustion of 100 liters of methane.

Solution: Assuming enough oxygen for complete methane combustion in Equation (3.14), and assuming gaseous water (vapor) as the product the amount of heat released is then calculated as the difference between the formation enthalpy sum and the broken bond enthalpy sum, with the appropriate values specified in the appendix tables, and also as in example 3.3 is:

$$\Delta H_{\text{Combustion}}^{CH_4} = \left[4\left(410 \text{ kJ/mol}\right) + 2\left(494 \text{ kJ/mol}\right)\right] - \left[2\left(799 \text{ kJ/mol}\right) + 4\left(460 \text{ kJ/mol}\right)\right]$$

$$= -810 \text{ kJ/mol}$$

A mol of substance has 22.4 liters, so the heat released by the complete combustion of 100 liters methane is then:

$$\text{Heat} = \frac{100 \text{ l}}{22.4 \text{ l}} \times 810 = 3616.071 \text{ kJ}$$

The cultivation of plants with the goal of producing biogas is not practiced, though there are plant species that are suitable in their fresh or in their ensiled forms for biogas production. However, through bio-methanization, several green plants could be used as energy feedstocks, because in contrast to the combustion, the raw material in its natural moist state is used. To put into practice the biogas crops, several fundamental conditions must be met. The plant species suitable for the biogas production are those that are rich in easily degradable carbohydrates, such as sugar and protein matter. For example, the maize, reed canary grass and perennial rye grass methane yields, after siltation and fermentation were identical. Raw materials from lignocellulose-containing plant species are hardly suitable for the biogas production. Biogas production from green plants is quite complicated, requiring a quite expensive and complex control of the fermentation process. In top of that, an acceptable yield of biogas from vegetable raw materials, production on a continuous, long-term basis and a homogeneous substrate is required.

The gasification process can convert most biomass feedstocks or residues to a clean synthesis gas. Once such a gas is obtained, it is possible to access and leverage the process technology developed in the petroleum and chemical industries to produce a wide range of liquid fuels and chemicals. There are several widely used process designs for biomass gasification: (1) staged reformation with a fluidized-bed gasifier, (2) staged reformation with a screw auger gasifier, (3) entrained flow reformation, or (4) partial oxidation. In staged steam reformation with a fluidized-bed reactor, the biomass is first pyrolyzed in the absence of oxygen. Then the pyrolysis vapors are reformed to synthesis gas with steam, providing added hydrogen as well as the proper amount of oxygen and process heat that comes from burning the char. With a screw auger reactor, moisture (and oxygen) is introduced at the pyrolysis stage, and process heat comes from burning some of the gas produced in the latter. In entrained flow reformation, external steam and air are introduced in a single-stage gasification reactor. Partial oxidation gasification uses pure oxygen with no steam, to provide the proper amount of oxygen. Using air instead of oxygen, as in small modular uses, yields produce gas (including nitrogen oxides) rather than synthesis gas. Biomass gasification is also important for providing a fuel source for electricity and heat generation for the integrated bio-refinery. Virtually all other conversion processes, whether physical or biological, produce residue that cannot be converted to primary products. To avoid a waste stream from the refinery, and to maximize the overall efficiency, these residues can be used for combined heat and power production. In existing facilities, these residues are combusted to produce steam for power generation. Gasification offers the potential to utilize higher-efficiency power generation technologies, such as combined cycle gas turbines or fuel cells. Gas turbine systems offer potential electrical conversion efficiencies approximately double those of steam-cycle processes, with fuel cells being nearly three times as efficient. A workable gasification process requires development of some technology: for example, feed processing and handling, gasification performance improvement, syngas cleanup and conditioning, development of sensors, analytical instruments and controls, process integration, and materials used for the systems. Biogas, also termed methane or gobar gas, comprises a mixture of gases. It is a fuel of high caloric value resulting from anaerobic fermentation of organic matter called biomass. Composition of this gas varies with the type of organic material used. Its basic composition is listed in Table 3.3. Methane is a colorless, odorless gas with a wide distribution in nature. It is the principal component of natural gas, a mixture containing about 75% methane (CH_4), 15% ethane (C_2H_6), and 5% other hydrocarbons, such as propane (C_3H_8) and butane (C_4H_{10}). The firedamp of the coal mines is mainly methane, which is the main component of biogas too. It is colorless and odorless, highly flammable and in combustion presents a lilac-blue flame and small red stains. It does not leave soot and produces minimal pollution. The caloric power of biogas depends on the amount of methane in its composition, reaching 5000–6000 kcal/m³. Biogas can be used for stove heating, campaniles, water heaters, torches, motors and other equipment. Biomass can be considered as all materials that have the property of being decomposed by biological effects, that is, by the action of bacteria. Biomass can be decomposed by methanogenic bacteria to produce biogas, in a process depending on factors such as temperature, pH, carbon/nitrogen ratio and the quality of each. Usable and accessible organic matter includes animal residue, agricultural residue, water hyacinth (*Eichornia crassipes*), industrial residue, urban garbage and marine algae.

Gasification is a technique that heats biomass, converting it into combustible gas, volatiles and ash. The technology behind gasification may vary based on the gasification agent or the reactor, but it is often more demanding because of feedstock specifications. Municipal solid waste and agricultural residues are the common feedstock. Gasification occurs in two endothermic steps. Biomass is first heated to over 700°C, which vaporizes volatiles such as hydrogen, CO, CO_2 and other hydrocarbon gases. The by-products are charcoal and ash. In the second step, the charcoal is gasified by reacting with oxygen, steam and hydrogen at high temperatures. The gasification products include syngas, bio-charcoal and tar. The specific amount of each depends on the feedstock, oxidizing agent, and the process conditions.

TABLE 3.3

Typical Composition of the Common Combustible Gases

Gaseous Fuel	Type	Percentage (%)
Methane	CH_4	60–90
Carbon Gas	CO_2	40–10
Hydrogen	H	Traces
Nitrogen	N	Traces
Sulfhydric Gas	H_2S	Traces

3.3.6 Principles of Anaerobic Digestion

Conversion of fresh biological materials into hydrocarbons or hydrogen can be achieved by quite a few anaerobic fermentation processes, i.e., processes performed by suitable microorganisms, without oxygen. Such *anaerobic digestion* processes, with biogas as output, are working on the basis of fresh biological materials, wood excepted, provided that the proper conditions are maintained (temperature, population of microorganisms, stirring, etc.). Biological materials in forms improper for storage or use are converted into liquid or gaseous fuels that can be utilized in the same ways, like oil and natural gas. In anaerobic digestion, the decomposition of the organic materials in the absence of the air (oxygen) by bacteria down to the organic matter produces methane (~65%), and carbon dioxide (~35%) with some traces of other gases. Anaerobic digestion consists of a series of biochemical processes with a set of bacteria to break down organic materials into biogas, methane, and a combination of solid and liquid effluents, the so-called digestate. It is the biological methane-genesis, an anaerobic process responsible for degradation of much of the carbonaceous matter in natural environment, where organic accumulation results in depletion of oxygen for aerobic metabolism. This process, which is carried out by a consortium of different microorganisms, is found in various environments, including sediments, flooded soils and landfills. In process diagram of anaerobic digestion feedstock is harvested, shredded and placed into a reactor, which an active inoculum of microorganisms has required for methane fermentation, as shown in Figure 3.5. The major types of farm anaerobic digester are covered lagoons, completed-mix digesters, plug-flow digesters and fixed-film digesters. A covered lagoon digester is large, in-ground or lined lagoon with a flexible or floating gas-tight cover. Covered lagoons are not heated, being best used in warmer climate to maintain digester temperature and have a hydraulic retention of 30–45 days. Complete-mix digesters are above-ground or under-ground cylindrical tanks, often with flexible cover with hydraulic retention from 10 to 25 days. A plug-flow digester consists of a long, narrow, insulated and heated cylindrical tank, partial or fully underground where the gas and by-products are pushed out by new manure and wastes. It is suitable for mechanically operated livestock facilities. A fixed-film digester is a column packed with media (wood chips or small plastic rings) supporting a thin-film bacteria (biofilm), methane-forming microorganisms are growing on this media. They have a very short hydraulic retention,

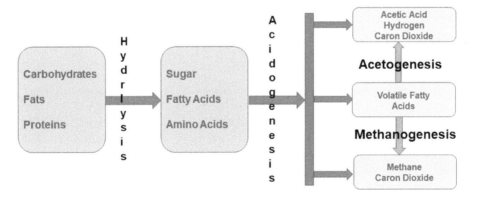

FIGURE 3.5
Flow diagram of anaerobic digestion process.

2–6 days, are smaller in size compared with other digester options, with the major draw-back that manure and residues can plug the media, and removing the manure solids from these digesters is reducing the potential biogas production.

The produced gas has a caloric value of 15–25 MJ/m³ (at standard temperature and pressure), with conversion efficiencies in the range of 40%–60%. Organic materials are composed of several organic compounds resulting from the remains or decomposition of the living organisms such as plants, and animals and their waste products. Organic material sources include diary manure, food processing waste, plant residues, food waste, fats, oils and grease. The end product biogas is composed of about 60%–70% methane, and CO_2 the remaining 40%–30%, and very small amount of trace gases. This biogas can be combusted to generate electricity and heat or processed into renewable natural gas, used in industrial processes and transportation fuels. Separated digested solids can be processed and used for diary bedding, crop lands, or covered into other products. Digested liquids, containing fewer pathogens and weed seeds, are rich in crop nutrients and can be used as agricultural fertilizers. The major benefits of anaerobic digestion include among others: improving air quality, by reducing odor and pollutant emissions, producing and harvesting biogas (methane) that is used for electricity and heat generation or in industrial processes, killing weed seeds, reducing fertilizer and other associate farm and agricultural costs. Associate biochemical processes of biogas production are quite complex, but anaerobic digestion systems are relatively simple to be described. Biogas is producing in four stages by different bacteria with specific functions that are break-down the complex organic compounds until biogas is produced. However, these bacteria are requiring a strictly oxygen-free (anaerobic) environment. Anaerobic digestion process can be split, as shown in Figure 3.5, into four main conversion steps or phases:

1. Hydrolysis: in which hydrolytic bacteria break down the wet biomass (biodegradables, municipal solid waste, organic waste, sewer sludge, processing waste, etc.) into sugars and amino acids;
2. Acidogensis: in which fermentative bacteria convert sugars and amino acids into organic acids;
3. Acetogensis: in which acidogentic bacteria convert the organic acids into hydrogen, carbon dioxide and acetate; and
4. Methanogenesis: in which methanogenic bacteria (methane forming) produce biogas from hydrogen, carbon and acetic acid. Staring from the cellulose from the wet biomass, the overall process can by summarized by this biochemical reaction:

$$\left(C_6H_{10}O_5\right)_n + nH_2O \Rightarrow 2nCO_2 + 3nCH_4 + 19 \text{ nJ/mol} \tag{3.16}$$

Anaerobic digestion is taking place in a digester (airtight chamber/container), which can operate in batch mode or in continuous mode (as shown in Figure 3.6), producing a steady biogas flow and are usually associated with large-scale operation. The digestion process is quite sensitive to the temperature range, requiring minimum 20°C (68°F) to 66°C (150°F), the higher the temperature the shorter the process and the smaller the digester size, but also higher temperatures are more difficult to maintain and are requiring closer monitoring. However, in order digestion to proceed, a number of conditions must be fulfilled. The bacterial action is inhibited by the presence of metal salts, penicillin, soluble sulfide, or ammonia in higher concentrations. However, reasonable nitrogen

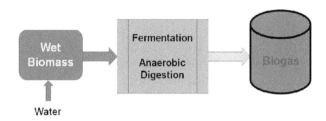

FIGURE 3.6
Schematic diagram of an anaerobic digester.

presence is essential for the growth of the microorganisms. If there is too little nitrogen relative to the carbon-containing material to be transformed, then bacterial growth will be insufficient and biogas production low. With too much nitrogen (carbon to nitrogen atom ratios less than 15), *ammonia poisoning* of the bacterial cultures may occur. When the carbon-nitrogen ratio exceeds 30, the gas production diminishes, but in some systems carbon-nitrogen values as high as 70 have prevailed without problems. If digestion time is not an issue, almost any cellulosic material can be used even pure straw. One initially may have to wait for several months, until the optimum bacterial composition has been reached, but then continued production can take place, and despite low reaction rates an energy recovery similar to that of manure can be achieved with properly prolonged reaction times. The anaerobic digestion process residue has a higher value as a fertilizer than the feedstock. Insoluble organics in the original material are, to a large extent, made soluble and nitrogen is fixed in the microorganisms. Sludge pathogen populations are significantly reduced, which a reason why the anaerobic fermentation has been used extensively as a cleaning operation in city sewage treatment plants, either directly on the sludge or after sludge growing algae to increase fermentation potential. In most of the new plants, the production of biogas for fuels is an integral part of the treatment process, but with proper management, sewage plants may well be net energy producers. If the digestion process is operating correctly then it, is almost pollutant-free, but gas cleaning, such as H_2S removal, may lead to emissions. The methane gas shares many of the hazards of other gaseous fuels, being asphyxiating and explosive at certain concentrations in air, about 7%–14% by volume. Notice that methane, the major AD generated biogas if mixed with air is highly explosive. In addition, biogas is heavier than air, displacing the oxygen near the ground, if leaks from a digester and can accumulate in unconventional spaces. Further, the biogas can act as deadly poison if H_2S is present, which is occurring in the biogas from the manure anaerobic digestion. Compared with the current mix of coal, oil or natural gas plants, biogas plants have 2 to 3 times lower SO_2 emission but higher NOx emissions. Higher ammonia content in the digested residue calls for greater care in using fertilizer from biogas plants, in order to avoid loss of ammonia. This is also true as regards avoiding loss of nutrients from fertilizer to waterways. Compared to spreading manure not refined by the biogas production process, a marked gain in fertilizer quality has been noted, including a much better defined composition, which will contribute to assisting correct dosage and avoiding losses to the environment. The dissemination of biogas plants removes the need for landfills, which is seen as an environmental improvement.

In the first stage of hydrolysis, or liquefaction, fermentative bacteria convert the insoluble complex organic matter, such as cellulose, into soluble molecules such as sugars, amino acids and fatty acids. The complex polymeric matter is hydrolyzed to monomer, e.g., cellulose to sugars or alcohols and proteins to peptides or amino acids, by hydrolytic enzymes, (lipases, proteases, celluloses, amylases, etc.) secreted by microbes. The hydrolytic activity is of significant importance in high organic waste and may become rate limiting. Some industrial operations overcome this limitation by the use of chemical reagents to enhance hydrolysis. The application of chemicals to enhance the first step has been found to result in a shorter digestion time and provides a higher methane yield. Hydrolysis or liquefaction reactions consist of:

$$\text{Lipds} \Rightarrow \text{Fatty Acids}$$

$$\text{Proteins} \Rightarrow \text{Amino Acids}$$

$$\text{Nucleic Acids} \Rightarrow \text{Purines \& Pyrimidines} \tag{3.17}$$

$$\text{Polysaccharides} \Rightarrow \text{Monosaccharides}$$

In the second stage, acetogenic bacteria, also known as acid formers, convert the first phase products to simple organic acids, carbon dioxide and hydrogen. The principal acids produced are acetic acid (CH_3COOH), propionic acid (CH_3CH_2COOH), butyric acid ($CH_3CH_2CH_2COOH$), and ethanol (C_2H_5OH). The products formed during acetogenesis are due to a number of different microbes. Other acid formers are anerobus, lactobacillus, or actinomyces. An acetogenesis reaction consists of:

$$C_6H_{12}O_6 \Rightarrow 2 \cdot C_2H_5OH + 2 \cdot CO_2 \tag{3.18}$$

Finally, in the third stage methane is produced by bacteria called methane formers (methanogens) in two ways: either by means of cleavage of acetic acid molecules to generate carbon dioxide and methane, or by reduction of carbon dioxide with hydrogen. Methane production is higher from reduction of carbon dioxide but limited hydrogen concentration in digesters results in that the acetate reaction is the primary producer of methane. The methanogenic bacteria include *Methanobacterium*, *Methanobacillus*, *Methanococcus* and *Methanosarcina*. Methanogens can also be divided into two groups: acetate and H_2/CO_2 consumers. *Methanosarcina* spp. and *Methanothrix* spp. (also, *Methanosaeta*) are considered to be important in AD both as acetate and H_2/CO_2 consumers. The methanogenesis reactions are expressed as:

$$CH_3COOH \rightarrow CH_4 + CO_2 \tag{3.19}$$

$$2C_2H_5OH + CO_2 \rightarrow CH_4 + 2CH_3COOH \tag{3.20}$$

$$CO_2 + 4H_2 \rightarrow CH_4 + 2H_2O \tag{3.21}$$

Usually the overall AD process can be divided into four stages: Pretreatment, waste digestion, gas recovery and residue treatment. Most digestion systems require pretreatment of waste to obtain homogeneous feedstock. The preprocessing involves separation of non-digestible materials and shredding. The waste received by AD digester is usually source separated or mechanically sorted. The separation ensures removal of undesirable or recyclable materials such as glass, metals, stones etc. In source separation, recyclables are removed from the organic wastes at the source. Mechanical separation can be employed if source separation is not available.

Example 3.9: How much methane gas can be generated through complete anaerobic degradation of 1 kg chemical oxygen demand (COD) of the anaerobic digestion from wastewaters at STP? Notice that chemical oxygen demand, acronym COD is a measure of the total organic carbon in a material derived from both biodegradable and recalcitrant molecules.

Solution: First step, in the COD estimate consists of finding the CH_4 equivalent of COD, which it is calculated from the complete methane combustion reaction:

$$CH_4 + 2 \cdot O_2 \Rightarrow CO_2 + 2 \cdot H_2O$$

In gram-equivalent: $16\,g\,CH_4 \sim 64\,g\,O_2\,(COD) \rightarrow 1\,g\,CH_4 \sim 64/4 = 4\,g\,COD$

Next is the conversion of CH_4 mass to equivalent volume, based on the ideal gas law, 1 mole of any gas at STP (Standard Temperature and Pressure) occupies volume of 22.4 L (liter).

$$1\,mol\,of\,CH_4 \sim 22.4\,L\,CH_4$$

$$16\,g\,of\,CH_4 \sim 22.4\,L\,CH_4 \rightarrow 1\,g\,of\,CH_4 \sim 22.4/16 = 1.40\,L\,CH_4$$

Finally, the CH_4 generation rate per unit of COD removed is calculated as:

$$1\,g\,CH_4 \sim 4\,g\,COD \sim 1.4\,L\,CH_4 \rightarrow 1\,g\,COD \sim 1.4/4 = 0.35\,L\,CH_4$$

or

$$1\,Kg\,COD \sim 0.35\,m^3\,CH_4$$

Theoretically, removal of 1 kg of COD from the wastewater results in the production of 0.35 m^3 of methane (at STP). This computation of methane production is *theoretical* because it assumes that all the COD removed appears as methane (since carbon dioxide, the other component of biogas, has zero COD). In actual practice, the process generates 0.27–0.30 m^3 methane per kg COD removed. Pure methane has a calorific value of 39,800 $kJ \cdot m^{-3}$. Biogas has 55%–70% methane (the rest is carbon dioxide and traces) depending on the type of reactor and reactor operating conditions. In fact, the maximum energy production through wastewater treatment occurs through anaerobic treatment of the wastewater.

3.4 Further Aspects of Hydrogen Economy

Hydrogen is viewed by many as the most promising fuel for light-duty vehicles for the future. Hydrogen can be produced through a large number of pathways and from many different feedstocks (both fossil and renewable). As hydrogen can be produced from water, it is one of the lightest, most efficient, cost-effective and cleanest fuel on the planet, if the mature and commercial viable technologies are developed. This is a realistic assumption since over 72% of the globe is covered with water and by-product again is water. In other words hydrogen economy starts and ends with water. It can avoid all harmful gases, acid rains, pollutants, ozone depleting chemicals and oil spillages due to conventional fuels. A biomass to hydrogen pathway and its use in fuel cells has only recently been included in transportation and energy research and development studies. The analysis is based on those studies which include biomass-derived hydrogen, comparing it to gasoline/diesel internal combustion engine vehicles and fuel cells which are utilizing the hydrogen from other feedstock (natural gas and wind-generated electricity). Since hydrogen is not commercially produced from biomass, all of these studies utilize process and emissions data based on research results and extrapolations to commercial scale. We find that direct comparison of results between studies is challenging due to the differences in study methodologies and assumptions concerning feedstock, production processes, and vehicles. However, the vast majority of hydrogen today is produced using fossil fuels, including natural gas and coal. On the other hand, there are quite intensive research studies to create an entirely natural and renewable method for producing hydrogen to generate electricity which could drastically reduce the dependency on fossil fuels in the future. The breakthrough means ethanol, which comes from the fermentation of crops, can be completely converted to hydrogen and carbon dioxide for the first time, through special stable catalyst assisted reactions which can generate hydrogen using ethanol produced from crop fermentation at realistic conditions. As with traditional methods of hydrogen production, carbon dioxide is still created during these processes. However, unlike fossil fuels which are underground, they are using ethanol generated from an above-the-ground source, by plants or crops, meaning that any carbon dioxide created during the process are back into the environment and are then used by plants as part of their natural cycle of growth.

The hydrogen use can afford the development of clean and adequate energy for sustainable development of all. Ever growing energy demands and the rising concern caused by the use of conventional fossil fuels, call for new and clean fuels. Among all kinds of energy sources, hydrogen is the best choice as a clean fuel. The main advantage of hydrogen as energy source lies in the fact that its by-product is water, and it can be easily regenerated. Hydrogen, the simplest element, consists of an atom of hydrogen, only one proton and one electron, being the most plentiful element in the universe. Despite its simplicity and its abundance, the hydrogen is not occurring naturally as a gas on the Earth, being always combined with other elements. Water, for example, is a combination of hydrogen and oxygen. Hydrogen is also found in many organic compounds, notably the "hydrocarbons" that make up many of our fuels, such as gasoline, natural gas, methanol and propane. In this presentation we would like to highlight the production, storage, transportation and applications of hydrogen energy are important issues. We have focused on the storage of hydrogen through carbon nanotubes. In its pure form, hydrogen is colorless and odorless gas. It is an energy carrier, not an energy source. Hydrogen is currently stored in tanks as a compressed gas or cryogenic liquid. The tanks can be transported by truck or the compressed gas is transported across distances of 50 miles or less by pipelines. **Safety** is essential in the entire energy conversion process. This begins with production, storage, transport, distribution and utilization. Each energy form poses its own specific risk, which should be taken care of. The safety of combustible energy carriers in their ignition, combustion, explosion and detonation behavior when mixed with air is still under study.

3.5 Environmental and Land-Usage in Biomass and Biofuels Applications

Production of biomass in any form requires the use of land, and it would require the involvement of rural people to do it. Chemical fertilizers, an important input required in agriculture, need large quantities of fossil fuel in their production. Our Institute is developing the concept of conducting agriculture without using chemical fertilizers. This concept is based on the assumption that soil microorganisms degrade the soil minerals to provide the green plants with the entire mineral nutrients that they need. If the soil microorganisms are adequately fed with organic matter, there is theoretically no need to apply chemical fertilizers to the soil. Traditional agricultural scientists recommend the application of organic matter in the form of compost. However, the nutritional value of composted organic matter is so low, that one has to apply 20–50 tons of compost per hectare. In practical terms, it means that one has to use the biomass produced in about 10 hectares for providing organic matter to one hectare. This is the reason why many planners and agricultural scientists consider organic farming to be impracticable. Land management associated with the production of biomass may result in decreased terrestrial carbon stocks in above-ground biomass, below-ground biomass, dead wood, litter and soil. For example, the production of biofuels from palm oil plantations causes large decreases in carbon stocks if the land was deforested to enable the establishment of the palm oil plantation. Similarly, a project that increases the collection of dead wood in an existing forest will lead to reduced carbon stocks if this practice depletes the carbon pool of dead wood in the forest. The planting of an annually bioenergy crop such as rapeseed on grassland is a third example: the annual tillage of the soil could cause a systematic decrease in the soil carbon stocks. On the other hand, bioenergy systems may also function as carbon sinks, or conversely afforestation, reforestation and re-vegetation can enhance carbon stocks in plants and soils, while at the same time contributing to a future biomass resource. The increases in carbon stocks could for instance result from establishment of short rotation tree plantations on cropland historically used for cereal production. It is not possible to assign a general ranking of land use options based on their contribution to climate change mitigation. The climate benefit of a specific option is determined by many parameters that are site-specific and can differ substantially depending on cultivation practice, conversion system configuration and the energy infrastructure context of its establishment (and the nature of direct and indirect land use).

A number of concerns with regard to the environmental impacts of the ethanol fermentation energy conversion chain must be considered. First of all, the biomass being used may have direct uses as food or may be grown in competition with production of food. The reason is, of course, that the easiest ethanol fermentation is obtained by starting with a raw material with as high a content of elementary sugar as possible, i.e., starting with sugar cane or cereal grain. Since sugar cane is likely to occupy prime agricultural land, and cereal production must increase with increasing world population, neither of these biomass resources should be used as fermentation inputs. However, residues from cereal production and from necessary sugar production (present sugar consumption is in many regions of the world too high from a health and nutrition point of view) could be used for ethanol fermentation, together with urban refuse, extra crops on otherwise committed land, perhaps aquatic crops and forest renewable resources. Concerns about proper soil management, recycling nutrients and covering topsoil to prevent erosion are very appropriate in connection with the enhanced tillage utilization that characterizes combined food and ethanol production schemes. The issues of competition with food and other (industry feedstock) uses of the biomass used for ethanol production could be reduced if only biomass residues were employed, and if nutrients were returned to the fields after energy extraction. This requires further enzyme additions for degradation of lignocellulosic components and may lead to slightly higher ethanol production costs. On the other hand, the biomass resource potential is greatly enlarged, as some 90% of harvested biomass does not end up as food. The enzymes and catalysts needed for producing ethanol from agricultural residues and household waste have already been developed.

Increased bioenergy use does not necessarily leads to an increased competition for food and feed crops. There are several conversion options that generate energy from biomass, which are using different feedstocks other than food or feed crops. Under strategies that shift demand to alternative, such as lignocellulosic feedstocks, bioenergy expansion could use other sources such as agriculture and forestry residues, no additional land or water required, although these could potentially cause negative effects if extraction rates are excessive. Lignocellulosic crops could also be grown on other land types, such as marginal lands, pastures and grasslands, not very suitable for first-generation biofuels, becoming an additional resource for feedstock production under sustainable management practices. The perennial energy crops are also presenting an opportunity for increasing water productivity, by decreasing the proportion of rainfall lost through evaporation. Marginal and degraded areas could also be used for lignocellulosic feedstock production. However, marginal lands have also alternative uses, implying that the current land users must be involved to ensure positive local socioeconomic development, requiring approaches other than monocultures, such as agro-forestry systems integrating bioenergy production with food crops and cattle production. Furthermore, biomass production on marginal or degraded land may not be the automatic outcome of increasing biomass demand. As bioenergy use increases and farmers adopt bioenergy crops, they may consider developments in both the food and bioenergy sectors when planning their operations. The economic realities at farm level can still lead to bioenergy crops competing with food crops, since it is the good soils are resulting in higher yields. Biomass plantations may eventually be pushed to marginal and degraded land due to increasing land costs following increased competition for prime croplands, the competition likely to be reflected in increasing food prices.

3.6 Waste-to-Energy Recovery

Industry and human activities produce solid, liquid and gaseous waste. Waste disposal is one of major problems being faced by all nations across the world. Generation of wastes both in the solid and liquid forms is associated with almost all human activities, residential, commercial and industrial processes. It is generally generated in urban, municipal, agricultural, commercial and industrial sectors. Municipal solid waste (MSW), sewage sludge (a by-product of wastewater treatment), and livestock manure can be sources of biogas energy, and it is unlikely that they will deplete, as there will always be waste generated across any civilization. The solid wastes include human excretes, household wastes, city garbage, commercial wastes and industrial wastes. The liquid form includes domestic sewage and effluents from community, institutional and industrial activities. Refuse or municipal solid waste can viewed as a fuel, refuse, or sold wastes, suffer from a lack of homogeneity, each large or small batch differing from another on a seasonal, daily, or location basis. It is dependent to some degree on weather and is often an index of the living standard. The refuse nature is often dependent on local regulations regarding recycling. MSW is typically a volatile fuel with a low heating value, not unlike many low-grade fossil fuels, being a cellulosic type fuel. The MSW disposal in many areas is laying it on the ground, while advancement in today landfills consists in the use of techniques that prevent seepage of acids and other liquids into the ground. Many landfills require daily covers of soil to contain debris and odors or taping the evolving methane for fuel use or even allowing the methane and other products of decomposition such as carbon dioxide to be released. Both of these compounds are major contributors to the *greenhouse effect*. As landfill sites fill up and as available land becomes scarcer, the problem of disposal becomes more acute. As the debates over *the not in my backyard syndrome* rage on as to where to locate landfills or treatment plants, the waste continues to grow. Over the last three decades, attempts have been made to develop processes that partially combust MSW to produce a gaseous fuel. There are seven major technologies involving gasification, or other innovative thermal processing technologies for processing municipal solid waste. However, many manufacturers were restricting their efforts to biomass. Often due to improper waste management and operation of these facilities and poor and/or incomplete waste treatment, these wastes find their way back into the environment, being an important source of pollutants and land degradation. Treatment of liquid and gaseous waste is widely practiced, to various extents determined by local regulations, sensitivity to the problem and pollution concerns. Liquid and gaseous waste is classified on the basis of pollutant content. The energy content in liquid and gaseous emissions can be partially recovered by means of heat exchangers, to be used in process or in cogeneration plants, depending on the temperatures and pollutant content. Solid waste from industry is classified in each country according to local regulations and nevertheless, the various classifications have large similarities. In an industrialized country, with a gross energy consumption of 160 MTOE/year (of which 50 MTOE/year is the share of industry), the amount of industrial waste is roughly 35 million tons per year, roughly speaking, 0.7–1.0 ton of industrial waste per TOE of gross industry energy consumption. In addition, in an industrialized country each person disposes of about 1.5–2 kg/day (or 0.5–0.7 ton/year) of urban waste,

The solid or liquid wastes are increasing to a quite large extent, while the existing landfills are filled near or beyond their capacities. Search for new sites or expiation of the existing ones are difficult tasks due to economic and regulatory aspects and strong public opposition. Even after, finding the new dumping yards or the approval for the expansion of the existing ones, the distance from the city increases the transport cost in the top of increased operation costs in order to comply with regulations for environment protection or other governmental requirements. In near future the municipal corporations are needed to pay heavy amounts to private parties to lift the garbage from cities. To overcome above serious problems, waste-to-energy recovery and biogas technology has been suggested. This will serve not only to treat waste but produce energy and manure reducing the effective volume of waste. Here again, decentralized installation system is advisable and recommended instead of centralized dumping yard. This will also solve many problem of transportation. Industrial solid waste is commonly broken down into four categories:

1. Inert waste from extractive, brick and similar industries, are also assimilated to the debris from building demolition and can easily be used for road construction and public works.
2. Urban waste includes plastics, paper, rubber and in limited quantities, glass.
3. Industrial waste includes all organic and inorganic materials, liquids, sludge and solids with limited concentrations of pollutant with the exceptions of the above-mentioned categories (inert and urban waste) and hazardous waste, involving risks for the environment.
4. Hazardous and infectious waste, industrial waste with high concentrations of toxics or pollutants.

Urban waste is commonly landfilled, but this is no longer an acceptable disposal method because of environmental problems and the scarcity of space. Utilities are very interested in burning waste to produce electricity and steam, but air pollution must be avoided. Urban waste treatment is not discussed here in detail, our main concern being industrial waste and its on-site utilization.

Non-hazardous industrial waste can be treated as urban waste, but the quantity produced by a single medium-sized factory is generally too small for on-site reuse in heat-recovery plants. Waste-management strategies comprise a few main sections, some of which are the same as in energy management: analysis of historical data, audits and accounting, analysis of local environmental regulations, engineering analyses and investment proposals based on feasibility studies, personnel training and information. Several factors must be considered:

- Incidence of waste costs on turnover and added value;
- Waste as a percentage of production;
- Identification and quantification of waste from the raw materials and energy flows inside the facility;
- Selection of significant indexes valid for the whole factory and for single production lines;
- Possibility of recycling inside the factory or outside; and
- Level of pollution and compatibility with local regulations.

Waste-to-energy (WTE) or energy from waste refers to any waste treatment that transforms waste resources into electricity, steam or heat energy. These include, for example, anaerobic digestion (discussed before), incineration or direct combustion, pyrolysis, gasification, refuse-derived fuel (RDF), etc. WTE technologies usually reduce the original waste volume (up to 90%), depending on the waste composition and the derived energy type. A waste management hierarchy follows the pattern of waste reduction, reuse, recycling, recovery treatment, followed by disposal. An integrated approach to WTE that practices waste separation and pretreatment of waste does not by-pass the waste hierarchy but precedes or replaces the disposal step which is a more sensible approach to WTE recovery than simply burning or converting raw unsorted waste. Nonetheless, the choice of WTE technology is important and the conversion plant itself may incorporate waste pretreatment units to facilitate this approach. Energy recovery from waste is classified on the basis of the kinds of treatment and of waste available. Heat recovery from hot liquid and gaseous streams is performed usually by means of heat exchangers, being the simplest and most economical recovery method, but requires that the recoverable stream match the demand profile of the end user in quantity and in quality. In addition, the pollutant content of the stream, which might provoke corrosion, must be considered carefully. Many types of solid and liquid waste can be stored when they are produced and then transformed into energy as liquid, solid, and gaseous fuels according to the end user demand. These systems are subject to the following considerations:

- Urban and industrial solid waste can be burned in incinerators to produce hot water, steam, and electricity. So can industrial and urban sludge. Solid waste can be prepared as RDF or as solid fuel pellets. Net heating values range from 16,000 to 21,000 kJ/kg.

- Liquid and solid waste can be treated by means of anaerobic digestion to produce biogas. The main parameter to be considered for liquid waste is the COD (chemical oxygen demand) content to which the production of biogas is generally referred.
- Other processes, such as pyrolysis, distillation or gasification are possible in particular applications, and must be investigated individually and evaluated carefully from the economic and technical points of view.

Industrial waste can be reused inside a factory as raw material and in energy-recovery plants. The main aspects are the following: the chemical composition of waste, the potential reuse for internal process, the technology available for treatment, waste produced by treatment plants and the consequent air and water pollution, and also the economic viability, which should take environmental, energy, and operating costs into account. Treatment of liquid and gaseous waste is widely practiced, to varying extents determined by local regulations, sensitivity to the problem and the kind of pollution concerned. Liquid and gaseous waste can be classified on the basis of pollutant content; specialized literature and practical experience are available to solve most cases either on-site or outside the factory. The energy content in liquids and gases, and pollutant emissions can be partially recovered by means of heat exchangers, to be used in process or in cogeneration plants, depending on the temperatures and pollutant content. Solid waste from industry is classified in each country according to local regulations; nevertheless, the various systems have much in common. Industrial solid waste is commonly broken down into four categories:

- Inert waste from extractive, brick and similar industries, to which can also be assimilated debris from building demolition or construction. These can easily be reused for road construction and public works.
- Urban waste: This includes plastics, paper, rubber, and, in limited quantities, glass.
- Industrial waste includes all industrial waste with the exceptions of the above-mentioned categories (inert and urban waste) and hazardous waste, involving risks for the environment, but it can be made less harmful by appropriate treatment or reused in processes. It includes organic and inorganic materials, liquids, sludge and solids with limited concentrations of pollutant.
- Hazardous and infectious wastes are industrial waste with high concentrations of toxics or pollutants.

Urban waste is commonly landfilled, a no longer an acceptable disposal method because of environmental problems and the scarcity of space. Utilities are very interested in burning waste to produce electricity and steam, but air pollution must be avoided. Urban waste treatment is not discussed here in detail, our main concern being industrial waste and its on-site utilization. Non-hazardous industrial waste can be treated as urban waste, but the quantity produced by a single medium-sized factory is generally too small for on-site reuse in heat-recovery plants. As a general rule, on-site treatment is economically viable if the quantity of waste is from about 500 tons per year to 1000 tons per year or better higher. This depends on the type of waste and particular situations, but it can be assumed as a general indication.

3.6.1 Waste Management

Waste-management strategies comprise a few main phases, often the same as in energy management: analysis of historical data, audits and accounting, analysis of environmental regulations, engineering analyses, and investment proposals based on feasibility studies, personnel training and information. Several factors need to be taken into consideration: waste costs on turnover and added value, production percentage of waste, waste identification and quantification, following the facility raw materials and energy flows, selection of significant indexes valid for the whole facility and for single-production lines, recycling inside or outside the factory, pollution levels and compatibility with regulations. Development of waste-management strategies at different levels depends on the importance of waste cost in comparison with other production factors and on the regulations, requiring specific procedures, regardless of the waste quantity involved. Waste management is performed and implemented in three approaches: (1) waste reduction by clean technologies, (2) waste elimination and (3) waste recycling as raw material or energy.

The clean-technologies approach is the most appealing, the most expensive and time-consuming approach, requiring significant changes in all the process phases, from raw-material input to packaging and delivery. It is important not to be overlooked are the packaging phase effects on urban and industrial waste, due to the content of plastic and wood waste. In addition, the use of natural gas, liquid and solid fuels with low sulfur concentrations is part of this area of waste management the so-called clean technology in facilities, gaseous pollutants from combustion are reduced without abatement plants at the combustion process ending. Solid and liquid waste elimination by employing an external company is the simplest but the most expensive policy. Quite often it is the only feasible one because the produced waste quantity is too small to justify an independent plant. Nevertheless, local regulations may require storage and transportation in accordance with strict rules, often increasing the final disposal cost. It must be pointed out that the difficulties in finding suitable sites for this operation are growing, particularly in crowded areas such as the industrialized part of the world. In the case of liquid waste, water discharge to external agencies costs according to the pollutant content. Otherwise, liquid waste could be treated before discharge by means of mechanical, chemical, and aerobic systems, which are consuming energy. In a similar way, the treatment of gaseous streams at high temperature must be performed in the factory, depending on local regulations, by burning the pollutant streams and by keeping them for a given time at temperatures higher than 900°C–1000°C (1650°F–1832°F), then cooling them to a lower temperature before discharge. Notice that the cooling phase can be economically performed by recovering heat from the exhaust through heat exchangers, thus reducing the energy consumed in treatment. Whenever possible, recycling as raw material or energy is the most attractive approach.

The use of waste as raw material in the process remains the best waste management form the waste does not have to leave the factory. The use of waste coming from other factories and processes is equally successful in a wider waste-management approach. It is a method often practiced in foundries and cement industries by using plastics, rubber, liquid waste, and RDF, but it requires separation of material flows in order to facilitate the recycling at different process steps. Although it is not always practicable because of environmental pollution issues and the equipment corrosion by substances such as polyvinyl chloride, incineration of waste (RDF, sludge, etc.) inside the factory is an attractive option. Pyrolysis, distillation and anaerobic treatment of solid waste, sludge, and liquids intended to produce fuel are good options. These treatments, such as anaerobic digestion processes, are often confined to small plants. Energy recovery in these plants is low and the recovered energy often only equals the input energy for auxiliaries. Nevertheless, unlike other energy-consuming systems, these treatments allow some energy saving. It is important in the WTE facility design to take into consideration the potential variations in both physical and chemical composition of MSW. One of the most troublesome areas in and WTE plant is the materials handling systems. To successfully select materials handling system components and design an integrated process, requiring adequate information on the variability and extremes of the physical size and shape the solid waste facility must handle, such as the bulk density, angle of repose of the material, and the variation in noncombustible content. In the design of the furnaces or boilers of WTE facilities, the refuse characteristics of interest are the calorific value, moisture content, proportion of noncombustibles, and other components whose presence during combustion are resulting in the need for flue gas cleanup.

3.7 Summary

As population increases and technology development begin to result in significant resource depletion and environmental deterioration, we must take a universal view on the ground rules for sustaining our species in a manner that is compatible with preservation of biosphere, requiring the production of feed, food and energy by technologies that are sustainable, have minimal environmental impacts, involving a major shift to renewable resources of energy, sustainable agricultural practices for production of food, feed and energy and recycle of all non-renewable resources. Energy from biological materials addresses a number of key energy and environmental issues, including climate change, energy security, and replacement of carbon-intensive energy sources. Biomass is one of the most important sources to increase the energy production based on renewable energy sources. In this chapter, various biomass conversion methods such as thermochemical, biochemical and chemical conversion have been covered. Biological materials refer to substances derived from living organisms that can be harnessed to produce bioenergy. Bioenergy can be generated through biomass (solid), biogas (gas) or biofuel (liquid). Derivation of methane from energy crops and organic wastes has an important role toward achieving this objective. Biomass energy or bioenergy is energy derived from organic matter, having the potential to greatly reduce pollutant emissions and to contribute to the world energy demand. The energy in organic materials comes from sunlight harvested via photosynthesis, the process where light energy is used to convert water and carbon dioxide into oxygen and organic compounds. Organic components from municipal and industrial waste, plants, agricultural and forestry residues, home waste and landfills can be used very efficiently in our society. In other word, the biomass as energy from plants, animal, feedstock and by-products of industrial, agricultural and human activities is storable, transportable, convertible, available and affordable, always with positive energy balance. Biomass represents the storage of solar energy in chemical form in plant and animal materials, being one of the most commonly used material, precious and versatile resources on the Earth. Biomass has been used for energy purposes ever since man discovered fire, and is a sustainable, environmentally benign and economically sound energy and material source. Different sources of bioenergy (i.e., solid biomass, biofuels, biogas, etc.) require distinct technical methods to convert the raw material to electricity and/or heat. Waste management is a strictly energy-related function, which concerns all factories, facilities, industrial, commercial, agricultural and the activities of every person at work and at home. Renewable fuels or biofuels are available in liquid form as bio-alcohols, vegetable oils, biodiesel and in gaseous form as biogas, syngas, and synthetic fuels. Biodiesel and bio-oils processed from biological materials such as vegetable oils and/or recycled cooking oils, animal fats, plant waste and forest waste products can be blended with non-renewable petroleum fuels to use in transport engines, boilers, space heating and industrial processes. Biofuels blended with commercially available fossil fuels promote energy efficiency and reduce pollutant emission. These fuel types have also the potential to reduce dependency on conventional fuels or fluctuating energy prices, through market and supply diversification. If the energy must be considered, taking into account the limited available resources and the search for new energy sources must always be continued, waste must be viewed in the perspective of its huge presence everywhere in the world. On the other hand, waste means energy: (1) energy can be produced from disposal at the end of a production process, and (2) energy is consumed both during the transformation of raw materials into the end product and during waste treatment. Cost of waste treatment is one of the factors to be considered in the economic analysis of investments. However, there are increased needs to generate electric energy and safely manage the MSW generated by modern society, particularly in the major metropolitan areas, together with the performance of modem WTE plants, indicate that this technology can be utilized to dispose of a part of the waste and provide electricity.

Questions and Problems

1. List the three U.S. states and the EU countries with the highest ethanol production. Do the same for biodiesel.

2. What are the three EU countries that are using the most biodiesel?

3. List the most common biomass sources.

4. Define the bioenergy and its importance in the world energy mix.

5. A forested area has an area of 1000 ha and is located at 45° N latitude. The annual solar radiation intensity is 180 W/m², if the trees are growing in average 7.5 months per year and are dormant the rest of the year. If in 20 years period the increases in the tree mass are about 1000 kg, what is the overall efficiency of converting solar radiation into carbohydrate stored chemical energy?

6. List the major advantages and disadvantages of biomass as a renewable energy source.

7. List the main advantages for the waste-to-energy recovering processes.

8. In your opinion is the biomass best used to produce transportation fuels or to generate heat and power?

9. What are the adverse effects of biomass direct combustion?

10. Can the biofuels play an important role in improving the energy security of a country? Briefly discuss the pros and cons.

11. Briefly describe the biomass to ethanol conversion.

12. What are gasification and pyrolysis?

13. What ingredients are needed to produce ethanol?

14. What ingredients are needed to produce biodiesel?

15. Briefly describe the biodiesel production process. Do the same for the ethanol production process.

16. What are the environmental and socioeconomic problems with biomass and biofuels?

17. List and briefly describe the major issues and problems with the municipal solid waste.

18. Estimate the area needed in your region or state to grow trees to generate 500 MW of power. You must find the annual average solar energy density for your area in order to make the estimate.

19. Sugar, starch and cellulose are raw materials in the production of ethanol. Name three plants that each produce sugar, starch and cellulose.

20. List the pros and cons whether your region should increase or not its biomass supply for combined heat and power generation or for bioethanol production.

21. Estimate the CO_2 emission by burning 1 liter of biodiesel and compared it with the emission of burning conventional diesel fuel.

22. Is biodiesel really green? Explain at least two arguments in support of the idea that biodiesel is a *greener* fuel. Also present one argument that biodiesel is not a *greener* fuel as it is usually considered.

23. Estimate the amount of ethanol that can be produced from 10 kg of glucose.

24. If 2.50 moles of CO_2 are produced, during glucose fermentation, how much ethanol was produced?

25. Balance the following chemical equations by writing whole number coefficients in front of the chemical formulae as needed.

$$| \quad |\cdot CO_2 +| \quad |\cdot H_2O + Light \Rightarrow C_6H_{12}O_6 +| \quad |\cdot O_2 \text{ (Photosynthesis)}$$

$$C_6H_{12}O_6 \Rightarrow | \quad |\cdot C_2H_5OH +| \quad |\cdot CO_2 + Heat \text{ (Fermentation)}$$

$$C_2H_5OH +| \quad |\cdot O_2 \Rightarrow | \quad |\cdot CO_2 +| \quad |\cdot H_2O + Heat \text{ (Combustion)}$$

26. The fermentation of sucrose ($C_{12}H_{22}O_{11}$) with water results in 4 ethanol molecules and 4 carbon dioxide molecules. Estimate how much ethanol can be produced from 2.5 ton of sucrose.

27. How much ethanol can be produced from 100 kg of glucose? What is the conversion efficiency?

28. Estimate the electricity amount in kWh that can be generated from a power unit, having a thermal efficiency of 0.42 that is burning the biomass grown on a 50 km², assuming the biomass yield similar to the one in Example 3.2.

29. Determine the amount of the thermal energy that can be generated from the ethanol produced through fermentation process from 300 kg of sucrose.

30. Determine the thermal energy released during the direct complete combustion of butanol (C_4H_9OH).

31. For 10 kg of carbon monoxide calculate the needed hydrogen and products resulting for the ethylene synthesis.

32. Find the heat amount released during the reactions of the complete combustion of 1 kg of methanol to produce carbon dioxide, gaseous water (vapor) and heat.

33. Repeat the above problem, but for 1 kg of ethanol.

Further Readings and References

1. A.W. Culp, Jr., *Principles of Energy Conversion* (2nd ed.), McGraw-Hill, New York, 1991.
2. R.E. Katofsky, *The Production of Fluid Fuels from Biomass*, Princeton University/Center for Energy and Environmental Studies: Princeton, NJ, 1993.
3. D.L. Klass, *Biomass for Renewable Energy, Fuels and Chemicals*, Academic Press, San Diego, CA, 1998.
4. G. Boyle (ed.), *Renewable Energy*, Oxford University Press, Oxford, UK, 2004.
5. V. Quaschning, *Understanding Renewable Energy Systems*, Earthscan, Routledge, UK, 2006.
6. A. Ristinen and J. J. Kraushaar, *Energy and Environment*, Wiley, Hoboken, NJ, 2006.
7. F.A. Farret and M.G. Simões, *Integration of Alternative Sources of Energy*, Wiley-IEEE Press, Hoboken, NJ, 2006.
8. P. Kruger, *Alternative Energy Resources: The Quest for Sustainable Energy*, John Wiley & Sons, New York, 2006.
9. J. Andrews and N. Jelley, *Energy Science, Principles, Technology and Impacts*, Oxford University Press, Oxford, UK, 2007.
10. E.L. McFarland, J.L. Hunt, and J.L. Campbell, *Energy, Physics and the Environment* (3rd ed.), Cengage Learning, Boston, MA, 2007.
11. F. Kreith and D.Y. Goswami (eds.). *Handbook of Energy Efficiency and Renewable Energy*, CRC Press, Boca Raton, FL, 2007.
12. A. Thow and A. Warhurst, *Biofuels and Sustainable Development*, Maplecroft, Bath, UK, 2007.
13. D. Pimentel (ed.), *Biofuels, Solar and Wind as Renewable Energy Systems, Benefits and Risks*, Springer, Berlin, Germany, 2008.
14. M. Himmel (ed.), *Biomass Recalcitrance: Deconstructing the Plant Cell Wall for Bioenergy*, Wiley-Blackwell, Hoboken, NJ, 2008.
15. C. Drapcho, J. Nghiem, and T. Walker, *Biofuels Engineering Process Technology*, McGraw-Hill, New York, 2008.
16. FAO Report, *State of Food and Agriculture—Biofuels: Prospects, Risks and Opportunities*, Food and Agriculture Organization of the United Nations, Rome, Italy, 2008.
17. S. Van Loo and J. Koppejan, *The Handbook of Biomass, Combustion & Co-firing*, Earthscan, Routledge, UK, 2008.
18. A. Vieira da Rosa, *Fundamentals of Renewable Energy*, Academic Press, Cambridge, MA, 2009.
19. IEA Bioenergy Report, *Bioenergy: A Sustainable and Reliable Energy Source, A Review of Status and Prospects*, ExCo IEA Bioenergy, Rotorua, NZ, 2009.
20. B.K. Hodge, *Alternative Energy Systems and Applications*, Wiley, Chichester, London, UK, 2010.
21. A.F. Zobaa and R.C. Bansal (eds.), *Handbook of Renewable Energy Technology*, World Scientific Publishing Co., Hackensack, N.J, 2011.
22. F.M. Vanek, L.D. Albright, and L.T. Angenent, *Energy Systems Engineering—Evaluation and Implementation* (2nd ed.), McGraw-Hill, New York, 2012.
23. E.E. Michaelides, *Alternative Energy Sources*, Springer, Berlin, Germany, 2012.
24. M. Kaltschmitt, N.J. Themelis, L.Y. Bronicki, L. Söder, and L.A. Vega (eds.), *Renewable Energy Systems*, Springer, Berlin, Germany, 2013.
25. G. Petrecca, *Energy Conversion and Management: Principles and Applications*, Springer, New York, 2014.
26. R.A. Dunlap, *Sustainable Energy*, Cengage Learning, Boston, MA, 2015.
27. V. Nelson and K. Starcher, *Introduction to Renewable Energy (Energy and the Environment)*, CRC Press, Boca Raton, FL, 2015.
28. M. Martín (ed.), *Alternative Energy Sources and Technologies*, Springer, Cham, Switzerland, 2016.

4

Electric Utility Integration of Renewable Energy Systems

4.1 Electrical Power System Basics

The electric power system is a complex technical system with main function to deliver electricity between generation, consumption and storage. A power system is an interconnected network of components converting nonelectrical energy into the electrical form and transferring the electrical energy from generating units to the loads or end-users. The next chapter subsections give a brief but comprehensive discussion of the electric power system, the AC systems, major system components, power system operation and management, and a conceptual description of the technical fundamental characteristics and performance of the electric power systems. The main function of any power system is to deliver electricity between generation, consumption and storage. A sustainable energy system involves key components, such as increased renewable energy sources, higher efficiencies and optimum electricity use in all sectors. Some of the critical issues for the electric power systems due to the extended renewable energy applications are: the intermittent electricity generation (e.g., wind and solar energy) integration to the electrical distribution systems and different energy storage solutions. Power systems are made up of three major distinct components or sub-systems: generation, transmission and distribution, as shown in Figure 4.1. The generation subsystem includes generation units, such as turbines and generators. The energy resources used to generate electricity in most of the power plants are fossil fuels, nuclear, or hydropower. In the last three decades was in increasing trend to include renewable energy sources for electricity energy generation, especially in the distribution section. Nuclear power and hydroelectric power are non-polluting energy sources, while the last one is also renewable energy source. Currently in the U.S. hydropower are accounting for about 8% and nuclear power 20% of the electricity generation.

The generated electricity is transmitted through a complex network consisting of transmission lines, transformers, control, monitoring and protective equipment. Transmission lines transfer the electrical energy from power plants to load centers. Transformers are stepping up the voltage at power plants to very high values (200–1200 kV) to reduce currents and losses, and reducing the size of transmission wires and implicit reducing the cost of the transmission system. The transmission lines carry power to load centers, where transformers lower the voltages up to 35 kV at bulk power substations. At load centers, the voltages are reduced by step-down transformers to lower values (4.5–35 kV) for power distribution networks. Large industrial customers are usually supplied from these substations. This is known as the sub-transmission power systems or sections. Distribution of power to commercial and residential users takes place through a power distribution

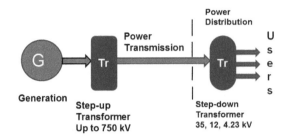

FIGURE 4.1
Basic structure of a simplified electrical power system.

system consisting of substations where step-down transformers lower the voltages to a range of 2.4–69 kV. Power is carried by main feeders to specific areas where there are lateral feeders stepping it down to customers levels. At customer sites, the voltage is further reduced to values, such as 120 V, 208 V, 220V, 280/277 V, etc. as required by users. Electrical power grids are divided into three main voltage levels, defined by the IEC 2009 standard, in low-voltage (LV) levels, 1 kV and lower, medium-voltage (MV), 1–52 kV and high-voltage (HV) section, above 52 kV. The highest voltage networks, operating above 220 kV, serve as transmission networks. Power distribution networks are serving restricted areas or regions, feeding local grid stations and costumers. Power systems are extensively monitored, controlled and protected. These complex electricity transmission and distribution networks encompass larger areas and millions and costumers. Each power system has several levels of protection to minimize the effects of any damaged system component on the system ability to provide safe reliable electricity to all customers. A power system serves one important function and that is to supply customers with electricity as economically and as reliably as possible. In summary the main functions of three sub-systems of the power systems are:

1. Generation subsystem, the grid components that are generating or are sources of electrical energy.
2. Transmission and sub-transmission subsystems are sections designed for transferring electrical energy from sources to load centers with voltages at 115 kV and higher to reduce minimize the power losses.
3. Power distribution networks, the most complex subsystem are the grid sections designated to distribute the electricity from substations, operating in the voltage range 44–12 kV to end users or customers at much lower voltages, 600–120 V, usually in single-phase connections.
4. Consumers or users, the utilization subsystems are the grid end who are using the electricity.

The basic power system structure and components is shown in Figure 4.1. In this structure, electro-mechanical systems are playing a key role in generating, transferring and using electricity. One of the essential power system components are the three-phase alternative current (AC) synchronous generators (alternators) and transformers. The electric generator converts nonelectrical energy provided by the prime mover, usually steam or hydro turbines to electrical energy. The turbine function is to rotate electrical generators by converting the steam thermal energy or water kinetic energy of the water into rotating mechanical energy. In thermal power plants, fossil fuels or nuclear reactions are used to produce high temperature steam, passed through the turbine blades causing turbine to rotate. Typical hydroelectric plants consist of dam, holding water upstream at high elevations with respect to the turbine. When electricity is needed, the water flows through the hydro turbines through penstocks rotating the generators. Since the generators are mounted on the turbine shaft, the generator rotates with the turbine generating electricity. At load level, the bulk of the energy is consumed by electrical motors, mostly induction type, appliances, specific equipment and devices. The synchronous generators have two synchronized rotating magnetic fields, one produced by the rotor driven at synchronous speed and excited by a DC circuit, and the other one produced into the stator windings by the three-phase armature currents. The excitation system maintains the generator voltage and control the reactive power flow. Due their structure and construction, the alternators can generate higher power and voltages, typically 30 kV. Alternators have magnetic field circuits mounted on the rotor, connected to the turbine shaft. The alternator stator has windings wrapped around its core in a three-phase configuration. Insulation requirements and other practical design issues limit the generated voltage to values, up to 30 kV.

The generator voltage (5–35 kV) is not high enough for efficient power transmission, being stepped-up by a transformer, connecting generators to transmission subsystems. The step-down transformers are connecting the transmission subsystems to power distribution networks. The transformers transfer electricity with high efficiency from one voltage level to another. Their main functions are stepping up the lower generation voltage to the higher transmission voltage and stepping down the higher transmission voltage to the lower distribution voltage. The main advantage of higher voltages in the transmission system is to reduce the grid losses. Since transformers operate at constant power, when the voltage is higher, then the current has a lower value. Therefore, the losses, a function of the current square, are lower at a higher voltage. However, when the electrical energy is delivered to the load centers, the voltage is stepped down for safer distribution and usage requirements. When the electrical power reaches customers' facilities is further stepped down to the required levels depending on the various standards worldwide. The electricity in an electric power system may undergo four or five transformations between generation and consumers. A power system is predominantly in steady-state operation or in a state that could with sufficient accuracy be regarded as steady state. In a power system there are always small load changes, switching actions and other transients occurring so that in a strict mathematical sense most of the variables are varying with the time. However, these variations are most of the time so small that an algebraic, i.e., not time varying model of the power system is justified. The electricity is delivered from the generation ends (power plants) to the loads (consumers) transmission lines and transformers. The bulk power is transmitted to the load centers over long distance high voltage transmission lines, operating at very high voltages, 220–1200 kV. The lines that distribute the electrical power within an area are called medium voltage distribution lines. There are several other categories such as sub-transmission and high voltage distribution line or power distribution networks, which will be discussed in later chapters or sections of this book. The transmission lines are high voltage conductors (wires) mounted on tall towers to prevent them to be in contact with humans, trees, animals, buildings, equipment or ground. Notice that other types of electrical generators, such as induction and DC generators, are often found in stand-alone, back-up power units and low-power generation networks.

The vast majority of today energy, including electricity, is provided by fossil fuels, such as coal, oil and natural gas. However, the fossil fuels are not the same in terms of availability and their environmental impacts. Fossil fuels are relatively easy to extract, transport and combust to meet the energy needs and demands. Fossil fuels are varying in terms of energy density (the energy content per unit of mass or volume), consumption rates, remaining resource availabilities and pollutant emissions per unit of energy produced. Since the time when coal surpassed wood as the leading energy source, fossil fuels are becoming the dominant primary energy source, counting for over 85% of all energy production. Fossil fuels are very important for meeting our energy needs, sustainable energy approach is considering high-efficient fuel combustion technologies, energy efficiency and conservation, distributed generation and the move toward the use of renewable and alternative energy sources. It is commonly accepted that the fossil energy resources are somewhat limited, and sometimes in the future their production begin to decline. There also is a quite strong opposition against nuclear power in many countries. In this scenario renewable energy resources tend to become a stronger contributor to the ever rising energy needs. The major renewable energy resources are the direct or indirect solar energy, with some forms attributed to the moon and the earth. Notable for their contribution to the energy demand are water, wind and solar energy, geothermal sources and biomass. Renewable energy is increasingly used in the electricity generation due to the technological advances in wind turbine, photovoltaic, energy storage, power electronic and control technologies. Renewable energy sources are providing clean energy, emitting fewer pollutants. For this reason, many experts and scientists advocate renewable energy sources over traditional fossil fuels.

Today power grids are operating unidirectional in the power flow from generation to end-users. The one-directional power flow and interactions are insufficient to facilitate increasingly complex and dynamic power demands from expanding consumer energy uses increasingly reliant on the non-essential electric appliances, life quality enhancement, and equally demanding of the grid communication. In achieving a balance between supply and demand, grid operators modulate the output of dispatchable electricity generation sources, using forecasting algorithms to follow consumer demand profiles on per second basis. These dispatchable methods function in idle modes, always spinning and burning fuel in preparation of being requested at short notice, while their output is a carbon intensive procedure. An idealized scenario is to replace the convention of supply follows demand, to a more efficient paradigm introducing demand side management (DSM) where demand follows the supply. By the load dynamic redistributions, it is possible to flatten the supply profile and using fast response mechanisms, enabling easier grid integrations of intermittent low carbon renewable energy resources. Such scenario requires the introduction of a new electrical grid through a series of retrofit measures to progressively replacing today grid architecture. The smart grid is such a modernized electrical grid, currently somewhat in a conceptual and small-scale demonstration phase, implementation phase, introducing functionalities for two-directional power flow and increasing communication between the utility and consumers. The smart grid is conceptualized as an extensive cyber-physical system, supporting and facilitating significantly the enhanced controllability and responsiveness of highly distributed energy resources within electric power systems. Electricity information is autonomously exchanged on a per second basis. Through utility control and monitoring capabilities, the consumer is either smartly requested or pervasively interrupted in their use of non-essential electrical equipment. The power industry has adopted *smart grids (SGs)*, using information and communication technologies, to make the electric power systems more reliable and efficient. To increase the transmission capacity, to reduce the transmission losses and to adopt the energy conservation practice and methods the *smart grids as a new power system model and structure* is adopted by most of the utilities at transmission and distribution levels. To enhance the generation capacity and to improve the performance of the power distribution systems, the local power generation is included, the renewable energy is a key player. The ability to increase the uses of intermittent energy sources by balancing them across large areas provided an electrical market which removes the congestion. For long time, the electric utilities are vertically integrated, generation, transmission and distribution being combined in a single system. Such centralized structure ensures that the energy supply equals the demand, lower the production costs, optimizing the generator dispatch and transmission capacity allocation. Power and money flows are unidirectional from consumer to utility, respectively, from utility to users, while the information flow is mainly between transmission and generation stations with limited information changes into the power distribution or with customers. Drawbacks economical aspect and lack of coordination between different utility components resulted into power system deregulation. The smart grid signifies digital upgrade of transmission and distribution subsystems in order to optimize operations, to reduce cost and to increase reliability and supply security. In summary, the main benefits of the smart grids are:

1. Improved connection and operation of all power generators, regardless the size and technology;
2. Allow the consumers to participate into the system operation optimization;
3. Provide consumers with better information and options for the supply choice, while significantly reducing the environmental impacts of the whole electricity supply system;
4. Improve the existing levels of system reliability, quality and security of supply;
5. Maintain and improve the existing services efficiently; and
6. Foster market integration.

4.1.1 Three-Phase Systems

The generation, transmission and distribution of electric power are accomplished by means of three-phase voltages and currents, as shown in Figures 4.2 and 4.3. An AC generator designed to develop a single sinusoidal voltage for each rotation of the rotor is referred to

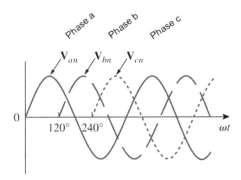

FIGURE 4.2
The diagram of a three-phase voltage system.

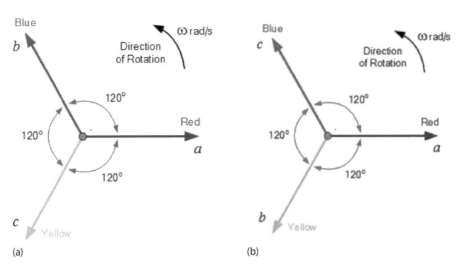

FIGURE 4.3
Three-phase voltage sequences: (a) positive and (b) negative sequences.

as a single-phase AC generator. If the number of coils on the rotor increases in a specified arrangement sets a poly-phase AC generator, which develops more than one AC phase voltage per rotation of the rotor. At the generating station three sinusoidal voltages are generated with the same amplitude but displaced in phase by 120°, resulting in the so-called a balanced source. If the generated voltages reach their peak values in the sequential order *abc*, the generator is said to have a positive phase sequence (Figure 4.3a). If the phase order is *acb*, the generator is said to have a negative phase sequence (Figure 4.3b). In a three-phase system, the instantaneous power delivered to the external loads is constant rather than pulsating as it is in a single-phase circuit. The three-phase motors, having constant torque, can start and run better than single-phase motors. This three-phase power feature, coupled with the inherent efficiency of its transmission compared to single-phase (less wire for the same delivered power), accounts for its universal use. In general, three-phase systems are preferred over single-phase systems for the power transmission for reasons, such as:

1. Thinner conductors can be used to transmit the same kVA at the same voltage, reducing the copper or aluminum amount required (typically about 25% less).
2. The lighter lines are easier to install, and the supporting structures are less massive and farther apart.
3. Three-phase equipment and motors run smoothly, having better running and starting characteristics than single-phase equivalents because of even power flows than is delivered by a single-phase supply.
4. Larger motors are usually three-phase type because they are essentially self-starting and do not require a special design or additional starting circuitry.

Three-phase sinusoidal voltages and currents generated with the same magnitude but are displaced in phase by 120° are setting a balanced source or generator, and the three voltages, V_{an}, V_{bn}, and V_{cn}, with the same magnitude and separated by 120° phase angles are produced:

$$V_{an} = V_m \sin(\omega t) = V_m \langle 0^\circ$$

$$V_{bn} = V_m \sin(\omega t - 120^\circ) = V_m \langle -120^\circ \qquad (4.1)$$

$$V_{cn} = V_m \sin(\omega t - 240^\circ) = V_m \sin(\omega t + 120^\circ) = V_m \langle -240^\circ$$

where V_m is the peak value or the magnitude of the generated voltage. The sum of the three waveform voltages, by using trigonometric identities, is

$$V = V_{an} + V_{bn} + V_{cn} =$$
$$V_m \left[\sin(\omega t) + \sin(\omega t - 120^\circ) + \sin(\omega t - 240^\circ) \right] = 0 \qquad (4.2)$$

Three-phase systems may be labeled by *a, b, c,* or by using the three natural colors, *red, yellow* and *blue.* The phase sequence is important for distribution and use of electrical power. If the generated voltages reach their peak values in the sequential order *abc,* the generator is said to have a positive phase sequence (Figure 4.3a). If the phase order is *acb,* the generator is said to have a negative phase sequence, as shown in Figure 4.3b. The three single-phase voltages can be connected to form practical three-phase systems in two ways: (1) star or wye (Y) connections (circuits), or (2) delta (Δ) connections (circuits). In the Y-connection, one terminal of each generator coil is connected to a common point or neutral n and the other three terminals represent the three-phase supply. In a balanced three-phase system, knowledge of one of the phases gives the other two phases directly. However, this is not the case for an unbalanced supply. In a star connected supply, it can be seen that the line current (current in the line) is equal to the phase current (current in each phase). However, the line voltage is not equal to the phase voltage.

In a three-phase system, the instantaneous power delivered to the external loads is constant rather than pulsating as it is in a single-phase circuit. Three-phase motors, having constant torque, start and run much better than single-phase motors. This feature of three-phase power, coupled with the inherent efficiency of its transmission compared to single-phase (less wire for the same delivered power), accounts for its universal use. A power system has Y-connected generators and usually includes both Δ- and Y-connected loads. Generators are rarely Δ-connected, because if the voltages are not perfectly balanced, there will be a net voltage, and consequently a circulating current, around the Δ loop. Also, the phase voltages are lower in the Y-connected generator, and thus less insulation is required. In Y-connected circuits the phase voltage is the voltage between any line (phase) and the neutral point, represented by V_{an}, V_{bn}, and V_{cn}, while the voltage between any two lines is called the line or line-to-line voltage, represented by V_{ab}, V_{bc}, and V_{ca}, respectively. For a balanced system, each phase voltage has the same magnitude, and we define:

$$|V_{an}| = |V_{bn}| = |V_{cn}| = V_P \qquad (4.3)$$

Here V_P denotes the effective magnitude of the phase voltage. We can show that

$$V_{ab} = V_{an} - V_{bn} = V_P\left(1 - 1\langle -120°\right) = \sqrt{3}V_P\langle 30° \tag{4.4}$$

Similarly relationships, we can obtain

$$V_{bc} = \sqrt{3}V_P\langle -90°$$
$$V_{ca} = \sqrt{3}V_P\langle 150° \tag{4.5}$$

In a balance three-phase Y-connected voltage system, the line voltage V_L whose magnitude is related to the phase voltage magnitude through:

$$V_L = \sqrt{3}V_P \tag{4.6}$$

> **Example 4.1:** A three-phase generator is Y-connected, and the magnitude of each phase voltage is 220 V RMS. For *abc* phase sequence, write the three-phase voltages, and calculate the line voltage magnitude.
>
> **Solution:** The expressions of the phase voltages are:
>
> $$V_{an} = 220\langle 0° \text{ V}$$
> $$V_{bn} = 220\langle -120° \text{ V}$$
> $$V_{cn} = 220\langle 120° \text{ V}$$
>
> While the magnitude for the
>
> $$V_{LL} = \sqrt{3} \times V_P = \sqrt{3} \times 220 = 380.6 \text{ V}$$

The line voltage is leading, by 30° the nearest phase voltage, and the magnitude of the line voltage is √3 times the phase voltage. A current flowing out of line terminal, I_L, the effective value of the line current is the same the phase current I_p, the effective value of the phase current for the Y-connected circuits, thus:

$$I_L = I_P \tag{4.7}$$

In delta connection, the line and the phase voltages have the same magnitude:

$$\left|V_L\right| = \left|V_P\right| \tag{4.8}$$

Similarly in the case of a delta connected supply, the current in the line is √3 times the current in the delta. In a manner similar as for the Y-connected sources we can easily prove:

$$I_{ab} = \sqrt{3}I_P\langle 0°$$
$$I_{bc} = \sqrt{3}I_P\langle -90°$$
$$I_{ca} = \sqrt{3}I_P\langle 150° \tag{4.9}$$

Balanced three-phase currents in a delta connection yields a corresponding set of balanced line currents, where I_L is the magnitude of any of the three line currents, related as:

$$I_L = \sqrt{3}I_P \tag{4.10}$$

4.1.2 Power Relationships in Three-Phase Circuits

A three-phase Y-connected voltage sources is supplying a three-phase balanced load, as shown in Figure 4.4, with the sinusoidal phase voltages:

$$v_a(t) = \sqrt{2}V_P \sin(\omega t)$$
$$v_b(t) = \sqrt{2}V_P \sin(\omega t - 120°) \qquad (4.11)$$
$$v_c(t) = \sqrt{2}V_P \sin(\omega t + 120°)$$

The phase currents flowing through the load are given by:

$$i_a(t) = \sqrt{2}I_P \sin(\omega t - \phi)$$
$$i_b(t) = \sqrt{2}I_P \sin(\omega t - 120° - \phi) \qquad (4.12)$$
$$i_c(t) = \sqrt{2}I_P \sin(\omega t + 120° - \phi)$$

Here ϕ is the phase angle between the voltage and current in each phase. The instantaneous power supplied to one phase of the load is expressed as:

$$pt) = v(t) \cdot i(t) \qquad (4.13)$$

Therefore, the instantaneous power in each of the three phases of the load is:

$$p_a(t) = v_a(t) \cdot i_a(t) = 2VI \sin(\omega t)\sin(\omega t - \theta)$$
$$p_b(t) = v_b(t) \cdot i_b(t) = 2VI \sin(\omega t - 120°)\sin(\omega t - 120° - \theta) \qquad (4.14)$$
$$p_c(t) = v_c(t) \cdot i_c(t) = 2VI \sin(\omega t - 240°)\sin(\omega t - 240° - \theta)$$

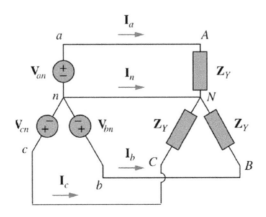

FIGURE 4.4
A Y-connected generator supplying a Y-connected load.

The total instantaneous power flowing into the load is the expresses as:

$$p_{3\phi}(t) = v_a(t)i_a(t) + v_b(t)i_b(t) + v_c(t)i_c(t) \text{ W} \tag{4.15}$$

By substituting expressions for phase voltages and currents, as given in Equations (4.11) and (4.12), into the three-phase power equation (4.15) and using additional trigonometric identity:

$$\cos(\alpha) + \cos(\alpha - 120°) + \cos(\alpha - 240°) = 0$$

Equation (4.15) can be re-written as:

$$p_{3\phi}(t) = 3|V||I|\cos(\phi) = 3P \text{ W} \tag{4.16}$$

Here: $|V| = \sqrt{2}V_P$ and $|I| = \sqrt{2}I_P$ are the peak magnitude (amplitude) of the phase voltage and current. Equation (4.16) represents an important result. *In a balanced three-phase system the sum of the three individually pulsating phase powers adds to a constant, non-pulsating total active power of magnitude three times the phase active power, as* shown in Figure 4.5. The single-phase power equations are applied to each phase of Y- or Δ-connected three-phase loads. The real, active, and apparent powers supplied to a balanced three-phase load are:

$$P = 3V_\phi I_\phi \cos(\theta) = 3ZI_\phi^2 \cos(\theta)$$

$$Q = 3V_\phi I_\phi \sin(\theta) = 3ZI_\phi^2 \sin(\theta) \tag{4.17}$$

$$S = 3V_\phi I_\phi = 3ZI_\phi^2$$

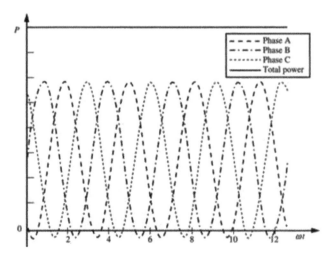

FIGURE 4.5
Power in a three-phase balanced system.

The angle θ is again the angle between the voltage and the current in any of the load phase, and the power factor of the load is the cosine of this angle. We can express the powers of Equation (4.17) in terms of line quantities, regardless the connection type (wye or delta) as:

$$P = \sqrt{3}V_{LL}I_L\cos(\theta)$$

$$Q = \sqrt{3}V_{LL}I_L\sin(\theta) \tag{4.18}$$

$$S = \sqrt{3}V_{LL}I_L$$

We have to keep in mind the angle θ in the above equations is the angle between the *phase voltage* and *the phase current*, not the angle between the line-to-line voltage and the line current.

> **Example 4.2:** The terminal line-to-line voltage of three-phase generator equals 13.2 kV. It is symmetrically loaded and delivers an RMS current of 1.350 kA per phase at a phase angle of 24° lagging. Compute the power delivered by this generator.
>
> **Solution:** The RMS value of the phase voltage is
>
> $$|V| = \frac{13.2}{\sqrt{3}} = 7.621 \text{ kV/phase}$$
>
> The per-phase active (real) and reactive powers are given by:
>
> $$P = 7.621 \cdot 1.350 \cdot \cos(24°) = 9.399 \text{ MW/phase}$$
>
> $$Q = 7.621 \cdot 1.350 \cdot \cos(24°) = 4.185 \text{ MVAR/phase}$$
>
> The instantaneous powers in phase a, b and c are pulsating and are given by:
>
> $$p_a(t) = 9.399\left(1 - \cos(2\omega t)\right) - 4.185\sin(2\omega t)$$
>
> $$p_b(t) = 9.399\left(1 - \cos(2\omega t - 120°)\right) - 4.185\sin(2\omega t - 120°) \tag{4.19}$$
>
> $$p_a(t) = 9.399\left(1 - \cos(2\omega t - 240°)\right) - 4.185\sin(2\omega t - 240°)$$
>
> The total (constant) three-phase power is:
>
> $$P_{3\phi} = 3 \times 9.399 = 28.197 \text{ MW}$$
>
> The fact that three-phase *active (real) power* is constant tempts us to believe that the *reactive power* in a three-phase is zero (as in a DC circuit). However, the reactive power is present in *each phase* as Equation (4.19) is showing. The reactive power per phase is 4.185 MVAR.

Example 4.3: A three-phase load draws 120 kW at a power factor of 0.85 lagging from a 440 V bus. In parallel with this load, a three-phase capacitor bank, rated 50 kVAR is inserted, find:

    a. The line current without the capacitor bank.
    b. The line current with the capacitor bank.
    c. The P.F. without the capacitor bank.
    d. The P.F. with the capacitor bank.

Solution:

    a. From the three-phase active power formula, the magnitude of the load current is:

$$I_{Load} = \frac{P}{\sqrt{3}V_L \times PF} = \frac{120 \times 10^3}{\sqrt{3}\,440(0.85)} = 185.25 \text{ A}$$

$$I_{Load} = 185.25 \left\langle -\cos^{-1}(0.85) = 185.25 \left\langle -31.8° \text{ A} \right.\right.$$

    b. The line current of the capacitor bank (a pure reactive load) is:

$$I_{Cap} = \frac{50 \times 10^3}{\sqrt{3}\,440} = 65.6 \left\langle 90° \text{ A} \right.$$

    The line current is:

$$I_L = I_{Load} + I_{Cap} = 160.6 \left\langle -11.5° \text{ A} \right.$$

    c. The PF without capacitor bank is PF = 0.85
    d. The PF with capacitor bank is

$$PF = \cos(11.5°) = 0.98$$

4.1.3 Power System Components and Structure

The device converting mechanical energy to electrical energy is called a generator. Synchronous machines can produce high power reliably with high efficiency, and therefore, are widely used as generators in power systems. A generator serves two basic functions. The first one is to produce active power (MW), and the second function, frequently forgotten, is to produce reactive power (MVAR). The discussion on generators will be limited to the fundamentals related to these two functions. More details related to the dynamic performance of the synchronous generators can be found in the references at the end of this chapter or elsewhere in the literature. The mechanical structure of generators is out of the scope of this material. A simplified turbine-generator-exciter system is shown in Figure 4.6. The turbine, or the prime mover, controls the active power generation. For instance, by increasing the valve opening of a steam turbine, more active power can be generated and vice versa. The exciter, represented as an adjustable DC voltage source, controls the field current that controls the internal generated voltage source, the so-called excitation voltage, E_f. In this way, the generator terminal voltage, V, is controlled.

The steady-state synchronous generator equivalent circuit consists of an internal voltage source and its (direct-axis) synchronous reactance in series (Figure 4.6). The system

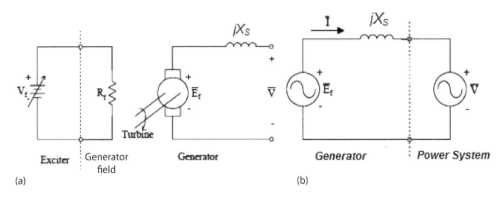

FIGURE 4.6
(a) An electrical representation of a simplified turbine-generator-exciter and (b) a per phase steady-state equivalent circuit of a synchronous generator and the system.

is represented with an infinite bus, which holds a constant voltage. The generator terminal voltage, or system voltage, is usually chosen as the reference, therefore, a zero degree angle. Then, the generator internal voltage can be obtained as:

$$\bar{E}_f = \bar{I}\left(jX_d\right) + \bar{V} = E_f < \delta \tag{4.20}$$

where the angle δ is called the power angle (see Figure 4.7 for the phasor representation). Notice that the power angles are corresponding to physical angels between the generator total magnetic flux contributors and also to the phase angles between the sinusoidal quantities considered as phasors. Figure 4.7 is showing the graphical representation of the quantities of Equation (4.20), together with the current and voltage, in the phasor notation of the internal generated voltage. Such representations are very useful in the calculations of the generator power and characteristics. The per phase analysis of the complex (apparent) power injected into the power system, by a synchronous generator is then calculated with:

$$\bar{S} = \bar{V} \cdot \bar{I}^* = \frac{VE_f}{X_S} \sin\delta + j\left[\frac{VE_f}{X_S}\cos\delta - \frac{V^2}{X_d}\right] = P + jQ \tag{4.21}$$

Here, X_S is the alternator synchronous reactance (the armature circuit resistance is neglected, being about one order of magnitude or even lower, the armature is the circuit where the voltage is induced by the internal magnetic field), P and Q are the generator

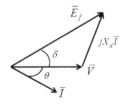

FIGURE 4.7
Phasor representation of the induced generator voltage.

active (real) power and reactive power, respectively. The generator active (real) power is then equal to the real term of Equation (4.21), expressed as:

$$P = \frac{3V \cdot E_f}{X_S} \sin(\delta) \tag{4.22}$$

However, the performance analysis is always conducted for simplicity in per-phase rather than three-phase circuits. The maximum value of the active power, P_{max}, is referred to as the steady-state stability limit and can be calculated as the limit for a power (torque) angle of 90°, and sin (90°) is equal to 1. It is worth mentioning that when the generator active power increases, the power angle also increases. However, at the peak power, P_{max}, the power angle is 90°, and this angle cannot be increased any further, since the generator is losing the synchronism with the rest of the power system. The generator electrical frequency is set by the number of poles, p, and the shaft speed of rotation, n, expressed by:

$$f_e = \frac{p \cdot n}{120}, \text{ Hz (cycles/second)} \tag{4.23}$$

The frequency is determined by the speed of the mechanical driver, prime mover and the generator construction (structure). The voltage is controlled in part by the magnetic field, set by the generator field circuits, and the generated power is controlled by the torque (power) angle, ultimately controlled by the generating unit prime mover controller, while the sign and magnitude of the reactive power is controlled by the DC field current. Among other factors the simpler voltage and power control are main reasons for the choice of the synchronous generators as the solely modern electricity generating units.

Example 4.4: A synchronous generator, connected to an infinite bus, has a negligible armature reactance and a synchronous reactance of 2.75 Ω. The generator excitation is adjusted to set the induced voltage, E_f to 24.5 kV and the input power is 110 MW. If the bus voltage is 23 kV, compute the power (torque) angle, δ and the generator maximum power.

Solution: From Equation (4.22) the torque (power) angle is calculated by:

$$\sin(\delta) = \frac{P \cdot X_S}{3 \cdot E_f \cdot V} = \frac{110 \times 10^6 \times 7.2}{3 \cdot \frac{24.5}{\sqrt{3}} \times 10^3 \cdot \frac{23.0}{\sqrt{3}} \times 10^3} = 0.537 \text{ and } \delta = 29.3°$$

Maximum power is obtained for a torque (power) angle of 90°, by:

$$P_{max} = \frac{3V \cdot E_f}{X_S} = \frac{3 \cdot \frac{23.0}{\sqrt{3}} \times 10^3 \cdot \frac{24.5}{\sqrt{3}} \times 10^3}{2.75} = 204.9 \times 10^6 \text{ W} = 204.9 \text{ MW}$$

The second power systems component are the three-phase and single-phase power transformers. A transformer consists of two or more windings, magnetically coupled through a ferromagnetic core. For a two-winding transformer, the winding connected to the AC supply is referred to as the primary while the one connected to the load is referred to as the secondary. A time-varying current passing through the primary coil produces a time-varying magnetic flux density within the core. According to Faraday law, the variable magnetic flux induces a voltage into the secondary. In order to understand how a transformer operates, the two inductors placed in close proximity to one another are magnetically coupled. The purpose of transformers is to change the primary voltage of the electric system to the required utilization voltage. Power transformers are static devices constructed with two or more windings used to transfer power from one circuit to another at the same frequency with different voltages and currents.

An ideal transformer is one with negligible winding resistances and reactances and no exciting losses, infinite magnetic permeability of the core and all magnetic flux remains into the transformer core. It consists of two conducting coils wound on a common core, made of high grade iron. There is no electrical connection between the coils, coupled to each other through magnetic flux. The coil on input side is called the primary winding (coil) and that on the output side the secondary. The essence of transformer action requires only the existence of time-varying mutual flux linking two windings coupling between the windings is made more effective through the use of a core of iron or other ferromagnetic material because most of the flux will be confined to a definite, high-permeability path linking the windings. Such a transformer is commonly called an iron-core transformer. Most transformers are of this type. The following discussion is concerned almost wholly with iron-core transformers. When an AC voltage is applied to the primary winding, time-varying current flows in the primary winding and causes an AC magnetic flux to appear in the transformer core. The primary is connected to AC voltage sources, resulting in an alternating magnetic flux whose magnitude depends on the voltage and the number of turns of the primary winding. The magnetic flux links the secondary winding and induces a voltage in it with a value that depends on the number of turns of the secondary windings. The transformer operation is subject to Faraday's and Ampere's Laws. If the primary voltage is $v_1(t)$, the core flux $\phi(t)$ is established such that the counter-emf $e(t)$ equals the impressed voltage (neglecting winding resistance), as:

$$v_1(t) = e_1(t) = N_1 \frac{d\phi(t)}{dt} \tag{4.24}$$

Here N_1 is the number of turns of the primary winding. The emf $e_2(t)$ is induced in the secondary by the alternating core magnetic flux $\phi(t)$, expressed as:

$$v_2(t) = e_2(t) = N_2 \frac{d\phi(t)}{dt} \tag{4.25}$$

Taking the ratio of Equations (4.24) and (4.25), we obtain:

$$\frac{v_1}{v_2} = \frac{N_1}{N_2} = a \tag{4.26}$$

Here, a is the transformer turns ratio. If $a > 1$ the transformer is step-down, if $a < 1$, the transformer is step-up, while if $a = 1$, the transformer is the so-called impedance transformer, used to separate electric tow circuits. If a load is connected across the secondary terminals results in current i_2. This current will cause the change in the mmf in the amount $N_2 i_2$. Ohm's law for magnetic circuits must be satisfied. The only way in which this can be achieved is for the primary current to arise, such as:

$$N_1 i_1 - N_2 i_2 = \Re \phi$$

Here $\Re$ is the core magnetic reluctance. The reluctance or magnetic resistance for a magnetic core is simply calculated as:

$$\Re = \frac{l_c}{\mu A} = \frac{l_c}{\mu_r \mu_0 A} \tag{4.27}$$

Here, l_c is the mean (average) core length (m), A is the core cross-sectional are a (m^2), μ and μ_r are the magnetic permeability and relative magnetic permeability of the core, respectively, and $\mu_0 = 4\pi \times 10^{-7}$ H/m is the magnetic permeability of the free space. For an ideal transformer we assumed infinite permeability or $\Re = 0$. Neglecting losses (ideal transformer), the instantaneous power is equal on both transformer sides (power conservations):

$$v_1 i_1 = v_2 i_2 \ \text{ or } \ V_P I_P = V_S I_S$$

Combining above relationship with Equation (4.26), we get:

$$\frac{i_1}{i_2} = \frac{N_2}{N_1} = \frac{1}{a}$$

$$\frac{I_P}{I_S} = \frac{1}{a} \tag{4.28}$$

If all variable are sinusoidal this equation applies also to the voltage and current phasors:

$$\frac{V_P}{V_S} = \frac{I_S}{I_P} = a \tag{4.29}$$

Example 4.5: A 220/20-V transformer has 50 turns on its low-voltage side. Calculate:

  a. The number of turns on its high side.
  b. The turns ratio a, when it is used as a step-down transformer.
  c. The turns ratio a, when it is used as a step-up transformer.

Solution: Transformer, used as the step-down one, the turns ratio is

$$a_{SD} = \frac{220}{20} = 11$$

The number of turns in the high-voltage side is then:

$$N_P = aN_S = 11 \cdot 50 = 550 \text{ turns}$$

The turns ratio when the transformer is used as step-up unit is:

$$a_{SU} = \frac{1}{a_{SD}} = \frac{1}{11} = 0.091$$

Depending on the transformer turns' ratio, the RMS secondary voltage can be greater or less than the RMS primary voltage. Equations (4.26) and (4.29) are showing that any desired voltage ratio, or ratio of the transformation, is obtained by adjusting number of turns of the transformer windings. Transformer action requires a magnetic flux to links two windings, obtained more effectively if an iron (or iron based) core is confining the magnetic flux to a definite path linking the windings. However, any magnetic material undergoes energy losses of energy due to the alternating voltage in the magnetic flux density and magnetic field intensity (B-H) loop, composed of two parts, the *eddy-current loss*, and the *hysteresis loss*. Eddy-current loss is basically an I^2R loss due to induced current in core magnetic materials due to the alternating magnetic flux linking the windings. To reduce these losses, the core is made by a stack of thin iron-alloy laminations. For analyzing an ideal transformer the following assumptions are made. The winding resistances can be neglected and the core reluctance is negligible. All the magnetic flux is linked by all the turns of the coil and there is no leakage of flux. The equation for sinusoidal voltage of the ideal transformer of the primary winding of turns N_1 is supplied by a sinusoidal voltage is:

$$v_1(t) = V_{1m} \cos(\omega t) \tag{4.30}$$

Maximum value of voltage and the RMS value are related through:

$$V_1 = \frac{V_{1m}}{\sqrt{2}} = 0.707 V_{1m} \tag{4.31}$$

In the case of sinusoidal excitation with supply frequency f Hz, the RMS value of the primary emf, from Equation (4.31) is given by:

$$V_1 = 4.44 f N_1 \Phi_m \tag{4.32}$$

Here Φ_m is the peak value of the magnetic flux. Similarly, the secondary emf RMS value is given by:

$$V_2 = 4.44 f N_2 \Phi_m \tag{4.33}$$

Example 4.6: Suppose a coil having 100 turns is wound on a core with a uniform cross-sectional area of 0.25 m². So 5 A, 60 Hz current is flowing into this coil. If the maximum magnetic flux density is 0.75 T, find the mmf and the voltage induced into the coil.

Solution:

$$mmf = NI = 100x5 = 500 \text{ At}$$

The induced voltage is computed by using Equation (4.32).

$$V = 4.44 \cdot 60 \cdot 100 \cdot 0.75 \cdot 0.25 x 10^{-4} = 0.4995 \approx 0.5 \text{ V}$$

Consider now an arbitrary load (Z_2) connected to the secondary terminals of the ideal transformer as shown in Figure 4.8.

The input impedance seen looking into the primary winding, using the transformer relationships for voltage and current, is given by:

$$Z_1 = \frac{V_1}{I_1} = \frac{aV_2}{\frac{I_2}{a}} = a^2 Z_2 \tag{4.34}$$

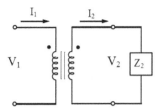

FIGURE 4.8
Load connected to an ideal transformer secondary.

The impedance seen by the primary voltage source of the ideal transformer is the secondary load impedance times the square of the transformer ratio. Using this property, the secondary impedance of an ideal transformer can be reflected to the primary side, expressed by Equation (4.34). In a similar fashion, a load on the primary side of the ideal transformer can be reflected to the secondary.

$$Z_2 = \frac{V_2}{I_2} = \frac{\dfrac{V_1}{a}}{aI_1} = \frac{Z_1}{a^2} \tag{4.35}$$

Another important property of an ideal transformer is the power conservation, stating that the primary and secondary apparent powers (volt-amperes) are equal in an ideal transformer:

$$V_1 I_1 = V_2 I_2 \tag{4.36}$$

The electric current is transmitted in the utility grids, as well as distributed locally to each load site by means of conducting wires and cables. Electricity use is dominated by alternating current, as far as utility networks are concerned, and most transmission over distances up to a few hundred kilometers or miles is by AC. For transmission over longer distances (e.g., by ocean cables), conversion to direct current (DC) before transmission and back to AC after transmission is common. Cables are either buried in the ground (with appropriate electric insulation) or are overhead lines suspended in the air between poles (masts), without electrical insulation around the wires. Insulating connections are provided at the tower fastening points, but otherwise the low electric conductivity of air is counted on, which implies that the losses are comprising conduction losses depending on the *instantaneous weather condition*, in addition to the ohmic losses connected with the conductor resistance R of the current, I as given in Equation (4.37). The leakage current between the elevated conductor and the ground depends on the potential difference as well as on the integrated resistivity, such that the larger the voltage, the further the wires must be placed from the ground. Averaged over different meteorological conditions, the losses in a standard AC overhead transmission line (138–400 kV at elevations up to 40 m) are 1% or lower per 100 km of transmission. However, the overall transmission losses of utility networks, including the highly branched power distribution networks in the load areas of the existing grids can be up to 10% of the power production.

$$\text{Ohmic Losses} \propto R \cdot I^2 \tag{4.37}$$

A conductor is one or more wires suitable for carrying electric current. Often the term wire is used to mean conductor. Conductors used in electrical power systems are made of copper or aluminum, solid or stranded, depending on the size and the required flexibility. They are packaged in several ways to from electric power cables. The most common cable construction type for low-voltage (600 V or less) power is the single-conductor cable covered with single layer insulation, or by an outer nylon jacket, and typically installed in conduit or other raceway systems. Insulation materials are usually extruded, providing insulation between the conductors and for the cable. Common voltage classes for cables are 600 V, 2 kV, 5 kV, 15 kV, 25 kV and 35 kV. A multi-conductor cable is an assembly of two or more conductors (often four conductors), in outer jacket and insulation for easy installation or replacement. Utilities are using aluminum for almost all overhead installations. However, the aluminum alloys are used only in larger conductors, provided that it is an approved alloy. Copper has very low resistivity and is widely used as a power conductor, although use as an overhead conductor are rare because copper is heavier and more expensive than aluminum. Copper has very good resistance to corrosion. Different conductor sizes are specified with gage numbers or area in circular mils. Smaller wires are usually referred to the American wire gage (AWG) system. The cable resistance and reactance are often provided by the manufacturers or is found in the reference book tables. In modeling, each phase conductor of a cable is usually represented by its resistance, R_{cd}, in series with the leakage reactance, X_{cd}, and the total cable impedance is:

$$Z_{cd} = R_{cd} + jX_{cd} \quad (\Omega) \tag{4.38}$$

The ampacity is the conductor maximum designed current, expressed in amperes. A conductor has several ampacities, depending on its applications and the assumptions used. It also is affected by the weather conditions for overhead cables. When higher current flows results in higher conductor and insulation temperatures as the conductor power losses increase. One limiting factor of the conductor ampacity is the current need to bring the conductor to a certain temperature. The power transmission capacity of an insulated cable system is the product of the operating voltage and the maximum current transmitted. Power transmission systems operate at fixed voltage levels so that the delivery capability of a cable system at a given voltage is set by its current carrying capacity. The delivery capability is defined as the cable *ampacity*. The operating voltage determines the cable dielectric insulation requirements, while the conductor size is set by the ampacity rating. These two cable system independent parameters (insulation and conductor size) are related by thermal considerations, a larger conductor size (less I^2R losses) results in higher ampacity, while insulation material increases, lower heat dissipation results in lower ampacity. The parameters influencing the ampacity are cable size, insulation characteristics, ambient temperature and the horizontal spacing between the circuits, while the soil thermal resistivity, burial depth are for underground cables. The basic characteristics of the cable physical construction include: conductor materials, shapes, types, surface coating, insulation types, number of conductors, installation conditions and arrangements (underground or above the ground, cable bunching and spacing). Current flowing through a cable generates heat through the conductor resistive losses, dielectric losses through the insulation and resistive losses from current flowing through any cable screens, shields or armoring. The components that make up the cable (e.g., conductors, insulation, bedding, sheath, etc.) must be capable of withstanding the temperature rise and cable heat. The current carrying capacity of a cable is the maximum current that can flow continuously through a cable without damaging the cable's insulation and other components (e.g., bedding, sheath, etc.). It is sometimes also referred to as the continuous current rating or ampacity of a cable. Cables with larger conductor cross-sectional areas have lower resistive losses and are able to dissipate the heat better

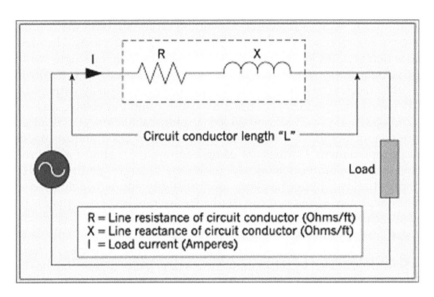

FIGURE 4.9
Simple circuit for explaining load-generator settings or voltage drop calculations.

than smaller cables. If the conductor current is known an important cable parameter, using a similar circuit, as shown in Figure 4.9, the voltage drop can be computed:

$$V_{drop} = I_{cd} \times Z_{eff} = I_{cd} \times \left(R_{cd} \cdot \cos\left(\theta_{PF}\right) + X_{cd} \cdot \sin\left(\theta_{PF}\right) \right) \qquad (4.39)$$

Here, θ_{PF} is the power factor angle, positive when the current is lagging the voltage and negative for current leading the voltage. The cable size is usually selected in the 20%–30% margin, limiting also the voltage drop usually in the range of 3%–5%. All calculations are usually in volts, amperes and ohms per phase, or in percent or per-unit values. Often, cable manufacturers are listing the effective cable impedance, Z_{eff} as the cable impedance at typical 0.85 lagging power factor, calculated as:

$$Z_{cable} = 0.85 \cdot R_{cable} + 0.527 \cdot X_{cable}, \quad \Omega/\text{phase} \qquad (4.40)$$

> **Example 4.7:** A three-phase feeder, Y-connected, rated 4.16 kV line-to-line voltage has a resistance of 54 mΩ and a reactance of 36 mΩ, both per 1000 ft and is supplying a 1.2 MW load. Determine the voltage drop at 0.85 power factor.
>
> **Solution:** The feeder phase voltage and current are:
>
> $$V_{\phi} = \frac{4160}{\sqrt{3}} = 2400 < 0° \text{ V}$$
>
> $$I = \frac{1.2 \times 10^{6}}{\sqrt{3} \times 4160} = 166.74 \text{ A/phase}$$
>
> For 1000 ft run, the voltage drop, Equation (4.39) is:
>
> $$V_{drop} = I_{cd} \times Z_{eff} = 166.7 \times \left(0.054 \cdot 0.85 + 0.036 \cdot \sqrt{1 - 0.85^{2}} \right) = 10.815 \text{ V}$$

4.1.4 Power Transmission and Distribution

Power distribution networks, since earlier grid beginnings, have a minimal role in the power grid operation management and control. Utilities often employ demand management schemes that are switching off non-critical loads to reduce load during emergency situations or peak demand periods. However, the offered controllability by such schemes is rather limited. The lower power distribution involvement is largely a consequence of the limited consumer communication capabilities. Smart grids promise cost-effective technology that overcomes such limitations, allowing consumers to respond to power system conditions and hence actively participate in power system operations. The power distribution networks transfer needed electricity from power plants to homes, industrial and commercial facilities. In order to transfer electrical power from an AC or a DC power supply to the loads, specific power distribution networks are used. Power distribution subsystems are divided into three components: *distribution substation*, *primary* and *secondary distribution networks*. At a substation, the voltages are lowered as needed and the power is distributed to the customers, usually one substation supply thousands customers. The transmission lines in any power distribution system are several times those of the transmission subsystems. Furthermore, most customers are connected to only one of the three phases in the power distribution. Therefore, the power flow on each of the lines is different and the system is unbalanced. Primary distribution lines are medium-voltage networks, with voltage ranging from 600 V to 35 kV. At a distribution substation, the transformers step-down the transmission voltages (35–230 kV), distributing power to several primary circuits, fanning-out from the substation. From power distribution transformers, the secondary distribution circuits connect the end-users at their service entrance. Power distribution employs equipment such as transformers, circuit breakers, monitoring equipment, metering and protective devices in order to deliver a safe and reliable power. The power distribution in the United States is usually in the form of three-phase 60-Hz AC current. The electric energy is changed in several ways, during the transfer through electrical special circuitry and devices. The distribution of electricity involves a system of interconnected transmission lines, originating at the electrical power-generating stations located throughout the country, with the ultimate purpose to supply the electricity needed for industrial, residential, and commercial uses. A typical electrical power distribution system is shown in Figure 4.10.

Electricity distribution involves a very complex system of interconnected transmission lines, power transformers, monitoring, control and protective equipment and devices. Electrical power systems are interconnected with one another in parallel circuit arrangements for supply security and power quality reasons. These power system interconnections are monitored, controlled and operated by computerized control centers. The control centers provide means for the data collection, recording, analysis, system monitoring, frequency and voltage control, and signaling. The electrical energy transmission requires long transmission lines from generation ends to the end-users, requiring optimum planning to assure best use of land and minimal environmental impacts, while minimizing the overall cost and investment. The location of transmission lines is limited by zoning laws and by the populated areas, highways, railroads, waterways, topographical and environmental factors. Overhead power transmission lines operate at voltage levels from 12 up to 750 kV or even higher, with the most common range from 50 to 350 kV. Underground power distribution is used primarily in urban and suburban areas, where the right-to-way for overhead power lines is limited. Underground distribution lines range from simple

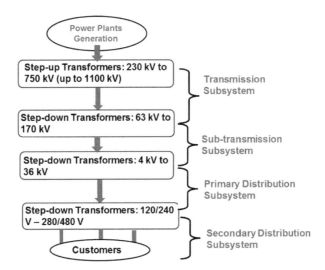

FIGURE 4.10
Electricity infrastructure from generation to power distribution.

coaxial cables to more sophisticated constructions insulated by a compressed gas. Major advantages of overhead transmission lines are their ability to dissipated heat, reduced costs, easier installation and maintenance. Underground cables are confined to short distances, being much more expensive than the overhead ones. To improve underground cables power-handling capacity, the research is focusing on forced-cooling techniques, such as circulating-oil and with compressed-gas insulation or on the cryogenic cables or superconductors operating at extremely low temperatures and having large power-handling capabilities. Notice that facilities using large electricity quantities are direct connected to transmission power systems.

Primary distribution lines are medium-voltage (MV) circuits, in the range of 600 V to 35 kV. At a distribution substation, transformers take the incoming transmission-level voltage (often in the 35–230 kV range), stepping it down to the voltage levels of primary circuits, which are fanning out from the substation. Close to end-users, power distribution transformers step-down the primary-distribution voltages to lower secondary circuit voltages, commonly 120/240 V, other needed voltage levels. However, the choice of voltage to be used on any particular distribution section is influenced by: decisions associated with voltage drops resulting from large current loads, capital cost of transformers used to change voltage levels and construction of distribution lines and associated switchgear to operate at the chosen voltage, and environmental aspects of the system installation. Urban distribution networks are often underground, while the rural constructions are mainly overhead type. Suburban structures are a mix, with a good deal of new construction going underground. The power distribution systems are capital-intensive businesses, with about 10% of the all utility capital investments. Cost lowering, simplification and standardization are all important power distribution design characteristics. However, few components or installations are individually engineered on power distribution circuits. Standardized equipment and designs are used wherever possible and recommended by engineering practices, codes and standards. Power distribution planning is the study of future power delivery needs. Its goals are to provide service at lowest cost and highest reliability and power quality possible.

Planning requires a mix of geographic, engineering and economic analysis skills. New circuits must be integrated into the existing distribution system within a variety of economic, political, environmental, electrical and geographic constraints.

Power distribution infrastructure has to be delivered to customers concentrated in cities, suburbs, rural areas and even in remote places, while few places in the industrialized world do not have electricity. Industrial facilities are using about 50% of all the electrical energy. The three-phase power is distributed directly to most large industrial and commercial facilities. Electrical substations use three-phase power transformers and associated equipment (circuit breakers and insulators) to distribute the electricity. Power distribution networks are divided into two base structures, unmeshed or meshed grids, further separated into: *radial networks*, in which the users or suppliers are connected by means of spur lines, radially fanning-out from a substation or to a node point, *ring networks*, the most common used, consisting of the two-sided feed lines, as the lines at the ends are returned to the feed-in point, and *meshed grids*, where the lines are running and connecting between nodes. Meshed grids offer the highest supply reliability, but are the most expensive and prone to increased short-circuit flows. Facilities must provide maximum supply safety, consuming fewer resources during construction or operation, having the flexibility to adapt to any future power requirements. Supply reliability is a measure of how well the limited power supply is fulfilling the given supply tasks, i.e., the covering of the load or the transmission and ensuring the power transfer even under disadvantageous operating conditions. The customer supply reliability is defined by the interruption frequency and duration. The supply reliability depends on the power consumer connection point, voltage level, position and the required power availability. The smart integration of all service installations offers an optimum to be attained for safety, flexibility, energy efficiency and environmental compatibility, offering in the same time maximum comfort. Power distribution networks are found along most of the secondary roads and streets. Urban construction is mainly underground power distribution, while the rural power distribution is mainly as overhead transmission type. Suburban structures are a mixture, with a large part of the new construction going underground.

Power distribution feeder circuits consist of overhead and underground transmission line networks in a mix of branching circuits, the so-called laterals from the station to customers. Each circuit is designed around several requirements such as the required peak load, voltage levels, distance to customers and local conditions (terrain configurations, street layout, visual and environmental regulations, or user requirements). The North America secondary voltage consists of a split single-phase service, providing customers with 240 and 120 V, connected then to devices depending on their ratings, being served from a three-phase distribution feeder, usually Y-connected, consisting of a neutral center conductor and a conductor for each phase. In most other parts of the world, the single-phase voltage of 220 or 230 V is provided directly from a larger power distribution transformer, providing also a secondary voltage circuit often serving hundreds of customers. The various branching laterals are operated in radial or in looped (ring) configurations, where two or more feeder are connected together usually through an open distribution switch. Power distribution networks are overhead or underground, highly redundant and reliable complex networks. Overhead lines are mounted on concrete, wooden or steel poles arranged to carry power distribution transformers or other needed equipment, besides the conductors. Underground distribution networks use conduits, cables, manholes and needed equipment installed under the street surface. The choice between the two systems depends of factors, such as safety, initial, operation and maintenance costs, flexibility, accessibility, appearance, life, fault probability, location and repairs and interference with communication systems. Underground distribution systems compared to overhead systems are more expensive, requiring higher investment, maintenance and operation costs, lower fault probabilities, more difficult to locate and repair a fault, lower cable capacities and voltage drops. Each system has advantages and disadvantages, while the most important is the economic factors.

Renewable sources are typically distributed over large areas in the upper central and southwestern United States, including the Dakotas, Iowa, Kansas, Texas, Minnesota and Montana and far from large demand centers east of the Mississippi or on the West Coast regions. New large area collection strategies and new longer distance transmission capability are required to deliver large amounts of power thousand miles across the country. There similar issues are found in EU, China, Brazil, India, Canada or Australia and on lesser extent in many other countries. This long distance transmission challenge is exacerbated in the United States by a historically low investment in transmission generally: from 1988 to 1998 demand grew by 30%, while transmission grew by only 15%; from 1999 to 2010 demand grew by 20% and transmission by only 3%.While high voltage DC is the preferred transmission mode for long distances, the drawbacks of single terminal origin and termination, costly AC-DC-AC conversion, and the decade or more typically needed for approval for long lines create problems for renewable electricity transmission. Superconductivity provides new alternatives to conventional high voltage DC transmission. Superconducting DC lines operate at zero resistance, eliminating electrical losses for any transmission length, and operate at lower voltages, simplifying AC-DC conversion and enabling wide-area collection strategies.

4.1.5 Requirements and Services for the Electrical Systems

The electric grid supply quality and reliability are the sum of all quality-determining conditions reliability from the user point of view and is defined by the following three parameters: supply reliability, voltage quality and service quality. The supply reliability as a part of the supply quality is the electrical system ability to fulfill its energy supply tasks under prescribed conditions during a specific period. The supply reliability at the customer level is determined by frequency and duration of electrical supply interruptions. Supply reliability differs depending on the grid structure and the power customer position into the grid. To evaluate the entire grid, the average values of corresponding power customers are taken into account. Standards are defining, describing and specifying the supply voltage characteristics at the transfer point of the grid to the user in low-, medium- or high-voltage supply networks under normal operating conditions. The supply voltage is described as a product with fixed characteristics. Every connection, whether a user or a generator (feeder), affects the voltage quality, as the grid is a rigid structure. A rigid grid is supposing an infinitely high short-circuit power. The quality characteristics are also different depending on the grid level. Mutual interaction must be taken into account here. For this reason, the connection guidelines are divided according to the voltage levels, the specific requirements being too variable to be summarized in a single guideline. The standard requirements consider the following supply voltage characteristics, such as frequency, voltage level, voltage waveform and the symmetry of the conductor/phase voltages. Influences on these supply voltage characteristics have the following effects: harmonics, flickers, voltage changes, grid-signal transmission voltages, frequency stability, and uneven load distribution on the individual conductors. The supply reliability as a part of the supply quality is the ability of an electrical system to fulfill its supply tasks under prescribed conditions during a particular time period.

Electrical power must be generated at the same instant that it is needed by the user. There are specific requirements and constrains for power supply, such as the supply quality and power quality. The power system control ensures the supply to the users is exactly the required quantity of electricity even in the case of unforeseen events in the grid. For example, short-term power adaptations that are designated for power regulation are implemented, through fast-start-up power stations (e.g., gas turbine power stations) or pumped hydropower storage stations can be used. Alternatively, certain customers with load control can be separated from the grid on a short-term basis. Power range control is a part of the equalization services that are required within the scope of the power systems preparation for covering losses and for equalizing of imbalances between power supply and uses. In addition, the transmission grid operator, for particular operating conditions and for maintaining the system safety, can automatically or per switch command remove loads from the grid or allocate rated values to power stations. A balancing circuit usually consists of the current supplier and his customers. The suppliers are responsible for estimating the amount of power that balancing grid requires, through informing the power stations of the available supply estimates. This procedure is called the scheduled supply. The supplier can source his power from many different energy sources, such as hydro-power stations, coal-fired power stations, nuclear power plants, wind turbines, biomasses or photovoltaic, and makes the balancing circuit available. The power requirement of the user and thus also the transport of the current over the grid is dependent upon the time of the year and the time of the day. The use of current reaches peak values in the morning, midday and evening. A base-load power station is a power station that operates for technical and commercial reasons possibly uninterruptedly and as near as possible to its full capacity.

Another important grid requirement is the frequency stability. If the frequency in the grid sinks or rises, then the functions and partly the life of numerous electrical devices, such as clocks (in which the grid frequency is used as a reference of the time cycle), computers, motors or compensation installations are influenced. The generation and utilization of power must be equalized constantly so that the frequency does not deviate greatly from the rated value of 50 Hz or 60 Hz. The grid frequency is determined by the active power flow, while the voltage level is controlled by the reactive power flows. Depending on the primary active and reactive impedances, the user or generation plant reactive power behavior decides upon the amount of the change of voltage in the grid or the grid connection nodes. If the behavior of a classic user in the past led to component failure, then if the position includes wind turbines the direction of the power flow can change. With the reactive power region in which today the normal wind turbines are operated, the voltage at the grid connection nodes is usually increased. In the reactive power take-up of the wind turbine (inductive behavior) the rise in voltage is much lower than the reactive power feed (capacitive behavior). In the higher voltage level networks, the voltage stability is kept within permissible limits by the so-called reactive power bands of power stations. Due to their massive expansion of wind and solar energy must also take part in the reactive power exchange and thus in the system voltage stability. System services in electrical supply are designated as the aid services necessary for the functionality of the system, and are provided by the generator of electricity for his customers in addition to the transmission and distribution of electrical energy, and which thus determine the quality of the power supply. Included in the system services are: active power control, relevant for the stability of the frequency, voltage stability, compensation of the active losses, reactive power control, black start/island operation capability, system coordination and operational measurements.

Ancillary services enable us to ensure the network management, i.e., the adjustment of essential power system electrical quantities, i.e., the frequency and voltage. The good electrical system operation is highly dependent on keeping the frequency and voltage in prescribed ranges. Frequency control is associated with the active power control, whereas voltage control is in principle associated with the reactive power. The first contributor to these services is the synchronous generators (alternators) of the conventional centralized power plants (nuclear, thermal or hydraulic type). The alternator adjustment principle, whose control of the driving turbine torque and the speed enables to control the frequency, while the excitation current control (inducting the magnetic flux inside the alternator) enables the voltage level adjustment. Other power system components can participate in ancillary services, in particular for on-load voltage control with the help of adjustable transformers. The frequency variations are caused by an active power production and consumption imbalance. The basic power system operation principle is: the electrical power must be generated at the same instant that it is needed by the user. If the production is equal to the consumption, the frequency remains stable (50 or 60 Hz), while if the production is higher than consumption, the frequency increases, due to the fact that during the transient control regime, the surplus of energy is stored in the form of kinetic energy on the revolving field, whose speed is increasing so the frequency and if the generation is lower than consumption, the frequency decreases because the energy deficit is taken from the kinetic energy of the revolving group, decreasing its speed. Active and reactive power transits can cause voltage drops. Voltage control is usually carried out locally to avoid the reactive power transit, consisting of multi-level controls. The electric system behavior is largely influenced by the fact that most of the electric energy is produced by synchronous machines, directly coupled to the network and driven by thermal or hydraulic turbines. The electric network conventional structure is based on a centralized network management, the level of transport network to which the conventional power generation units are connected. In this structure, the power distribution only hosts consumers, thus only the power flows from high voltage levels, from the interconnection points with the transmission network toward the lowest voltage points, where the controls are very limited, enabling to adjust the voltage levels, and all the protections are based on the one-directional power flow. Ancillary services are then mainly ensured on the transmission levels by the connected generation groups. The development of decentralized production has considerably affecting the power systems. Indeed, because of its generally reduced power, DG and RES are often connected to the power distribution networks. Electricity market liberalization and the impact of the integration of such generation types, especially on the level of the power distribution network and on the transport network level, from the connection constraints of the production units to these networks are discussed in the next chapter sections and subsections.

Standards define, describe and specify the important characteristics of the supply voltage at the transfer point of the grid user in low-, medium- and high-voltage supply grids under normal operating conditions.

With the standards the supply voltage is described as a product with fixed characteristics. Every grid connection consumer, whether user or generator (feeder), influences the voltage quality, as the grid is not rigid. The quality characteristics are different depending on the grid level. The standard requirements are respectively considered, in that the following supply voltage characteristics are defined: frequency, voltage level, voltage curve form, and symmetry of the phase voltages. Influences on these characteristics of the supply voltage have the following effects: harmonics, grid flicker, voltage changes, grid-signal transmission voltages, frequency stability and uneven load distribution on the individual conductors. Harmonics are constant, periodic deviations of the grid voltage from the sine wave form (voltage distortion) are caused by additionally superimposed fluctuations whose frequency is an integer multiple of the grid frequency. The permissible harmonic voltage in the respective grid levels are used up to a great extent already by the connected user devices. The additional harmonic voltage values generated by wind turbines, PVs or other RES units must therefore be limited to permissible values. Voltage changes can occur due to, for instance, the switching-on of larger loads (e.g., motors, capacitors), controlled loads (fluctuation packet controls) and due to variable feeders (e.g., WTs or large PV arrays). In the case of wind turbines the fluctuations of the wind (e.g., gusts) can lead to a change in the feed-in power. The power change, for its part, affects the impedances and can lead to impermissible flicker values.

Grid (power systems) supply quality has been an important issue from the earlier power system begging and becoming even more important as customers of all scales from residential to large industrial facilities are using electronic equipment and systems that are often very sensate to supply and power quality disturbances and events that last even a fraction of second. The frequency and voltage of the electrical network are the most relevant characteristics from costumer point of view. Grid operator must ensure the control of power flows at the required standard into the acceptable limits for voltage and frequency. For example, reactive power flows can give rise to substantial voltage changes across the power system, meaning that it is need to control and maintain reactive power flow balance between the generator and demand points. System frequency is consistent throughout the interconnected power networks, while the voltages are experiencing the so-called voltage profiles at various pints into the system that are related to the local power generation and energy demand at any instant, being also affected by the power network configurations. Notice that the supply is also affected not only by the demand itself, but also by the *power supply demand nature*. For example, a large industrial facility with hundreds or thousands of electrical motors can have large fluctuations in the demand for active and reactive power. Similar effects can have the local large integration of RES and DG units. The generation system as a whole must be sized to accommodate the highest power demand level, as well as a reserve in the event of the unavailability of any generators or for any unforeseen extra demand peak. Notice that when a load or generation unit is added or removed from the power system it affects the supply producing power system disturbances. The size and the extension of such disturbances are determined by the load or generation unit size. Other events, such as short-circuits, can produce *transients* on transmission or distribution networks. Such disturbances have effects on the grid-connected equipment, loads, power electronics, communication and computer systems. Large disturbances in the grid can affect any generating plant, which beyond the set limits can be disconnected from the grid to limit the damages. Most of the loads and generators have the some ability to *ride through faults*, being less likely to disconnect and cause the cascade effects. For example, in the past the wind turbines were designed not to ride through the faults and were immediately disconnected to avoid turbine damages by the connection. However, for the modern wind farms, the modern electronic management systems can emulate the rite-through ability of a large electric generator.

4.1.6 Distributed Generation and Renewable Energy Sources in Modern Power Systems

As electricity demands keep increasing on daily basis causing grid system unbalance, resulting in various causes like load shedding, voltage unbalance, which ultimately are affecting the consumers. To avoid all such issues the only option is to meet the demands by increasing generation, however there are significant lags with the conventional energy sources so generating more power by conventional ways may not be an option. The current interest resurgence in the renewable energy use is driven by the needs to reduce the environmental impacts of fossil-based energy systems, to ensure supply security and improve overall grid efficiency. Harvesting renewable energy on larger scales is undoubtedly one of the main challenges of our time. Future energy sustainability depends heavily on how the renewable energy issues are addressed. Thus, RES and DG use is important for preserving and improving our living standards. Renewable energy sources can significantly enhance and diversify the energy resources and options. However, the electricity generation using renewable energy resources like hydropower, solar photovoltaic, biogas, biomass and wind turbine are often taking place in small-scale settings due to disperse resource nature and the capacity can vary from a few hundreds of kilowatts to several megawatts. Distributed generation or distributed resources (DR) are where such small-scale electricity generators are usually connected to the grid either at the primary or secondary distribution level. DERs include both renewable and non-renewable small-scale generation as well as the energy storage. These technologies enable higher integration levels of renewable and conventional energy sources. However, the renewable energy sources are not *dispatchable*, meaning that the power output cannot be controlled. Daily and seasonal effects and limited predictability result in an intermittent generation. Future energy sustainability depends heavily on how the renewable energy issues are addressed. Smart grids promise to facilitate the renewable energy integration, while providing other benefits as well. Industry must overcome a number of technical issues to deliver renewable energy in significant quantities. Control is one of the key enabling technologies for the deployment of renewable energy systems. Solar, wind and marine power systems require effective use of advanced control techniques. Other renewable energy sources, such as geothermal energy, biomass and hydropower are more predictable, and not suffering on the same issues. In addition, smart grids cannot be achieved without extensive use of control technologies at all levels.

Renewable energy resources are often used for power generation in stand-alone or isolated system configurations. However, their major benefits are significantly enhanced when they are integrated into the electric power grids. Each renewable energy resource is different from grid perspective and some are easier to integrate than others. With the smart grid advent, higher degrees and penetration rates can be achieved. Leading RES characteristics that impact their grid integration are their size (generation capacity as compared to the conventional power generation units), location (geographically and with respect to power network topology), and their variability (minute, daily, seasonally and intermittently). RES gird integration is reducing the dependence on the foreign fossil fuels by enabling the seamless integration of cleaner, greener energy technologies into power networks. Usually RES units are connected at the power distribution levels, while larger RES units, such as wind farms or solar power plants are connected at the transmission levels. The important RES grid integration benefits included among others. Energy sustainability, in that the renewable energy sources, can make a significant contribution to environment protection, diversify resources, easing the dependence on fossil resources, domestic energy carriers and contributing to regional value creation and help to secure employment. Empowering grid into peak hours, through the RES and energy storage integration, the smart grid, having real time information can substitute with RES generation whenever is possible. Increasing renewable energy into generation mix improves the operational efficiency and reduces peak demands. Smart metering and monitoring are helping to adopt demand side management and demand response usage leading to optimum utilization, resulting into saving of energy. Renewable energy systems included in microgrid configurations are independent systems able to operate isolated system during grid failure, reducing impact on customer. Industrial and commercial consumers adopt grid-connected renewable energy systems which helps to reduce power demand. Sometimes isolated system in residential areas conserves the energy. Upgrading electrical market, through power exchange, provides opportunities for electricity trading.

4.2 Grid Integration of Renewable Energy Systems

The modern electric power system is rapidly evolving from a centralized system largely based on fossil fuels, large hydropower and nuclear energy based generation, toward a more decentralized electric power system with high penetration of CHP power generation and RES-based electricity. According to Ackermann et al. (2001), decentralized or DG is defined as the electric power generation sources connected directly to the power distribution networks or on the customer side of the meter, regardless the size, fuel or technology used. Distributed generation or embedded generation term refers to smaller scale generation units, usually connected to power distribution networks that include: renewable energy generators, DG, CHP systems, standby generation units that are used when centralized generation is inadequate or too expensive. Distributed generation is not a new concept. The first power plants generated electricity locally to meet the demand of the neighboring customers. The economies of scale and technological advances paved the way for today larger and centralized power generation. However, during the last decades, the interest for DG and RES has been significantly increased in United States, Europe, Japan and Canada and in most of developed and developing countries. Aside from all expected benefits resulting from the connection of decentralized resources, several concerns arise due to the variable and unpredictable nature of some RES such as wind and solar generation. The RES intermittency has to be addressed in order to keep the optimal power system operation. Traditionally, conventional power plants are used as operational reserves to satisfy the electric power balance on the short time scale. In general terms, the reserve power task is to deal with unforeseen occurrences on the supply side and deviations from the expected demand. However, the large penetration of intermittent (i.e., largely variable and to an extent uncertain) RES requires an adaptation of the electric power system to this new reality.

Electricity generation using renewable energy resources is often taking place in smaller scale settings due to disperse nature of the resources. Examples include small hydropower, solar photovoltaics, residential solar-thermal energy, biogas, biomass and small wind turbine based electricity generation. These renewable energy based power generators up to several megawatts are often connected to the grid at the primary or secondary distribution level and are considered distributed generation or distributed energy resources (DER). The generation technology or the operational characteristics requires the use of some interface between the generator and the utility power distribution networks. For example, solar PV panels are generating DC electricity and therefore, a power electronics DC-AC converter is required between the grid and the PV generator. Some other RES-based technologies, such as small hydropower or small wind turbines are using induction generators that can be directly connected to the AC power grid. However, concerns such as starting transients, energy conversion efficiency and power quality issues make connecting them through a power electronics interface a better choice. Although renewable energy systems are clean and renewable sources of electric power, many challenges must be addressed. Two critical technical challenges are identified with a higher RES penetration: (1) managing variability and uncertainty during the continuous balancing of the system, and (2) balancing supply and demand during generation scarcity and surplus situations. Variable renewable energy sources are both much more uncertain and variable than the conventional power generators. Generally, PV power output is more variable than the wind turbine output (changing faster on a minute-to-minute basis), but less uncertain. With both wind and solar power, the system ability and generation operators to predict generation levels are improving. Distinct but related from the previous challenge is the issues of balancing the supply and demand, when higher renewable energy production is present, such as the case of low demand and high RES generation or low renewable energy production and high demand. Supply of the wind and solar are usually not coincident with the demand, introducing challenges at the bulk power system level and, if there is significant distributed PV generation, at the local power distribution system level. An illustration of this power distribution system challenge is the *reverse power flow* that can occur during the midday in the areas with large amounts of distributed PV, the residential electricity demand is low, larger generated electricity feeds back up through the transformer to the medium-voltage network. Power distribution systems have not designed to anticipate such situations because the generation has come from large-scale systems located on the transmission section, with power flowing predictably one way down to lower voltage systems. Challenges related to high peak load during periods of low variable renewable energy production are less technically demanding and are primarily economic challenges related to market solutions chosen to remunerate reserve capacity or demand response activities. A circuit similar to one of Figure 4.5b or Figure 4.9 is used the initial estimates of the voltage rise, for an RES-PCC Thévenin equivalent impedance, $Z_{th} = R + j \cdot X$, the voltage rise is expressed by:

$$\Delta V \approx \frac{P \cdot R + Q \cdot X}{V} \tag{4.41}$$

Here, P and Q are the generator active and reactive power, being positive if they are flowing for the generator to the network. For example, an induction generator is absorbing reactive power form the network, Q being negative in this case. The voltage rise level is determined by the electric network operation requirements, and usually a 1% voltage rise may be a concern for the grid operators. Voltage rise can be mitigated by extracting reactive power at PCC. Voltage rise can be often a major concern for wind farms, usually located in rural areas, connected by long and high impedance lines, putting strict limits on the generated power that can be connected to a specific rural area.

Example 4.8: A 3.5 MVA, 0.85 power factor WT permanent magnet synchronous generator is connected at PCC to 12.8 kV bus, by using a 100 m cable having impedance $Z_{cable} = 0.08 + j0.06 \ \Omega$. Determine the voltage rise level by connecting this RES generator.

Solution: The active and reactive powers injected by this generator are:

$$P = S \cdot PF = 3.5 \times 0.9 = 2.975 \text{ MW}$$

$$Q = S \cdot \sqrt{1 - PF^2} = 3.5 \times 0.527 = 1.844 \text{ MVAR}$$

From Equation (4.41) the percent voltage rise level at PCC is:
The percent voltage rise is 0.74%, which is less than 1%, being acceptable.

4.2.1 Generation Unit Power System Connection

The basic requirement for any grid-connected generator is that it not adversely affects the power supply quality or power system operation. All generators are connected to the power system through the so-called point of common coupling (PCC), defined as the closest power network point to the generator, where other customers are not connected. The PCC is the point where the connected generator can cause most disturbances. The PCC fault level is a power network strength measure, and an important design parameter for predicting the fault currents and performances under normal operating conditions, such as voltage rise or the overall generator effects on the network. When a part of network is disconnected from the grid and if the electric network is operated as independent power system supplied by few electric generators is called *islanded electric network* of a power network in *islanding operation mode*. Islanding may lead to abnormal voltage and frequency variations. If the islanding is to be permitted, the power system must be designed to operate in both grid-connected and stand-alone modes. The stand-alone operation is more difficult to operate since must include full generation-load matching in order to ensure that the voltage and frequency are kept into acceptable limits. However, the most of electric generators are designed to operate in both grid-connected and stand-alone mode. Standby generators usually operate by diesel or gas engines installed to provide emergency power to critical loads. If a generator is not designed to operate in stand-alone mode must be able to shut-down quickly the grid connection is lost for safety, power quality and protection reasons. RES-based generation units are unlikely to operate in both stand-alone and grid-connected modes. During a fault, if an auto recloser is opened then two independent systems are formed which are operated at two different frequencies. Disastrous results may occur when the auto recloser is made to reclose during out of phase of two systems. In addition to that the islanding operation may create an ungrounded system which depends on the transformer connection. Based on the synchronous generator conventional power system models, for the large-scale renewable energy generations such as hydropower units, steam turbines operated with biomass or geothermal energy, the conventional technologies and integration methods are used. However, for large-scale wind farms or large solar power plants the grid integration are posing completely different set of challenges and issues, due to their variability of the generator output power and irregular nature of wind and solar energy.

4.2.2 Issues Related to Grid Integration of Small-Scale Renewable Energy Generation

The operation of the generating units connected in low-voltage networks are based on the fundamental principle of minimizing the reactions of generating units on the power distribution grid and thus on maintaining the supply quality. These requirements are also requiring that a rapid uncoupling of the generating units from the grid occurs in the case of disturbances in the power distribution. The installed WTs, PVs or small-hydro units must also participate actively in the voltage and frequency stability. For instance, the consequences of a grid malfunction due to a failed power feeder must be limited to prevent uncontrolled fault spread, while a rapid uncoupling of the generating plant in the case of malfunctions in the higher voltage grids must not occur unselectively. Conventional electric distribution systems have a single voltage source on each distribution feeder, which is not the case when small-scale RES generators are interconnected to the power distribution. In order to ensure grid safe and reliable operation special requirements need to be satisfied during interconnection. Major protection-related problems are fault clearance, reclosing and inadvertent islanded operation. Usually the grid and DG or RES interfaces are based on power electronics inverters, power processing units or asynchronous generators. The utilities are concerned about their impact on power quality, which includes harmonics, voltage dips, over voltages and voltage flicker. Several identified technical requirements imposed by utilities to address these concerns are a major technical barrier for grid integration of DG and RES. The general technical requirements for interconnection of the small-scale generation units are based on the IEEE P1547 standards as guideline. The most serious issue related to the interconnection of distributed generation is the system protection. Usually radial power distribution network is used with time graded over current protection scheme. When a DG or RES is interconnected it may alter the coordination of the existing protection scheme and can lead to the protection equipment malfunctions.

The common power system protection equipment and devices include reclosers, circuit breakers, fuses and instrument transformers. A *circuit protection* refers to a scheme or device used to disconnect a section of a system component in the event of a *fault*. A fault means that an inadvertent electrical connection is made between the energized components and something or component at a different potential. The basic fault types are *phase-to-ground* and *phase-to-phase* faults. The object of circuit protection is to reliable detecting the fault when it happens and interrupt the power flow to clearing the fault. *Fuses* are the simplest protection device with limited use, *circuit breakers* have moving parts actuate by relays to operate. Switches and circuit breakers in power transmission and distribution are referred as *switchgears*. *Reclosers* are circuit breaker that are designed to clear immediately very-short term faults, such as one caused by lightning or wind induced momentarily contact between phases. The main parameter used in the selection of fuses, reclosers, circuit breakers and currents transformers, and the coordination between over current relays are based on *short-circuit current* level. The equivalent system fault point impedance and the expected fault current level are decided by the short circuit level. Power distribution systems are operating as passive networks but the interconnection of DG and RES generators makes the equivalent network impedance to decrease, resulting in increase in the fault current level. Therefore if fault occurs the existing circuit breakers may not withstand to the sudden high fault currents, leading also to current transformer saturation, and the changed fault current levels may upset the synchronization among the overcurrent relays which can lead to unacceptable protection system operation and behavior. During fault currents the recloser and fuses operate faster, the coordination can be lost, due to reduction in the required margin between the recloser fast curve and the fuse melting curve. Another major issue of the large grid DG and RES penetration is *the reverse power flow*. The power flows are unidirectional in radial power distribution systems, which are setting the design of the protection schemes. When a DG units and RES generators are connected, the coordination among the protection relays and device may be altered due to the power flow reversal. In the fault event the protection relays are detecting and discriminating the fault currents only if the measured fault current by the relays is significantly higher compared to the normal rated load current. When the DG or RES unit fault current contribution is limited, the effective detection of faults by the overcurrent protection relays is difficult and limited. Usually, the induction generators, small synchronous generators or power electronic converters are employed in RES units or systems. The supply of the fault currents to three-phase faults is not possible by the induction generators and fault current contribution to the asymmetrical faults is also limited, while the small synchronous generators cannot sustain the fault current supply, significantly at higher levels than the rated currents. Power electronic converters are designed to limit the output current internally since the power semiconductor devices cannot resist longer to significant overcurrent. The ability of the protection relays to detect faults is affected and even compromised by this lack of the sustained fault currents.

4.3 Power Electronics Circuits

Power electronics is the application of solid-state electronics, signals and systems for the electric power control and conversion, through the specialized electronic circuits, the power electronics converters. The main task of power electronics is the conversion of one form of electrical energy to another, which involves the conversion of voltage and current in terms of magnitude or RMS values, the change of frequency and the number of phases. Power electronic circuits or components doing such conversions are referred to as converters. Converters are used in many different power and voltage ranges, their spectrum of converting power ranges from a few mW to several 100 MW, the voltage range extends from a few volts to several 10 kV or even 100 kV and even current ratings ranges from mA to kA. Power electronic converters are found wherever there is a need to modify an electrical energy form (i.e., change its voltage, current, or frequency). For example, variable-speed wind generator systems or PV arrays need power electronic interfaces to the electrical grid. Even for fixed-speed wind generator systems, when the energy storage system is connected to the grid power electronics devices are essential. Power semiconductor devices that are used as switches in power electronic circuits. Most power semiconductor devices are used only in commutation mode and are therefore optimized for this operation. Common power devices are the power diode, thyristor, power metal-oxide-semiconductor field-effect transistors (MOSFETs), and insulated gate bipolar transistor (IGBT). A power diode or MOSFET operates similar to their low-power counterparts but are able to carry a larger current and typically are able to support a larger reverse-bias voltage in OFF state. Structural changes are made in power devices to accommodate the higher current density, higher power dissipation, or higher reverse breakdown voltage. There are newer trends and directions in modern power electronics for the integration of wind energy conversion systems, small hydropower, solar thermal energy systems and photovoltaic systems. There are several reasons for these developments, among others are: increasing number of renewable energy sources and distributed generators, into transmission and power distribution sections of the grid, new strategies for the operation and management of the electricity grids, improved the power-supply reliability and quality, and liberalization of the grids leads to these newer management structures. Power electronics technologies, together suitable protection configurations, methods and the modern and adaptive control schemes and procedures are envisioned to lay an important and critical role into the integration and extended use of the distributed generation and renewable energy sources into the future electrical grids. This chapter section provides a brief overview and basic understanding of various power electronics devices: rectifiers, inverters, DC-DC choppers, cyclo converters, pulse width modulation (PWM)-based voltage source converters (VSCs), and current source inverters (CSIs). For detailed study, readers are referred to books of power electronics, listed in the chapter reference section or elsewhere in the literature.

The power electronics application to perform power conversion with minimum losses usually eliminates the usage of resistive elements or other similar elements with high dissipative losses (at least not in the critical paths). Also electronics components such as transistors are not operated in their linear regions, because this would also incur high power losses. Only the components with minimum losses, which are essentially reactive elements such as capacitors, inductors and transformers, together with the joins to these elements, the switches. Switches are key components in power electronics because they are the only elements that can control the flow of currents and voltages selectively either actively (by gate pulses, externally driven) or passively (as results of external electrical behavior of load or network, i.e., load or source commutated). The simplest (idealized) switch is the two-pole switch, it can assume two states, ON and OFF. The power loss of the ideal switch is always zero, because either the voltage or the current is zero, so their product, the switch power. The circuits connected to input and output of the switch must be capable of such step-like switch parameter changes. The most important candidates of two-pole switches are diodes, MOSFET and bipolar transistors. Many devices can only conduct the current in one direction and withstand blocking voltage only in reverse direction. Exceptions are the MOS-FETs which can conduct the current in forward and reverse directions, and the thyristor which is capable both of forward and reverse blocking voltage. For many applications, these shortcomings of the real devices are no serious handicaps, because handling of both voltage and current polarities is often not required. If that is necessary in some cases, several devices must be combined in order to cover all necessary specifications.

4.3.1 DC-DC Power Converters

The use of DC-DC converters are extensive in renewable energy, energy storage devices and industrial electronics applications. Such DC-DC power converters are also used in cases where a DC voltage produced by rectification or supplied by a power sources is used to supply secondary loads, and the DC voltage needed to be changed as required by the loads. The conversion is often associated with stabilizing, i.e., the input voltage is variable but the desired output voltage stays the same. The converse is also required, to produce a variable DC from a fixed or variable source. Since the conversion function is achieved by switching the DC power *ON* and *OFF* (chopping the power) at high frequencies, these power converters are also known as *DC choppers* or *switch-mode power supplies*. There are three basic configurations of the DC-DC power converters are Buck, Boost and Buck-Boost converters and a few variants of them. This chapter subsection describes the circuit topologies and the operation characteristics of the main types of DC-DC power converters. The basic circuit of a Buck DC-DC converter is shown in Figure 4.10 connected first to a purely resistive load. By using the fact that the average voltage across the inductor is zero, assuming perfect filter, the voltage across the inductor is V_d during t_{ON} and $-Vout$ the remaining of the cycle. If the diode the output voltage $v_{out}(t)$ is equal to the input voltage V_d when the switch is closed and to zero when the switch is open, an average output voltage V_{out} is computed as:

$$V_{out} = \frac{1}{T_{SW}} \left[\int_0^{t_{ON}} V_d \cdot dt + \int_{t_{ON}}^{T_{SW}} 0 \cdot dt \right] = \frac{t_{ON}}{T_{SW}} V_d = D \cdot V_d \qquad (4.42)$$

Here, $T_{SW} = 1/f_{SW}$ is the converter switching period, f_{SW} is the switching frequency (Hz), and $D = t_{ON}/T_{SW}$ is the duty ratio. The output voltage of the Buck converter is always lower the input voltage, so it is a step-down converter. A low pass filter is used to attenuate the high frequencies (multiples of the switching frequency) and leaves almost only the DC component. Assuming ideal components and that the input and output powers are the same, power conservation then:

$$V_d \cdot I_d = V_{out} \cdot I_{out} \implies I_{out} = \frac{I_d}{D} \tag{4.43}$$

The output current is the same as the inductor current, the inductor, L being in series with load. Finally, the consideration on the output voltage ripples, which are assumed that the ripple current is absorbed by the converter capacitor, C, i.e., the voltage ripples are small, being given by:

$$\frac{\Delta V_{Out}}{V_{Out}} = \frac{1-D}{8f_{SW}^2 LC} \tag{4.44}$$

Example 4.9: A Buck DC-DC converter has the following parameters: $V_d = 50$ V, $D = 0.4$, $L = 400\ \mu H$, $C = 100\ \mu F$, $f_{SW} = 20$ kHz, and the load, $R = 20\ \Omega$. Assuming ideal components, calculate (a) the output voltage and (b) the output voltage ripple.

Solution:

a. The inductor current is assumed to be continuous, and the output voltage is computed from Equation (4.42), as:

$$V_{out} = D \cdot V_d = 0.5 \times 50 = 25 \text{ V}$$

b. The output voltage ripple is computed from Equation (4.44):

$$\frac{\Delta V_{Out}}{V_{Out}} = \frac{1-D}{8f_{SW}^2 LC} = \frac{1-0.5}{8 \times (20 \cdot 10^3)^2 \times 400 \cdot 10^{-6} \times 100 \cdot 10^{-6}} = 0.003906 \text{ or } 0.39\%$$

For the setup Boost converter, the average inductor current is the average output current the output voltage is always higher than the input, and its topology is shown in Figure 4.11a. Similarly, there are two different topologies, based on the switch conditions of the switch, and again, the way to calculate the relationship between input and output voltage we have to take the average current of the inductor to be zero, and the output power equal to the input power hence the output voltage and current are:

$$V_d \cdot t_{ON} + (V_d - V_{out}) \cdot (T_{SW} - t_{ON}) = 0$$

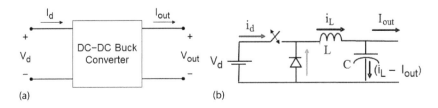

(a) (b)

FIGURE 4.11
(a) Block diagram of a Buck converter and (b) Boost converter circuit.

and

$$V_{out} = \frac{V_d}{1-D}, \quad \text{and} \quad I_{out} = I_d(1-D) \tag{4.45}$$

Following an analysis similar to that of a buck converter the output voltage ripples is given by:

$$\frac{\Delta V_{Out}}{V_{Out}} = \frac{D}{f_{SW} \cdot RC} \tag{4.46}$$

It is important to note that the operation of a Boost converter depends on parasitic components, especially for duty cycle approaching unity. These components limit the output voltage to levels. The Buck-Boost converter, having the topology as shown in Figure 4.12b, can provide output voltage that can be lower or higher than that of the input, depending on the duty ratio range. Again the operation of the converter can be analyzed using the two topologies resulting from operation of the switch. By equating the integral of the inductor voltage to zero, power conservation, the output voltage and current are estimated as:

$$V_d \cdot DT_{SW} + (-V_{out}) \cdot (1-D)T_{SW} = 0$$

and

$$V_{out} = \frac{D \cdot V_d}{1-D}, \quad \text{and} \quad I_{out} = I_d \frac{1-D}{D} \tag{4.47}$$

The output voltage ripples are calculated by using Equation (4.46), the one used for the Boost converters. The voltage ratios achievable by the DC-DC converters are summarized in Figure 4.13. Notice that only the Buck converter shows a linear relationship between the control (duty ratio) and output voltage. The Buck-Boost can reduce or increase the voltage ratio with unit gain for a duty ratio of 50%. All the DC-DC converters are transferring energy between input and output by the inductor, so the analysis is based on voltage balance across the inductor. In many DC-DC applications, multiple outputs may be required and output isolation may be needed. In addition, input to output isolation may be required to meet safety standards and/or provide impedance matching. The above discussed DC-DC topologies can be easily adapted to provide isolation between input and output. The Flyback converter can be developed as an extension of the Buck-Boost converter provides insulation to the inclusion of a transformer into its circuit. The Buck-Boost converter works by storing energy in the inductor during the ON phase and releasing it to the output during the OFF phase. With the transformer the energy storage takes place into the magnetization of the transformer core, while providing also insulation between converter input and output stages. To increase the stored energy a gapped core is often used.

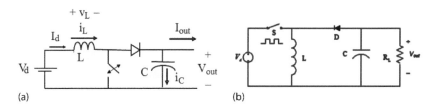

FIGURE 4.12
(a) Boost converter circuit diagram and (b) Buck-Boost converter circuit diagram.

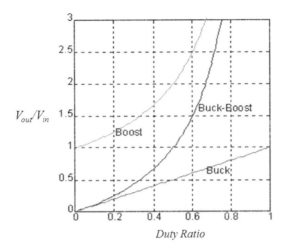

FIGURE 4.13
Voltage ratio vs. duty ratio for Buck, Boost and Buck-Boost converters.

4.3.2 Rectifiers and Other Power Electronics

A rectifier is a circuit that converts AC signals to DC, a process known as rectification. The simplest rectifier is the half-wave configuration, consisting of a diode, a single-phase AC sources and a purely resistive load (Figure 4.14a). If the load includes an inductance and a source (e.g., charging a battery), the diode continues to conduct even when the load voltage is negative as long as the current is maintained, changing the rectifiers waveforms. When the source voltage is positive, the current flows through the diode and the load voltage equal the sources voltage (Figure 4.14b). However, usually half-wave rectifiers are not common, rather a single phase diode bridge rectifier such as shown in Figure 4.15. The load can be modeled with one of two extremes: either as a constant current source, representing the case of a large inductance that keeps the current through it almost constant, or as a resistor, representing the case of minimum line inductance. The rectifier analysis, assuming that the instantaneous input voltage is $v(t) = V_m \cdot \sin(\omega t)$ including its waveforms,

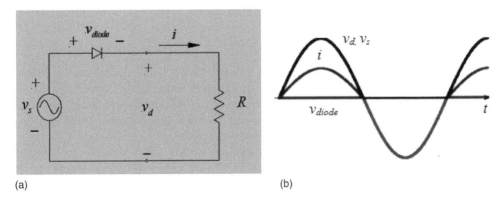

(a) (b)

FIGURE 4.14
(a) Half-wave rectifier circuit diagram and (b) current and voltage waveform.

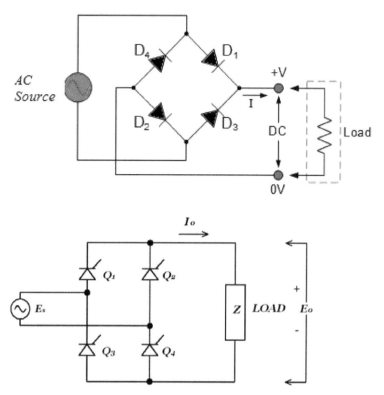

FIGURE 4.15
Single-phase full wave rectifier circuit diagram, upper panel diode-based and lower thyristor-based.

determines the following expressions for the output voltages and currents in the case of diode full-wave rectifier and the thyristor full-wave rectifier, as:

$$V_d \simeq \frac{2 \cdot V_m}{\pi}, \quad \text{and} \quad I_d = I_S \qquad (4.48)$$

and, respectively:

$$V_d \simeq \frac{V_m}{\pi}(1+\cos(\alpha)), \quad \text{and} \quad I_S = 0.816 I_d \qquad (4.49)$$

where V_m is the amplitude of the sinusoidal voltage, and I_S is the RMS value of the input AC voltage and current, while α is the delay angle, corresponding to the time we delay triggering the thyristors after they became forward biased. On the DC side, only the DC voltage component carries power, since there is no harmonic content, while on the AC side the power is carried by the fundamental, since there are no harmonics in the voltage. If the current on the DC side is sustained even if the voltage reverses polarity, then power is transferred from the DC to the AC side. The DC side voltage can reverse polarity when the delay angle exceeds 90°, and the current is maintained (e.g., a battery). Another important measure is the RMS voltage, more useful in estimating the effect on heating or incandescent lighting equipment. The two mean and RMS values normalized to their base when $\alpha = 0$. The RMS voltage is given by:

$$V_{RMS} = V_m \sqrt{1 - \frac{\alpha}{\pi} + \frac{\sin(\alpha)}{2\pi}} \qquad (4.50)$$

Example 4.10: Find the peak (amplitude) source voltage that a full-wave diode rectifier is producing 100 V (DC) across a 100 Ω. What is the power absorbed by this load?

Solution: The output rectifier DC voltage is computed by Equation (4.48), so the peak AC voltage is calculated as:

$$V_m \simeq \frac{\pi \cdot V_d}{2} = \frac{3.14 \times 100}{2} = 157 \text{ V}$$

The power absorbed by the 100 Ω resistive load is then:

$$P_{100-\Omega} = \frac{V_d^2}{R} = \frac{157^2}{100} = 246.5 \text{ W}$$

4.3.3 DC-AC Power Converters or Inverters

A DC-AC power converter, or inverter, is an electronic device or circuitry that changes a direct current (DC) into an alternating current (AC), more precisely, inverters transfer power from a DC source to an AC load or to grid. A power inverter can be entirely electronic or may be a combination of mechanical effects and electronic circuitry, while the so-called static inverters are not using moving parts in the conversion process. Inverters can generate single or poly-phase AC voltages from a DC supply, while into the class of poly-phase inverters, three-phase inverters are by far the largest group. These power converters are used in applications such as electric motor drives, uninterruptible power supplies, and utility applications such as grid connection of DC-type renewable energy sources or energy storage systems. Inverters are classified into two main categories: voltage source inverter (VSI), which has stiff DC source voltage, DC voltage has limited or zero impedance at the inverter input terminals, and current source inverter (CSI), which is supplied with a variable current from a DC source that has high impedance. The resulting current waves are not influenced by the load. Single-phase inverters are of two main types: full-bridge inverter and half-bridge inverter. Half-bridge inverter is the basic building block of a full-bridge inverter. It contains two switches, and each of its capacitors has a voltage output equal to $V_{dc}/2$. The switches complement each other, that is, if one is switched ON the other one goes OFF. Full-bridge inverter achieves the DC-to-AC conversion by switching in the right sequence. It has four different operating states which are based on which switches are closed. A three-phase inverter converts a DC input into a three-phase AC output. Its three arms are normally delayed by an angle of 120° so as to generate a three-phase AC supply.

The input voltage, output voltage and frequency, and the power handling depend on the specific device or circuitry design. Notice that he inverter power is provided by

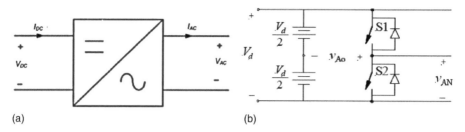

FIGURE 4.16
(a) Block diagram of an inverter and (b) one leg of an inverter.

the DC source. Figure 4.16a is showing the inverter block diagram. An inverter is usually able to set the converted AC signal characteristics at any required voltage and frequency levels with the use of appropriate transformers, switching and control circuits. Solid-state inverters have no moving parts and are used in a wide range of applications, from small switching power supplies in computers to large electric utility high-voltage direct current applications that transport bulk power. Inverters are commonly used to supply AC power from DC sources such as solar panels or batteries. A typical power inverter device or circuit requires a relatively stable DC power source capable of supplying enough current for the intended power demands of the system. The input voltage depends on the design and purpose and application of each inverter. The output voltage $v_{out}(t)$ can be theoretically any value in the range $-V_{DC}$ to $+V_{DC}$ (the DC source voltage), or zero, depending on the state of the inverter switches. The current waveform in the load depends on the load components. For the resistive load, the current waveform matches the shape of the output voltage, while inductive loads have a current that has more of a sinusoidal quality than the voltage because of the inductance filtering property. A separate class of inverters is the line commutated inverters for multi-MW power ratings that are using thyristors (silicon controlled rectifiers, SCRs). SCRs can only be turned *ON* or *OFF* on command. After being turned on, the current in the device must approach zero in order to turn the device off. Many other inverters are self-commutated. Such systems, converting DC into AC through the use of switching devices, such as GTOs, BJTs, IGBTs and MOSFETs, are allowing the transfer of power from the DC source to any AC load, and gives considerable control over the resulting AC signal. Line commutated inverters need the presence of a stable utility voltage to function.

Figure 4.16b shows the operation of the one leg of an inverter regardless of the number of phases. To illustrate its operation, the input DC voltage is divided into two equal parts. When the upper switch, S_1, is closed, S_2 is open, the output voltage is $+V_d/2$, and when the lower switch, S_2 is closed, S_1 is open, the output voltage is $-V_d/2$. To control the output voltage waveform the inverter switches are controlled by pulse width modulation (PWM), where the time each switch is closed is determined by the difference between a control waveform, and a carrier, usually a triangular waveform. When the control waveform (signal) is greater than the triangular waveform, the carrier, S_1 is closed, and S_2 is open. When the control wave is less than the triangular wave, S_1 is open, and S_2 is closed. In this way, the width of the output is modulated (hence the name, PWM). If the frequency of the triangular wave is f_{tri}, and the reference (sine) frequency is f_{ref}, then their ratio defines the frequency modulation index, m_f, while the ratio of the control voltage, V_{ref} to the triangular waveform voltage, V_{tri} the amplitude modulation index, m_a, is defined as:

$$m_f = \frac{f_{ref}}{f_{tri}} \qquad (4.51a)$$

and

$$m_a = \frac{V_{m,ref}}{V_{m,tri}} \tag{4.51b}$$

In full inverter, the diagonal switches are operating in tandem, such that S_1 and S_3 open and close together, and S_2 and S_4 open and close together, while the inverter output oscillates between $+V_d/2$ and $-V_d/2$. Applying the Fourier transformation of a PWM square wave shown, the amplitude of the fundamental is a linear function of the amplitude index $V_o = m_a \cdot V_d/2$ as long as $m_a \leq 1$, and the RMS output voltage is:

$$V_{o1} = \frac{m_a}{\sqrt{2}} \cdot \frac{V_d}{2} = 0.353 m_a \cdot V_d \tag{4.52}$$

When voltage modulation index becomes larger then 1, the output voltage increases also, but not linearly with m_a. The output voltage amplitude reaches a peak value of $(4/\pi) \cdot V_d$, when the reference signal becomes infinite and the output voltage is a square wave waveform. Under this condition, the RMS value of the output voltage is:

$$V_{o1} = \frac{\sqrt{2}}{\pi} \cdot V_d = 0.45 \cdot V_d \tag{4.53}$$

Assuming the ideal components and no losses, the DC side power is equal with the AC side power:

$$P = V_d \cdot I_{do} = V_{o1} \cdot I_{o1} \cdot PF \tag{4.54}$$

Here, PF is the power factor, thus for normal operation, the RMS output current is:

$$I_{do} = 0.353 m_a \cdot I_{o1} \tag{4.55a}$$

And in the limit for a square wave waveform is:

$$I_{do} = 0.45 \cdot I_{o1} \cdot PF \tag{4.55b}$$

Example 4.11: A PWM full-bridge inverter is producing a 50 Hz AC voltage across of a 10 Ω resistive load. The DC input to the bridge is 90 V, the amplitude modulation ratio m_a is 0.9, and the frequency modulation ratio m_f is 25. Determine the amplitude of the 50 Hz component of the output voltage and load current, and the power absorbed by the load (resistor).

Solution: The amplitude of the 50 Hz is computed from Equation (4.52):

$$V_1 = \sqrt{2} \cdot V_{o1} = \sqrt{2} \cdot 0.353 m_a \cdot V_d = \sqrt{2} \cdot 0.353 \cdot 0.9 \times 90 \approx 40.5 \text{ V}$$

For the fundamental frequency, being a resistive load the current is:

$$I_1 = \frac{V_1}{R} = \frac{40.5}{10} = 4.05 \text{ A}$$

The power absorbed by the load (assuming ideal conditions), and for a resistive load PF = 1, is:

$$P = V_1 \cdot I_1 = 40.5 \times 4.05 = 182 \text{ W}$$

A single-phase half-bridge DC–AC inverter is shown in Figure 4.17a. The analysis of the DC–AC inverters is taking into account the following assumptions and conventions. The switches S_1 and S_2 are unidirectional, i.e., they conduct current in one direction. The RMS output voltage is given by:

$$V_{out} = \frac{V_{in}(\text{DC source voltage})}{2} \tag{4.56}$$

The instantaneous output voltage $\nu_{out}(t)$ is rectangular in shape, not a purely sinusoidal waveform and its amplitude is set by the DC source voltage, V_{DC}. The instantaneous value of the output voltage $\nu_{out}(t)$ and output current, having angular frequency, ω, for the case of a resistive load, R can be expressed in Fourier series format as:

$$\nu_{out}(t) = \sum_{n=1,2,3,\ldots}^{\infty} \frac{4 \cdot V_{DC}}{n\pi} \sin(n\omega t) \tag{4.57a}$$

and

$$i_L(t) = \sum_{n=1,2,3,\ldots}^{\infty} \frac{4 \cdot V_{DC}}{R \cdot n\pi} \sin(n\omega t) \tag{4.57b}$$

A single-phase full bridge DC-AC inverter is shown in Figure 4.17b. The analysis of the single phase DC-AC inverters is taking in account following assumptions and conventions. The switches S_1, S_2, S_3 and S_4 are unidirectional, i.e., they conduct current in one direction. A single-phase square wave type voltage source inverter produces square-wave output voltage for a single-phase load. Such inverters have very simple control logic and the power switches need to operate at lower frequencies compared to switches in other inverter types. The first-generation inverters, using thyristor switches, were almost invariably square wave inverters because thyristor switches could be switched on and off only at lower frequencies.

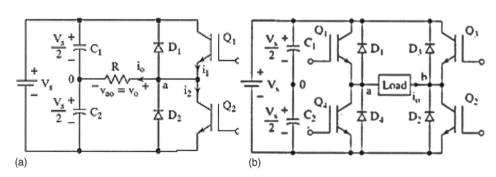

(a)   (b)

FIGURE 4.17
Half-bridge (a) and single-phase full bridge inverter schematics (b).

However, the present-day switches are much faster and used at switching frequencies of several kilohertz. Single-phase inverters mostly use half bridge or full bridge topologies. All inverter topologies, discussed here, are analyzed under the assumption of ideal circuit conditions, assuming that the input DC voltage is constant, while the switches and inverter components are lossless. With the ideal component assumption and constant DC voltage, the power supplied by the source must be the same as absorbed by the load. Power from a DC source (source current, I_S) is calculated by the well-known relationship:

$$P_{DC} = V_{DC} \cdot I_S \tag{4.58}$$

Far a single-phase full bridge inverter, the output voltages are computed by using Equation (4.57a), while the load currents for the case of a purely inductive load, having and inductance, L and for the case of a resistive-inductive (R-L) load are given by these relationships:

$$i_L(t) = \sum_{n=1,2,3,\ldots}^{\infty} \frac{1}{n \cdot \omega L} \frac{4 \cdot V_{in}}{n\pi} \sin\left(n\omega t - \frac{\pi}{2}\right) \tag{4.59a}$$

and

$$i_L(t) = \sum_{n=1,2,3,\ldots}^{\infty} \frac{4 \cdot V_{in}}{n\pi \sqrt{R^2 + (n\omega L)^2}} \sin\left(n\omega t - \theta_n\right) \tag{4.59b}$$

where the phase angle, θ_n, is given by:

$$\theta_n = \tan^{-1}\left(\frac{n\omega L}{R}\right)$$

Power absorbed by a load with a series resistance is determined from well-known relationship:

$$P_{\text{Load}} = R \cdot I_{\text{RMS}}^2 \tag{4.60}$$

where the RMS current can be determined from the RMS currents at each of the components in the Fourier series by

$$I_{RMS} = \sqrt{\sum_{n=1}^{\infty} I_{n,RMS}^2} = \sqrt{\sum_{n=1}^{\infty} \left(\frac{I_n}{\sqrt{2}}\right)} \tag{4.61}$$

Here $I_n = \frac{V_n}{Z_n}$, Z_n is the load impedance at harmonic n, for example in the case of a resistive inductive load, operating at frequency, f (Hz), and the load impedance amplitudes are expressed as:

$$Z_n = \sqrt{R^2 + (n\omega \cdot L)^2} = \sqrt{R^2 + (2\pi n f \cdot L)^2}$$

Notice that the calculation of the RMS component by dividing square root of 2 is valid only for sinusoidal waveforms, being only an approximation in the case of the square waves.

Equivalently, the power absorbed in case of the load resistor can be determined for each frequency in the Fourier series, through Ohm law. Total power is then computed through:

$$P = \sum_{n \geq 1} P_n = \sum_{n \geq 1} R \cdot I^2_{n,RMS} \qquad (4.62)$$

Example 4.12: A full-bridge inverter has a switching sequence that generates a square wave voltage waveform across a series *R-L* load. The switching frequency is 50 Hz, the DC source voltage is 280 V and the load resistance is 20 Ω, and load inductance is 20 mH. Determine the load voltage and current amplitudes, and the power absorbed by the load.

Solution: The load voltage amplitudes for a resistive-inductive load, Equation (4.57a) is:

$$V_n = \frac{4 \cdot V_{DC}}{n\pi} = \frac{4 \times 280}{n \times 3.14} = \frac{356.7}{n} \text{ V}, n = 1,3,5,...$$

For example: $V_1 = 356.7$, $V_3 = 118.9$ V, $V_5 = 71.3$ V

The load current amplitudes, Equation (4.59b) are:

$$I_n = \frac{V_n}{Z_n} = \frac{V_n}{\sqrt{R^2 + (n\omega L)^2}} = \frac{356.7}{n\sqrt{100^2 + (n \cdot 100\pi \times 20 \cdot 10^{-3})^2}}, n = 1, 3, 5,...$$

And for example $I_1 = 0.812$ A, $I_3 = 0.157$ A, $I_5 = 0.0515$ A

Power at each frequency is determined from Equation (4.56) as:

$$P = \sum_{n=1,3,5,...} R \cdot I^2_{n,RMS} = \sum_{n=1,3,5,...} R \cdot \left(\frac{I_n}{\sqrt{2}}\right)^2$$

And first three power terms are:

$$P_1 = 20 \times \left(\frac{0.812}{\sqrt{2}}\right)^2 = 6.59 \text{ W}, P_3 = 0.248 \text{ W, and } P_5 = 0.0265 \text{ W}$$

Basically only first term has a significant contribution to the load power, and an approximate value of the power delivered to the load is:

$$P \approx P_1 + P_2 + P_3 = 6.59 + 0.248 + 0.0515 = 6.863 \text{ W}$$

The voltage waveforms of practical inverters are, however, non-sinusoidal and contain certain harmonics. Square wave or quasi-square wave voltages are acceptable for low and medium power applications, and for high power applications low, distorted, sinusoidal waveforms are required. The output frequency of the inverter is determined by the rate at which the semiconductor devices are switched on and off by the inverter control circuitry and consequently, an adjustable frequency ac output is readily provided. By sequentially switching them on and off, the voltage across the load changes polarity cyclically and produces an alternating the voltage and current. Pulse width modulated (PWM) inverters are among the most used power-electronic circuits in practical applications. PWM is a technique which is characterized by the generation of constant amplitude pulse by modulating the pulse duration by modulating the duty cycle. PWM control requires the generation of both reference and carrier signals that are feed into the comparator and based on some logical output, the final output is generated. The reference signal is the desired signal output maybe sinusoidal or square wave, while the carrier signal is either a saw-tooth or triangular wave at a frequency significantly greater than the reference. These inverters are capable of producing ac voltages of variable magnitude as well as variable frequency. The quality of output voltage can also be greatly enhanced, when compared with those of square wave inverters. The PWM inverters are very commonly used in adjustable speed ac motor drive loads where one needs to feed the motor with variable voltage, variable frequency supply. For example the fundamental of the output voltage of single-phase, as function of the DC source voltage, V_{DC}, is expressed by this simple relationship:

$$V_{o1(\text{rms-fundamental})} = \frac{4V_{DC}}{\pi\sqrt{2}} = 0.90 \cdot V_{DC} \tag{4.63}$$

The three-phase bridge type VSI inverter with square wave pole voltages has been also considered. The output from this inverter is to be fed to a three-phase balanced load. Such circuits may be identified as three single-phase half-bridge inverter circuits put across the same DC bus. The individual pole voltages of the three-phase bridge circuit are identical to the square pole voltages output by single-phase half bridge or full bridge circuits. The three pole voltages of the three-phase square wave inverter are shifted in time by one-third of the output time period. The three-phase square wave inverter as described above can be used to generate balanced three-phase AC voltages of desired (fundamental) frequency. However, harmonic voltages of 5th, 7th and other non-odd multiples of fundamental frequency can severely distort the output voltage. In many cases such distortions in output voltages may not be tolerable and it may also not be practical to use filter circuits to filter out the harmonic voltages in a satisfactory manner. In such situations the inverter discussed in this lesson will not be a suitable choice. Fortunately there are some other kinds of inverters, namely pulse width modulated (PWM) inverters, which can provide higher quality of output voltage. The square wave inverter discussed in this lesson may still be used for many loads, notably AC motor type loads. The motor loads are inductive in nature with the inherent quality to suppress the harmonic currents in the motor. The example of a purely inductive load discussed in the previous section illustrates the effectiveness of inductive loads in blocking higher order harmonic currents. In spite of the inherent low-pass filtering property of the motor load, the load current may still contain some harmonics. These harmonic currents cause extra iron and copper losses in the motor. They also produce unwanted torque pulsations.

4.4 System Grid Integration Specifications, Issues and Requirements

Depending on the scale of generation, the renewable energy can be integrated into the utility grid either at the transmission level or at the distribution level. Wind farms or power plants, large renewable energy generation can be directly interconnected to the transmission system whereas the small scale DG units are usually interconnected to the medium- or low-voltage distribution systems. Before designing the system for both the types of interconnections a detailed analysis have to be made to face the different challenges in it. Leading characteristics of the RES and DG units that impact their integration into power grids are their size (generation capacity as compared to other sources of power generation on a system), their location (both geographically and with respect to network topology), and their variability (minute-by-minute, daily, seasonally and intermittently). Variable generation, provided by many renewable energy sources, is a challenge to electric grid operations. However, when integrated into smart grid as responsive distributed generation can be a profit to system operations if coordinated to relieve stress in the system (e.g., peak load, line overloads, etc.). RES and DG technology advancement and development have allowed extended grid integration. When such RES and DG systems or units are interconnected to the power grid or load centers can provide increased efficiency, reliability, better power quality and a broad set of other economic and power benefits. The RES and DG interconnection with power systems, often with power distribution networks are regulated by several standards and codes, addressing performance, safety and power quality issues, issued by the Institute of Electrical and Electronics Engineers (IEEE), the National Fire Protection Association (NFPA), International Electrotechnical Commission (IEC) and Underwriters Laboratory (UL) and other organization form different countries, Canada, France, Germany, Japan, United Kingdom, China, etc. IEEE, for example is developing voluntary consensus standards for electrical and electronic equipment and systems, with the participation of users, manufacturers, utilities and public interest groups. IEEE approved IEEE 547 Standard for Interconnecting Distributed Resources with Electrical Power Systems. This standard focuses on the technical specifications for testing and interconnection, providing relevant requirements for the performances, operation, testing, safety and maintenance of the interconnection, including general requirements for response to abnormal conditions, power quality, islanding, test specifications, design requirements, installation evaluation, production, commissioning and periodic tests. These requirements are universally needed for RES and DG interconnections and are sufficient for most of the installations and applications.

The definition of PCC varies, however in common words is defined as the point where a power generating unit is connected to the power system, its exact location depends on who owes the separation line between the generator and rest of the power network. PCCs are one of the points where many of the grid disturbances occur. PCC fault level is important, in the case of a generator connection because determines the effect of the generator on the electric network and in the protection circuit selection. Low fault level implies, higher network source impedance and a relatively large PCC voltage change, caused by the transfer in and out of active or reactive power. The RES generators impact on the electrical network depends on the PCC fault level and the RES generator size. The appropriate voltage level where an RES or DG unit is connected depends on its capacity rate. Units up to 250 kVA are connected on 400 V networks, the RES and DG units, from 1 kVA up to 10 kVA are connected on the 10–12.8 kV networks, while the larger RES and DG generation units usually are connected to the networks above 15 kV voltage levels. Notice that higher voltage connections tend to be more expensive than the lower voltage ones. RES and DG connections tend to raise the PCC voltage, leading to overvoltages to nearby users, while these voltages need to be limited rather than exceeding line thermal capacity limit. The effect being the limitation of the RES generator capacity connected at a specific location.

The significant increase in the wind energy development, a paradigm shift took place at the turn of the twentieth century into the grid connection conditions for wind turbines. Before unselective separations from the grid were applied in the case of grid malfunctions, the protection concept stability that the wind turbine connection conditions or for other

RES-based power generation forms changed fundamentally. The grid connection guidelines or user regulations are since the basis for the technical connection conditions, summarizing the important points that must be considered in the grid connection. They are defining the processing duties of the grid operators, system installers, planners or power customers. Wind turbines must be able to participate in the static and dynamic grid support in grids at all voltage levels, high or medium voltages, and even at lower voltage levels. With the static frequency stability in normal operating cases, the slow voltage changes are kept into tolerable limits by the reactive power conditions. The dynamic grid support prevents the undesirable switching-off of large power feeds that may result into the grid collapses in the case of component failures in the higher voltage grid sections. As wind turbines with connections in the medium-voltage grid have an immediate or longer influence on the high and highest voltage grid sections, these power units must also be in a position to fulfill system services as described before. In the future, wind or solar power plants feeding the low-voltage grids must participate in the static voltage stability. As they are required, during normal grid operations to contribute maintain the voltage level of the low-voltage grid section, having immediate effects on the design of such power plants. Dynamic grid support by generating plants that feed into the low-voltage grid is not required.

Solar and wind power plants exhibit changing dynamics, nonlinearities and uncertainties. Hence the grid, and especially the smart grid, requires advanced control strategies to solve effectively. The use of more efficient control techniques would not only increase the performance of these systems, but would increase the number of operational hours of solar and wind plants and thus reduce the cost per kilowatt-hour (KWh) produced. Wind behavior changes daily and seasonally, and sunlight is only available during daylight hours. Both wind and solar energy can be viewed as aggregate energy resources from the power grid point of view, with power levels that vary within a 10 minute to 1 hour time frame, so they are not representing the same form of intermittency as an unplanned interruption in a large base-load generator. The major functional components required for the DG and RES units and systems grid integration include power conversion and conditioning, protection devices and circuits, transfer switches and switchgear, RES and DG unit control and monitoring, and often coordination with the area electric power operators. These functions are not necessarily independent, discrete and separate objects, being often combined in specific application equipment or subsystems. The functions that are included into the interconnection subsystems may consist of power conversion (e.g., change the electricity types if needed or required to make the system compatible), power conditioning (e.g., providing basic power quality to supply clean power to the load), protection is monitoring the PCC and the RES and DG output and input power to ensure that the operating conditions are not exceeding the interconnection requirements, such as frequency, over- and under-voltages. RES and DG system functions can include: unit and load control, communication between RES and DG systems, loads or grid, metering, monitoring capabilities and ancillary services, such as voltage support, regulation or backup supply, etc.

4.4.1 Grid-Connected PV Systems

One of the greatest scientific and technological opportunities, faced today is the development at affordable costs of the efficient methods to collect, convert, store and utilize solar energy, the most abundant energy sources. Solar-based electricity can be generated either directly using PV cells or indirectly by collecting and concentrating the solar power to produce steam, which is then used to drive a turbine to provide the electric power. A grid-connected PV system is made up of an array or arrays of PV panels mounted on rack-type supports or integrated into a building. The PV panels are connected in series or parallel to achieve optimal voltage and current, and feed into an inverter transforming DC array current into AC current in phase and at the same voltage as the grid. The typical operating voltages of PV arrays are in the range of 150–400 V DC for small-scale systems (up to 3 kW), and 400–700 V DC for systems with power range from 10 to 500 kW. Maximum voltage is limited by the panel insulation to avoid any current leakage, and by the inverter maximum operating voltage. The inverters are equipped with a maximum power point tracking (MPPT) system that constantly adjusts the entry voltage to the characteristics of the PV modules, which vary according to temperature and solar radiation. As the system is linked to the grid, the rules and standards to be followed are those of small generating units of electrical energy, not controlled by the local utility, while the specific safety measures must be taken during installation and operation, as required by codes and standards. In essence, the AC voltage fed into the grid must be in phase with the grid voltage and comply with a number of regulations and safety requirements, which are more demanding than for a PV stand-alone system. All PV inverter subsystems connected to the grid incorporate an MPPT unit, and common characteristics such as automatic disconnection in the case of grid absence, minimal harmonics generation and a high frequency precision. The inverters of the grid-connected PV systems consist of three types, determined by the system size. First types are *PV module inverters*, which are the smallest models (100–200 W), usually being fixed in the back of the PV solar panel, producing directly 230 or 130 V AC. Their advantages are reduced cabling, only in AC, simplified connection to the building and reduced sensitivity to shading, since one shadow does not usually affect all the panels of the system. The second category is the *PV string inverters*, a sort of more powerful inverter type, connected to each series string of PV panels. Its advantages are cabling economy and DC protection. Its design is similar to first one with the advantage of working at a higher power and voltage, being so more efficiently. Central inverters are intermediate-sized models (1–5 kW), usually single-phase, and are intended for individual houses or small buildings. The solar power station type models are usually three-phase for power rages in excess of several hundred kilowatts. The advantages are clear separation of the DC and AC subsystems and simplified maintenance, while the disadvantages are more complex cabling and increased shading sensitivity.

Grid-connected PV system uses conventional panels, as those used in stand-alone systems, with the difference being that the number of PV cells is not tied to multiples of 36, the usual number used for the recharging of lead batteries. The PV panels for grid-connected systems have the sizes limited by the manufacturers, with the usually area of 1.7 m^2 for a typical output power of 200–300 W. This size corresponds to an approximate weight of 25 kg, which enables two people to carry out the installation. Larger (double surface) panels are also available, but their installation is more difficult, requiring a crane. Since PV modules are always installed outside, their supports must be resistant to corrosion, and therefore structures and fixings must preferably be in stainless steel, or aluminum, if the frames of the modules are themselves made of that material. The sizing of the PV module support structures must be in accord with the weight and size of the modules, wind resistance, and possibly the weight of snow in some areas. All cables, mechanical devices, fixings and electrical components must be installed according to the IEC and IEEE standards, codes and appropriate local regulations. Grid-connected inverters use two techniques to generate alternating current: the sine wave is produced by the device, using the grid zero point for the synchronization or the grid is used as signal and synchronization source. Some inverters use a transformer to ensure a galvanic separation from the grid, enabling the PV panels to be insulated from the grid, or transformer-less inverters, to improve efficiency (about 2%) and reduce cost. For these latter inverters, the cabling of the panels must be floating as the inverter continuously tests all current leaking in the direction of the earth. In case of leakage, the inverter cuts out to avoid all direct contact between the panel frame and the grid. All inverters use a certain amount of energy for their internal operation, either be supplied by the PV generator (not use current at night, but no continuous measurement possible), or from the grid (supply regularity and stability, but continuous consumption). Such continuous consumption by the inverter has a slight influence on the annual efficiency of the system. Inverters, in general, are equipped with devices to measure basic data, often having an interface for the data collection and storage. To ensure the safety of the system, prescriptions relating to the local grid and those relating to the inverter must be distinguished. Inverters do not have a constant efficiency: they are generally more efficient at three quarters of maximum power, and less efficient at low power. Inverters are sized about 85%–90% of the STC power of the PV generator, due the variability in the solar energy, higher operation temperature than 25°C, associated with higher insolation levels, reducing the generated power and the inverter and other system losses. Oversizing the generator allows one to improve the efficiency of the inverter, since reaching a high efficiency more quickly, which is improving the system annual output.

4.4.2 WECS Grid Integration

Wind turbines are complex machines, with large flexible structures working under turbulent and unpredictable environmental conditions, and are connected to a constantly varying electrical grid with changing voltages, frequency, power flow, and the like. Wind turbines have to adapt to those variations, so their efficiency and reliability depend heavily on the control strategy applied. Current growths of wind energy generation have results in the development of large wind farms with more than 100 MW capacities and are usually interconnected to the transmission grid. As wind energy penetration in the grid increases, additional challenges are being revealed: response to grid disturbances, active power control and frequency regulation, reactive power control and voltage regulation, restoration of grid services after power outages, and wind prediction, for example. With the increased wind energy integration in the power systems, it is necessary to require the wind farms to operate as much as possible like conventional power plants to support the network voltage and frequency not only during steady-state conditions but also during grid disturbances. In order to do that, the utilities in several countries have established and developed grid codes for the operation and grid connection of wind farms, covering all technical aspects relating to connection, operation and electricity transmission system use. The major requirements of the grid codes for operation and grid connection of wind turbines are:

1. Wind turbines are required to operate within range of required grid voltage variations.
2. Wind turbines are required to operate within typical grid frequency variation range.
3. Grid codes require wind farms to provide active power control to ensure a stable frequency in the system and to prevent overloading of lines.
4. Grid codes require that wind farms to provide frequency regulation capability to help maintain the desired network frequency.
5. Grid codes require that individual wind turbines control their own terminal voltage to a constant value by means of automatic voltage regulators.

6. Wind farms are required to provide dynamic reactive power control capability to maintain the reactive power balance and the power factor in the desired range.

7. Wind farms are required, by grid codes to provide the electric power with a desired quality, such as maintaining constant voltage or voltage fluctuations in the desired range or maintaining voltage–current harmonics in the desired range.

8. Wind farm operators are required to provide information corresponding to a number of parameters important for the system operator to enable proper operation of the power system. They also require installation of monitoring equipment to verify the actual behavior of the wind farm during faults and to provide the characteristics and data during such events to the grid operators.

Predictability is a key factor in managing wind power variability and fluctuations, and significant advances have been made, over last decade to improve wind velocity and power forecasting methods. Today the wind power prediction of the aggregate wind farms is quite accurate, through weather forecast, wind power generation models and statistical analysis, making possible to predict power generation from 5-minute to hourly intervals over 72-hour period with 20% or lower errors for a 36-hour horizon forecast. For regionally aggregate wind farms the output wind power forecast errors are in the order of 10% for a day ahead and less than 5% for 1–4 hours interval. Due to the random wind velocity nature, the wind generator output power, frequency and terminal voltage fluctuate, and in consequence the power quality deteriorates. However, the frequency, grid voltage and transmission line power must be kept constant, as is required by the standards and codes. To achieve these objectives controls are needed. Though there are many advances in short-term (and long-term) wind forecasting technology, accurate forecast of wind speed is still difficult. Hence, to cover the fluctuations in wind farm production a significant amount of spinning reserve and stand by capacity is required. Even though, there are some moderating effects in wide geographic area distribution of wind forms, the chance of simultaneous low energy generation from all wind farms cannot be completely neglected. Short circuit in high voltage transmission network causes short duration voltage dip which results in disconnection of several thousand megawatts of wind generation from the grid and leads to severe stability problems within the power system.

New connection regulations like fault ride through capabilities force wind turbine to have a supporting effect on grid operations. The requirement of fault ride through capability in large modern wind farms implies the guarantee that the system is maintaining stability after the fault clearance and the wind farm is capable and ready to supply power immediately. Serious challenges are faced by the wind turbine system designer through *strict* fault ride through standards are presented. Some of the requirements are:

1. The power transfer capability of the wind turbine generator (WTG) is reduced during fault, resulting in mechanical stress on blades and other rotating parts due to over speeding.

2. The WTG may draw increased amount of reactive power from the system during fault, depending on the employed generator technology which can result in a poor fault recovery response.

3. WTG technology is setting the nature and the magnitude of the fault current, while better issue knowledge helps in designing equipment ratings for protection and control settings.

4. The input mechanical power has to be regulated by the pitch control of turbine blades. During faults, the mechanical over speeding of rotating elements can be limited by the fast pitch control of blades.

5. The overall performance of the wind farm can be improved by applying the appropriate reactive power compensation device. One of the key issues related to wind power generation is the requirement of reactive power for voltage support. Induction generators are mostly used in wind generation systems. Induction machines do not have reactive power supply capability as synchronous machines. As per interconnection regulation the wind farms have to maintain the power factor within F0.95.

4.4.3 Balancing Power Systems with Large RES and DG Integration

Traditionally power system operation relays on the principle that at the user (load) energy demand is matched by the generation through the units of the system power plants. For such reasons, if either the demand, the generation or both are experiencing sudden changes, the occurred power unbalances result in the power system instability. Moreover the grid operators are not controlling the demand so their actions are focusing on the power generation. Power generation dispatch or unit commitment consists of defining which generation units are operated at any specific moment in time, as well as the generated power by each unit (the dispatch) to satisfy the power demands. Energy demands are varying in time due to several factors, such as seasonality, time of the day, weather conditions, geographical location, economic activity, etc. From operator point of view the uncertainties are only due to the demand side. However, the increasing integration of non-dispatchable and variable RES and DG units are resulting in uncertainties of the generation side. Moreover, besides the power unbalance induced by the RES and DG systems there is an additional issue that needed to be taken into account, the coincidence of higher availability of the renewable energy sources (e.g., wind energy) with lower energy demand periods, fact confirmed by data from various locations. Positive power unbalances correspond to an excess of generation, while negative unbalances correspond to a generation shortage or when the available power capacity is not enough to cover all power demands. The resulting power unbalance is characterized by such relationship:

$$P_{\text{Unbalance}}(t) = P_{\text{Generation}}(t) - P_{\text{Demand}}(t) \qquad (4.64)$$

The concept of flexibility of a power system is key factor in terms of balancing the RES and DG variable sources while keeping the lights on. On the supply side, flexibility arises from innovations into flexible coal, nuclear and natural gas power plants, energy storage units, and renewable energy systems. On the demand side, many of the distributed generation resources, flexible demand, energy storage and electric vehicle can significantly contribute, and likewise transmission and power distribution networks, grid operations and market designs. The grid integration challenges are encompassing several critical elements. Key among them is the *flexibility* concept of a power system, in terms of balancing variable wind, solar and other renewable energy resources in particular, and more general in terms of how all elements of a power system, on both supply and demand sides, can work together to ensure reliability (*keep the lights ON*) while minimizing the operation and electrical energy cost. Another key element is the design of the electricity markets, in ways that is aiding grid integration, while ensuring the most economically efficient operation. A further element is the planning and strengthening of transmission networks to balance geographical patterns of renewable energy resources and power demand. One final element is how power distribution systems can be transformed in their planning and operation, to support RES and DG grid integration and flexibility. Flexibility can come from both supply-side resources and demand side resources on a power system. Flexibility also arises from the design and operation of electricity markets, from transmission and power distribution networks, and from the technical operation of the grid itself. Long-term power system planning for flexibility incorporates all of these elements. Power system flexibility has existed since the early begging of the electric power systems over a century ago. However, the conventional flexibility concept was based primarily on being able to vary the power generation output to match the changes into the load or energy demands, and to respond to sudden unexpected changes into the power system components such as a transmission line or generator experiencing a fault, damage or a malfunction. The flexibility is needed were driven by the accuracy of load or energy demand forecasting and the probabilities of various discrete events. However, recently the flexibility needs and the meaning of power system flexibility is being reconsidered and redefined, in terms of being able to balance large shares of variable solar, wind and other renewable energy resources whose power output is not constant or dispatchable and to accommodate a variety of new technologies like advanced energy storage systems, electric vehicles, and demand side response, all of which are changing the nature of power systems. Renewable energy systems such as large and small hydropower, biomass and biogas, geothermal, and concentrating solar-thermal power units can contribute to power system flexibility. These types of power plants can be usually dispatched similarly to the conventional power plants, and can offer flexibility in terms of start-up time, ramping rates, and minimum power output limits. The 2014 IEA report categorizes such power systems as *firm* or *dispatchable* renewable energy. The additional costs of increasing the power system flexibility to accommodate higher shares of renewable energy systems are usually called *integration cost* or *flexibility cost*. Because each power grid is different, consequently the measures needed to increase flexibility are different, and due to the fact that such methods are not fully developed (including what counts as additional or incremental costs), the field is still under development, and controversy exists over how and what to count.

Power system operation practice requires that no power unbalances occur, in order to maintain perfect operation and the standard power quality levels. In the events when the power systems tend to be unbalanced, for example driven by an increased renewable energy generation, the power unbalances must be mitigated. In order to do that, individual or combined solutions to mitigate the power unbalance issues, such as curtailment of renewable energy generation, interconnections with other power systems or the use and addition of energy storage. The renewable energy generation curtailment consists in the event of power unbalance to cut-off partially or totally the RES and DG generation. RES curtailment implies a waste of the energy resources, and an increase into the fossil fuels consumptions, being recommended to be considered only in the extreme contingency cases. Strong power systems interconnections are an important advantage in terms of energy management and local power unbalances. Larger RES and DG integration into today grids is one of the drivers to reinforce the grid interconnection capacities in order to create a diversified mix and to extend the RES grid integration. However, there are some constrains in its application, such as the potential of the simultaneous power unbalances in the interconnected power systems. Energy storage offers an important benefit in the utility settings, decoupling the demand from the supply, thereby mitigating the power unbalances, allowing increased asset utilization, facilitating the RES and DG penetration, while improving electrical grid flexibility, reliability and efficiency. The energy storage option requires large investments, but can avoid the renewable energy curtailment disadvantage and/or the power exchange through grid interconnections constraints.

There two major categories of the energy storage systems for grid applications: short-term discharge energy storage devices and long-term discharge energy storage devices. Former devices have a very fast response to the power system needs, and they can supply power only for short periods up to minutes. Such devices are usually employed to improve the power quality, to compensate RES transient responses, or to cover load during start-up and synchronization of back-up generators. Long-term energy storage devices are able to provide power from seconds to hours, being characterized by slower responses to the power system needs. Such systems are usually applied on the energy management, large capacity renewable energy systems integration or power grid congestion management. Short-term energy storage devices, such as flywheels, supercapacitors and battery banks are now common in power systems with large RES penetration or weak interconnections and microgrids. The long-term discharge energy storage devices, such as compressed air or pumped hydropower systems are mature technologies and it is expected that their usage to increase with the larger penetration of the non-dispatchable renewable energy systems. Regardless, the long-term or short-term discharge energy storage types, all energy storage devices are requiring power electronics units for power conversion and management, as shown in Figure 4.18.

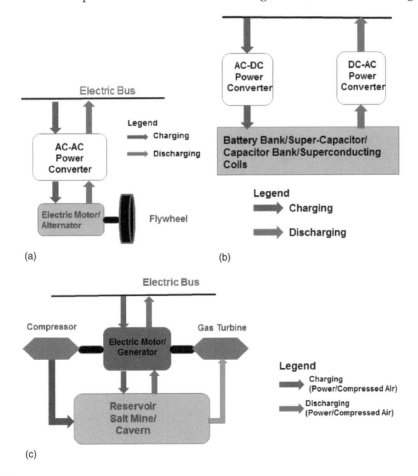

FIGURE 4.18
Grid interconnection of (a) flywheel, (b) battery bank, supercapacitor unit or superconducting coil and (c) compressed air storage.

The negative effects of the increased uses of such power electronics subsystems is the reduction in the overall grid power quality, requiring additional measures and costs to mitigate.

4.5 Summary

An electrical grid is an interconnected network for delivering electricity from suppliers to consumers. The electrical grid has evolved from an insular system that serviced a particular geographic area to a wider, expansive network that incorporated multiple areas. Most of the today power systems from generation, transmission and large part of the power distribution are in the form of three-phase systems. In the earlier days of the power generation tesla not only led the battle if the power system use DC or AC, but also proved that the three-phase electric power was the most efficient way to generate, transfer and use electricity. The essential feature of a three-phase system is although all currents are sinusoidal alternating waveforms if the system is balanced the total instantaneous power of the system is constant. Power systems are fundamentally reliant on control, communications and computation for ensuring stable, reliable, efficient operations. Today inflexible, unidirectional and rigid grids are less able to extract maximum value from RES and DG investments and can struggle to accommodate higher levels of variable RES. This struggle raises the likelihood of lost economic potential due to curtailed generation, challenges with maintaining a stable grid and lost opportunities in terms of meeting environmental, climate, economic development and energy access goals. The power industry has adopted smart grids that use information and communication technologies, which may make electric power systems more reliable and efficient and able to accommodate distributed generation and renewable energy into the transmission and power distribution networks. Smart grid concepts encompass a wide range of concepts, technologies and applications. Smart grid technology can control renewable resources to effect changes in the grid operating conditions and can provide additional benefits as distributed generation assets or when installed at the transmission level. Distributed generation can support weak grids, adding grid voltage and improving power quality. In certain circumstances, distributed generation can be used in conjunction with capacitor banks for management of power flows or to manage active and reactive power balance. A significant challenge associated with smart grids is the integration of renewable generation. Traditionally, power systems have addressed the uncertainty of load demand by controlling supply. With renewable energy sources, however, uncertainty and intermittency on the supply side must also be managed. These uncertainty and intermittency in generation are major complications that must be addressed before the full potential of these renewables can be reached. The smart grid represents an evolution of electricity networks toward greater reliance on communications, computation, and control, promising solutions to a broad range of energy issues.

Questions and Problems

1. List the major issues of the grind integration of wind and solar energy resources.
2. What are the current grid issues and problems?
3. What are the major benefits of the smart grids?

4. List and briefly describe the power grid sections.

5. What are the advantages of the three-phase systems?

6. List the advantages and disadvantages of high-voltage and low-voltage transmission lines.

7. What is the function of power plant generators and transformers?

8. List the main requirements of the wind farms grid integration.

9. Briefly describe the concepts of power system flexibility and balancing.

10. What are the criteria and inverter size level for the PV grid-connected systems.

11. List the main *firm* or *dispatchable* renewable energy sources and their role in the power system operation and management.

12. What are the reasons of the increased use of RES and DG in power and energy systems?

13. What are the major issues and problems of the small-scale RES grid integration?

14. List the briefly discussed main power electronics converters used in renewable energy.

15. What are the relationships between the phase and the line voltages and current for a Y-connection?

16. What are the relationships between the phase and line voltages and current for a Δ-connection?

17. The per-phase load impedance of a three-phase delta-connected load is $4 + j6\ \Omega$. If a 480 V, three-phase supply is connected to this load, find the magnitude of: (a) phase current; and (b) line current.

18. Calculate the RMS value, supply frequency and the phase shift in degrees for the AC voltage given by:

$$v(t) = 180\sin\left(300t + 0.866\right)\ \text{V}$$

19. A Y-connected balanced three-phase source is supplying power to a balanced three-phase load. The source phase voltage and current are given by:

$$v(t) = 420\sin\left(377t + 30°\right)\ \text{V}$$

$$i(t) = 90\sin\left(377t + 15°\right)\ \text{V}$$

Calculate: (a) the RMS and line-to-line voltages and currents; (b) supply frequency; (c) power factor at the source side; (d) three-phase active, reactive, and apparent powers supplied to the load; and (e) the load impedance if the load is balanced and Y-connected.

20. A synchronous generator, connected to an infinite bus has a negligible armature reactance and a synchronous reactance of $3.60\ \Omega$. The generator excitation is adjusted to set the induced voltage, E_f to 17.5 kV. If the bus voltage is 15 kV, compute the generator maximum power.

21. A three-phase, Y-connected synchronous generator, rated at 15 MVA, 13.5 kV has a negligible resistance and a synchronous reactance of $2.1\ \Omega$. Determine the generator induced voltage, E_f and the maximum power.

22. A three-phase feeder, Y-connected, rated 12.8 kV line-to-line voltage has a resistance of 0.2 Ω and a reactance of 0.16 Ω per 1000 m and is supplying a 13.5 MW load. Determine the voltage drop at 0.75, 0.85 and 0.90 power factor lagging.

23. A 600V cable, operating at 60 Hz has a resistance per 1 km of 1.37 Ω, and an inductance per 1000 m of 1.544 mH. The cable is connected an induction motor to AC source. If the RMS current flowing the cable is 95 A, what is the cable voltage drop?

24. A Buck converter has the following parameters: $V_d = 24$ V, $D = 0.60$, $L = 25$ μH, $C = 24$ μF, and $R = 10$ Ω. The switching frequency is 80 kHz. Assuming ideal components, determine (a) the output voltage, and (b) the output voltage ripple.

25. A Buck-Boost converter has the following parameters: $V_d = 12$ V, $D = 0.6$, $L = 12$ μH, $C = 20$ μF, and $R = 10$ Ω. The switching frequency is 120 kHz. Assuming ideal components, determine (a) the output voltage, and (b) the output voltage ripple.

26. A Boost converter has the following parameters: $V_d = 30$ V, $D = 0.75$, $L = 20$ μH, $C = 60$ μF, and $R = 15$ Ω. The switching frequency is 100 kHz. Assuming ideal components, determine (a) the output voltage and (b) the output voltage ripple.

27. A power transformer has 1000 turns in primary and is rated at 230 V and 50 Hz, determine the core magnetic flux.

28. A 12.5 kΩ resistor is connected to the secondary of a transformer that has turns ratio 6.0. What is the resistance seen in the primary?

29. Find the peak (amplitude) source voltage that a full-wave controlled rectifier is producing 100 V (DC) across a 25 Ω. What is the power absorbed by this load?

30. A single-phase inverter supplied by a 150 V DC source is delivering power to a resistive-inductive load, consisting of 15 Ω and 30 mH inductance, operating at frequency of 60 Hz), determine the amplitudes of the square wave load voltage, the amplitudes of the load current, and the power absorbed by the load.

31. A single-phase full-wave bridge rectifier is connected to a 130 V (RMS), 60 Hz AC voltage source is supplying a resistive load of 10 Ω. Calculate the load current and voltage.

32. A single-phase square-wave inverter is connected to a 120 V DC voltage source. Calculate the magnitude of the fundamental and the first three harmonics produced by this inverter.

33. For a Buck converter with input voltage 12.5 V, output voltage 5 V and switching frequency 20 kHz, having an inductance of 1 mH and a capacitor of 470 μF, determine the ripple in output voltage.

34. A full-bridge square-wave inverter has the DC source voltage 105 V, and the output frequency of 60 Hz, and a resistive load of 15 Ω. Determine the average and RMS values of output voltage and current.

35. A PWM full-bridge inverter is producing a 60 Hz AC voltage across of a series R-L load of a 10 Ω resistance and a 25 mH inductance. The DC input to the bridge is 100 V, and the amplitude modulation ratio m_a is 0.8. Determine the amplitude of the 60 Hz component of the output voltage and load current, and the power absorbed by the load resistor. Hint: The power absorbed by the load is computed by using: $P_R = R \cdot I_{RMS}^2$

36. A full-bridge inverter has a switching sequence that generates a square wave voltage waveform across a series R-L load. The switching frequency is 60 Hz, the DC source voltage is 270 V, the load resistance is 30 Ω, and load inductance is 25 mH. Determine the load voltage and current amplitudes, and the power absorbed by the load.

Further Readings and References

1. R.D. Schultz and R.A. Smith, *Introduction to Electric Power Engineering*, John Wiley & Sons, New York, 1988.
2. B.W. Williams, *Power Electronics: Devices, Drives, Applications and Passive Components*, MacMillan Press, London, UK, 1992.
3. C. Bayliss, *Transmission and Distribution Electrical Engineering*, Butterworth-Heinemann, Oxford, UK, 1996.
4. S. Heier, *Grid Integration of Wind Energy Conversion Systems*, John Wiley & Sons, Hoboken, NJ, 1998.
5. R. Messenger and J. Ventre, *Photovoltaic Systems Engineering*, CRC Press, Boca Raton, FL, 2000.
6. N. Jenkins, R. Allan, P. Crossley, D. Kirschen, and G. Strbac, *Embedded Generation*, The Institution of Electrical Engineers (IEE), London, UK, 2000.
7. T. Ackermann, G. Andersson, and L. Söder, Distributed generation: A definition, *Electric Power Systems Research*, Vol. 57(3), pp. 195–204, 2001
8. N.R. Friedman, *Distributed Energy Resources Interconnection Systems: Technology Review and Research Needs*, NREL/SR-560-32459, National Renewable Energy Laboratory, Golden, CO, 2002.
9. T.L. Skvarenina, *The Power Electronics Handbook*, CRC Press, Boca Raton, FL, 2002.
10. IEEE 1547, *IEEE Standard for Interconnecting Distributed Resources with Electric Power Systems*, 2003.
11. IEEE P1547.1, *Draft Standard for Conformance Test Procedures for Interconnecting Distributed Resources with Electric Power Systems*, IEEE Press, Piscataway, NJ, February 2005.
12. IEEE P1547.2, Draft Application *Guide for IEEE 1547 Standard for Interconnecting Distributed Resources with Electric Power Systems*, IEEE Press, Piscataway, NJ, July 2004.
13. G. Pepermans, J. Driesen, D. Haeseldonckx, R. Belmans, and W. D'haeseleer, Distributed generation: Definition, benefits and issues, *Energy Policy*, Vol. 33(6), pp. 787–798, 2005.
14. T. Ackermann, *Wind Power in Power Systems*, Wiley, London, UK, 2005.
15. A. von Meier, *Electric Power Systems: A Conceptual Introduction*, Wiley – IEEE Press, Hoboken, NJ, 2006.
16. S. Heier, *Grid Integration of Wind Energy Conversion Systems* (2nd ed.), Wiley, London, UK, 2006.
17. F.A. Farret and M. Godoy Simões, *Integration of Alternative Sources of Energy*, Wiley – IEEE Press, Hoboken, NJ, 2006.
18. L. Freris and D. Infeld, *Renewable Energy in Power Systems*, John Wiley & Sons, Chichester, UK, 2008.
19. J.R. Cogdell, *Foundations of Electric Power* (2nd ed.), Prentice Hall, Upper-Saddle River, NJ, 2008.
20. P. Schavemaker and L. van der Sluis, *Electrical Power System Essentials*, John Wiley & Sons, Chichester, UK, 2008.
21. M. Crappe, *Electric Power Systems*, ISTE, London, John Wiley & Sons, New York, 2008.
22. M.E. El-Hawary, *Introduction to Power Systems*, Wiley: IEEE Press, Hoboken, NJ, 2008.

23. M.A. El-Sharkawi, *Electric Energy: An Introduction* (2nd ed.), CRC Press, Boca Raton, FL, 2009.
24. N. Hadjsaid and J. C. Sabonnadière, *Power Systems and Restructuring*, ISTE, London, UK, John Wiley & Sons, New York, 2009.
25. IEC – 2010, Standard 50160, *Voltage Characteristics of Electricity Supplied by Public Distribution.*
26. J.L. Kirtley, *Electric Power Principles: Sources, Conversion, Distribution and Use*, Wiley, Hoboken, NJ, 2010.
27. B.K. Hodge, *Alternative Energy Systems and Applications*, Wiley, Hoboken, NJ, 2010.
28. A. Keyhani, M.N. Marwali, and M. Dai, *Integration of Green and Renewable Energy in Electric Power Systems*, Wiley, Hoboken, NJ, 2010.
29. A. Keyhani, *Design of Smart Power Grid Renewable Energy Systems*, Wiley: IEEE Press, Hoboken, NJ, 2011.
30. R. Teodorescu, M. Liserre, and P. Rodríguez, *Grid Converters for Photovoltaic and Wind Power Systems*, Wiley, Hoboken, NJ, 2011.
31. D.W. Hart, *Power Electronics*, McGraw-Hill, New York, 2011.
32. M.H.J. Bollen and F. Hassan, *Integration of Distributed Generation in the Power System*, Wiley-IEEE Press, Hoboken, NJ, 2011.
33. IEA Report, *Harnessing Variable Renewables: A Guide to the Balancing Challenge*, Paris, France, 2011.
34. P.T. Krein, *Elements of Power Electronics*, Oxford University Press, Oxford, UK, 2012.
35. R.G. Belu, Power electronics and controls for photovoltaic systems, in *Handbook of Research on Solar Energy Systems and Technologies* (eds. Sohail Anwar et al.), IGI, Global, pp. 68–125, 2012. doi:10.4018/978-1-4666-1996-8.ch004.
36. M.R. Patel, *Introduction to Electrical Power Systems and Power Electronics*, Taylor & Francis Group/ CRC Press, Boca Raton, FL, 2013.
37. B.M. Bernd and S. Zbigniew, *Smart Grids—Fundamentals and Technologies in Electricity Networks*, Springer, Heidelberg, Germany, 2014.
38. IEA Report, *The Power of Transformation: Wind, Sun and the Economics of Flexible Power Systems*, Paris, France, 2014.
39. IEA – RETD (Renewable Energy Technology Deployment), *RE-Integration: Integration of Variable Renewable Electricity Sources in Electricity Systems – lessons learnt and guidelines*, OECD, Paris, France, 2014.
40. L. Jones, *Renewable Energy Integration: Practical Management of Variability, Uncertainty, and Flexibility in Power Grids*, Elsevier, London, UK, 2014.
41. S.W. Blume, *Electric Power System Basics for the Nonelectrical Professional* (2nd ed.), John Wiley & Sons, New York, 2016.
42. J.E. Fleckenstein, *Three-Phase Electrical Power*, CRC Press, Boca Raton, FL, 2016.

5

Economics, Energy Management and Conservation

5.1 Introduction, Renewable Energy Potential and Economics

Over the past century, energy was the backbone of technology and economic developments. In addition to men, machines and money, energy is the fourth production critical factor. Without energy, no machine or equipment can run, the electricity is needed for everything in modern society. Hence, energy demands have dramatically increases in the years following the industrial revolution. This rapid increase in energy usage has created problems of demand and supply. If the growing energy demands are to be met with fossil fuels, there are higher probabilities significant declines in the energy producing in the future, so it is a today need to concentrate on renewable energy source to satisfy the demand and conserve our finite natural energy resources for future generations. Even more, the energy consumption and demands in the developing countries is increasing even at faster rates compared to the developed countries. Most of the capital stocks and infrastructure of modern economic systems are based on the fossil fuel energy use, and the transitions from fossil fuel uses are involving massive restructuring and huge investments. While private markets are playing a critical role in this process, major changes in policies are also needed to foster this transition. The considerable transition economic implications are justifying special attention on the renewable energy and on energy and environmental research. Fossil fuels, coal, oil and natural gas are the dominant energy sources in the modern society. However, the twenty-first century is already seeing the next energy transition, away from fossil fuels toward renewable energy and more environmental friendly energy systems. The transition is motivated by several factors, such environmental concerns, fossil fuel supply security, cost and significant technological advances. In order to produce or generate energy, energy carriers must be harvested and used. The main reason that current fossil fuel sources are widely used is that they have large amounts of conveniently concentrated energy, readily available to use and perform useful work. For example, early oil deposits were found near the ground surface, avoiding the need for deep drilling and pumping the oil great vertical distances. As time went on, the energy and cost required to find, extract and process crude oil, coal or natural gas increased significantly. This energy cost of acquiring fossil fuels is deducted from the harvested energy, reflecting the net available energy. Net energy, the return on the invested energy is expressed as a ratio of the available energy for consumption divided by the require energy to produce it. Net energy is a physical attribute of an energy source, and one important component of the energy cost, e.g., if a better technology, is reducing the required energy to make a solar PV panel, the PV panel cost falls and the cost of delivered

PV-based electricity is also fallen. A *large net energy ratio* means that large useful energy quantities are produced for smaller energy investments, as is the case with the exploitation of the original earlier oil or natural gas deposits. Net energy ratios for the same source vary significantly, depending on specific production technologies, local conditions and energy source characteristics.

Energy economics studies the energy resources and commodities, including drivers that are motivating companies, organizations and consumers to supply, convert, transport and use energy, to dispose and process the used energy residuals, to set market and regulatory structures, to analyze its distributional and environmental consequences, in order to make possible for the most economic and efficient energy uses. Energy economics recognizes the principles of physics: (1) energy is neither created nor destroyed but can be converted among forms; (2) energy comes from the physical environment and ultimately returns there. The industrial civilization history is in fact a history of the energy transitions and expansions. Energy is the primary and most universal measure of all work types performed by the humans and nature, a crucial input in the processes of economic, social and industrial activities. Energy economics requires that all decisions to develop a new facility, to buy new equipment are made on the cost analyses. However, the project cost calculations used in engineering are more appropriately specified as project finance. Energy demand is derived from preferences for energy services and depends on the properties of the conversion technologies and costs. About 80% of the modern society energy supplies are from fossil fuels because these sources provide energy at the lowest cost. However, the fossil fuels' cost advantages over renewable energy sources are decreasing, and certain renewables can already compete with fossil fuels solely on financial terms. Renewable energy costs are expected to further keep declining, while the fossil fuel prices are likely to rise. Thus even without policies to promote renewable energy, the economic factors are moving in that direction. Ultimately the economic analysis aims to select one system out of several alternatives that provides the required energy form at the lowest cost. Therefore different renewable energy types are compared with conventional generation systems, trying to take into considerations all benefits and costs. The result of an economic analysis is *one energy unit cost*, related to the $/kWh for electricity or heat. In an estimation of a specific energy cost, all the costs (e.g., plant construction, equipment, installation, operating and maintenance, fuel, upgrading and disposal costs) are divided to the total generated energy (kWh) by the power plant during its lifetime. All the costs included in such calculations must include the inflation rates and other adjustments in order to have accurate estimates.

Economic potential, one measure of renewable generation potential, is a metric that attempts to quantify the amount of economically viable renewable generation that is available at a location or within an area.

Economic potential may be defined in several ways. For example, one definition might be the expected revenues (based on the local market prices) minus generation costs, considered over the expected lifetime of the generation asset. Another definition might be generation costs relative to a benchmark (e.g., a natural gas plant) using assumptions of fuel prices, capital cost and plant efficiency. Economic potential is the subset of the technical potential that is available where the cost required to generate the energy, determining minimum revenue requirements for the resource development is below the available revenues. Finally, the market potential is the energy amount, expected to be generated through market deployment of renewable technologies after considering the impacts of current or future market factors, such as incentives and other policies, regulations, investor response and the economic competition with other generation sources. Economic analysis and evaluation methods are facilitating the comparisons among energy technology investments and the project viability. Usually, the same or similar methods can be used to compare investments in energy supply, energy development or energy efficiency. All sectors of the energy community and industries, including the renewable energy and distributed generation sectors need guidelines and comprehensive cost analysis in order to make economically efficient energy-related decisions. In general, in the power generation technologies, the electricity costs are primarily affected by three major factors: investment and capital costs, operation and maintenance (O&M) costs and the fuel costs. There are quite often incorrect cost and economic studies and analysis of renewable energy projects, due to the lack of the understanding of both technology and economics involved, while the misleading cost comparisons of different energy technologies are not uncommon.

The decision-making process of the energy projects is very similar to the decision-making processes in any engineering projects, involving the following steps: (1) the identification of the project needs, (2) project alternative solutions are formulated, analyzed and evaluated, and (3) a decision is made on the best and optimum alternative solution and then the project is identified and fully specified. These steps include: feasibility analysis and a detailed economic analysis of all the identified alternatives. In evaluating the projects, all factors must be examined and some of the proposed alternative solutions may be excluded for reasons other than economic or technical considerations. It is worth to notice here that the final decision must be specific enough to allow the facility planning, installation and construction, but also allows sufficient flexibility for the next project stages or future changes and adaptations. The decision-making process for engineering projects and especially for energy projects follows a well-defined need, being accomplished and performed by multi-step, structured procedures, simulations and mathematical techniques. The project or investment economic analysis should be made for a few carefully selected alternative solutions to the problem and the final decision takes into consideration the economic analysis as well as previous experience, environmental and social factors as well as engineering input.

In all engineering projects, it is imperatively necessary to justify the installation of new equipment, technology or new system implementation. Important questions for investors, planners and regulators are the project profitability and balance between the return and cost of an RES or DG project. However, there are similarities but also significant differences between the cost structure and investments of an RES or DG project and development and conventional power and energy projects. For example, the cost structures of PV or WT systems are different from that of conventional generation systems using fossil fuels such as coal, oil or natural gas. The RES and DG initial capital costs are higher than for the conventional generation units, because the basic components of a PV system (PV solar panels, power conditioning and control units, energy storage, support structures) or the ones of a wind energy conversion system (turbine, foundation, tower, drivetrain, generator and power conditioning and control units) are quite expensive. However prices of solar PV panels or wind turbines are dropping faster, for example the average one-off installation cost of solar PV panels has already dropped from more than $2 per unit of generating capacity in 2009 to about $1.00 in 2015. On the other hand, there are no fuel costs and emission costs during the lifespan of PV system or wind turbine operation of 20–30 years. The maintenance and operation costs of PV systems or wind energy conversion systems are also relatively low. In weighing the various alternatives, the economic analysis problems can be classified as fixed input type, fixed output type, or situations where neither input nor output is fixed. Whatever the nature of the problem or project, the proper economic criteria are to optimize the benefits and costs ratio.

5.2 Economic Aspects and Cost Analysis of Renewable Energy Systems

In all engineering projects and applications, including the renewable energy and distributed generation projects, it is necessary to justify the project development, implementation and/or installation of new equipment (Kaltshmitt, Streicher and Weise. 2007, Belu, 2018). In weighing the various alternatives, the economic analysis problems can be classified as fixed input type, fixed output type, or situations where neither input nor output is fixed. Whatever the nature of the problem, the proper economic criteria are to optimize the benefit-cost ratio. Keep also in mind that the lifetime of any RES or DG project typically spans on over 20 years. Therefore, it is important to compare savings and expenditures of various amounts of money properly over the project or development lifetime. In engineering economics, savings and expenditures of amounts of money during the project are usually called *cash flows*. To perform a sound economic analysis of an RES or DG development and project the most important economic parameters, data and project technical concepts must be well defined and known. The important parameters and concepts that significantly affect the economic decision making on the project and development include:

1. The time value of money and interest rates including simple and compound interests;
2. Inflation rate and composite interest rate;
3. Taxes including sales, local, state and federal tax charges;
4. Depreciation rate and salvage value;
5. Tax credits and incentives, local, state and federal.

In any project finance, there are two major costs: the *investment costs* and the *operating costs*. The investment cost relates to the fixed costs of materials and installation to deliver an RES or DG system to the client or stakeholders. The client only has to make this payment once, and the system will last for decades. In the case of systems like photovoltaics, wind turbine or solar hot water cannot be bought as a half of a module, meaning that the system size increases in lumps, rather than in smooth increments. Lumpy costs are common to energy systems deployment and projects, especially large-scale power systems for utilities. In comparison to a nuclear power plant the investment in incremental PV modules is fairly fine-grained and smooth. But from the perspective of a residential homeowner, the incremental investment in solar technologies appears to have a high fixed cost, and to be a lumpy cost. RES and DG investments, the total system costs (C_{Sys}) are divided into direct capital costs (C_{dir}) and indirect capital costs (C_{indir}), both of which can have a scale dependency, as shown here:

$$C_{Sys} = N(units) \cdot C_{dir}(\$/unit) + N(units) \cdot C_{indir}(\$/unit) \tag{5.1}$$

From an economics perspective, we would frame the scaled dependency of the size of the project in terms of the unit cost, that is the fixed and variable costs per characteristic unit of performance or area. Area dependency is one type of unit in residential solar hot water systems, and costs per unit area (C_{area}, in $/m^2$) are proportional to the area of the aperture for the hot water collector system (in m^2). For example, for a PV system, direct costs include the PV modules, inverters and the remaining balance of systems (BOS) equipment costs. Equipment BOS describes structural and electrical parts, mounting components, supports, and wiring. In broader PV evaluation, BOS is used to describe all costs outside of the fixed and variable costs per characteristic unit of performance or direct capital costs for the PV modules themselves. Extensive studies are progressing to reduce the costs of the PV modules and the BOS to the investment. In order to perform a realistic economic analysis, the effects of federal income tax, depreciation, tax credits and the inflation have to be included into the economic analysis. Federal, state and local taxes are part of the business expenditure. Income tax forms a major portion of the taxes. Federal and state governments provide tax credits in order to encourage investments in energy-related projects. However, these tax credits are in place for few years and may go. Some of the specific projects for tax credits are: (1) alternative energy sources, (2) wind or solar energy, (3) recycling equipment, (4) shale oil equipment, (5) cogeneration and DG projects and (6) hydroelectric generation equipment.

Depreciation is a term used in accounting, finances and economics to spread the cost of an asset or installation over the several years interval, for equipment or installation usually the lifespan. In simple words, the depreciation is the reduction in the asset or a good value due to the usage and factor as: time performance degradation, technological changes, depletion, inadequacy, etc. The depreciation method or approach is a useful method for the recovery of an investment through income tax over a period of time. Usually, the depreciation is accounted on capital assets such as a house, car, installation and industrial equipment. Depreciation is not applicable for items such as land, salvage value and interest amounts. Investment depreciation is defined as the decline in the capital value of the investment using the internal return rate as the discount factor. Tax and accounting depreciation are set mechanically by a specific set of rules. There are two types of depreciation, account (book) and tax depreciation. In book depreciation, a certain amount is off each year and credited to a depreciation reserve for the purpose of reinvesting. Tax depreciation is based on the existing tax laws and taxation codes and is used by the federal, state and local governments as a tool to encourage investments. The rate is decided by the government and can be changed at any time. The depreciation methods, used in engineering economic analysis are: (1) straight-line method (usually not used for tax purpose), (2) sum of year method, (3) declining balance method and (4) the accelerated cost recovery system. The recovery of depreciation on a specific asset may be related to the use rather than time, in which the depreciation is calculated by unit of production depreciation. This is not applicable to electrical and energy equipment and installation. If the investment amount and the number of years of depreciation are known, then the depreciation per years is calculated as the ratio of the investment over the lifetime, in years. In the sum of years of digits method, a greater depreciation amount is used in earlier years and there is lesser depreciation in the later years compared to the straight-line method. In the declining balance depreciation method, a constant depreciation rate is applied to the account (book) value of the property. This depreciation rate is based on the type of property and when it was acquired. Since the depreciation rate in this method is twice the straight line rate, the method is called the double declining balance method and the *double declining balance depreciation* in any year is given by the twice of book value divided by the number of years, N. The book value equals the cost depreciation to date, assuming depreciation zero in the first year, is expressed by:

$$\text{Double Declining Depreciation in Any Year} = \frac{2 \cdot (\text{Cost} - \text{Depreciation to Date})}{N} \quad (5.2a)$$

Straight-line depreciation is the simplest and most used calculation method, in which the real asset value at the end of the period that it generates revenue can be estimated, by expensing a part of the original cost into equal increments over the period. The real value of a good or asset, which can be zero or even negative, is its value at the time of disposal, removal or selling and is expressed by this relationship:

$$\text{Annual Depreciation Expenses} = \frac{\text{Original Cost} - \text{Real Value at the End}}{N} \left(\text{in \$/year}\right) \quad (5.2b)$$

5.2.1 Energy Cost Calculations without Capital Return

In any energy project, a payment of interest is expected for the invested capital, which depends on the economic factors and risks. The energy unit cost estimate without considering return is relatively simple. All cost over the generation system lifetime are added (the total cost) and divided by the operating period in order to calculate the annual cost, and then by diving the annual cost to the annual generated energy, the cost per unit of energy is obtained. If an energy generation system produces annually, E_{anl}, expressed in kWh with a total annual cost C_{anl}, then the *specific energy cost*, SE_{Cost} is calculated by:

$$SE_{Cost} = \frac{C_{anl}}{E_{anl}} \ (\$/kWh) \quad (5.3)$$

The annual cost is calculated by dividing the total cost C_{total} by N, the number of the power plant or energy generation system operating years, the plant or system lifespan. The total cost includes: initial investment, INV_0 and all payments P_k for every operational year k, for the entire lifespan of the system, N. For the power plant or generation system operating period of N years, the annual cost is calculated by:

$$C_{anl} = \frac{C_{total}}{N} = \frac{INV_0 + \sum_{k=1}^{N} P_k}{N} \quad (5.4)$$

The annual cost is also called the leveled electricity cost (LEC) in the case of power systems and leveled heat cost (LHC), in the case of thermal or heating systems. Before discussing other issues, it is worth to clarify the difference between *price* and *cost* of a product, often mistakenly as synonym. The product price is determined by the supply and demand for the product. The price of a product consists of its cost (production), the profit, taxes, installation, transportation and maintenance costs, from which are subtracted incentives and tax credits (if any), and resulting amount maybe corrected by a scarcity factor.

> **Example 5.1:** A 1.2 MW wind energy conversion system is generating an average of 2.85 millions of kWh per year. The investment cost of \$1450/kW of the installation and an average annual cost of 2.5% of the investment cost for the annual operation cost for this wind turbine. Assuming the wind turbine lifespan of 20 years, calculate the specific energy cost.
>
> **Solution:** The total cost of the wind energy system is:
>
> $$C_{total} = 1.2 \times 10^3 \cdot 1650 \cdot (1 + 20 \times 0.025) = \$2,610,000$$

The specific energy cost, computed with Equation (5.3) is:

$$SEC = \frac{C_{total}}{20 \times 2.85 \times 10^6} = \frac{2.61 \times 10^4}{5.7 \times 10^7} = 0.045789 \text{ \$/kWh} \simeq 4.6 \text{ Cent/KWh}$$

5.2.2 Basic Concepts, Definitions and Approaches

Before proceeding to the discussion of the economic and cost analyses and methods, it is important to provide key definitions that are used in this field. These definitions have been adapted from the literature, and the interest readers are directed to the reference and suggested readings of the chapter for additional information (Kaltshmitt, Streicher and Weise. 2007, Newnan, Eschenbach and Lavelle, 2011). The definitions of the most common and needed concepts and definition used in the fields of economics, cost analysis and management, which are critical in the decision making process for energy projects, are discussed here. The term *interest* can be defined as the money paid for the use of money. It is also referred to as the value or worth of money. Two important terms are *simple interest* and *compound interest*. Simple interest is always computed on the original principal (the borrowed money). Unlike simple interest, with compound interest, interest is added periodically to the original principal. The term conversion or compounding of interest simply refers to the addition of interest to the principal. The interest (conversion) period in the compound interest calculations is the time interval between successive conversions of the interest, while the interest period is the ratio of the stated annual rate to the number of interest periods in 1 year. The *present worth* is the current value of an amount of money due at a later time and the effect of the applied interest. *Average cost* represents the total of all fixed and variable costs calculated over a period of time, usually one year, divided by the total number of units produced, while the *average revenue* is the total revenue over a period of time, usually one year, divided by the total number of units produced. Average profit represents the difference between average revenue and average cost. *Fixed costs* are all costs, not affected by the level of business activity or production level, e.g., rents, insurance, property taxes, administrative salaries and the interest on borrowed capital. *Life cycle cost (LCC)* is the sum of all fixed and variable costs of a project from its initiation to its end of life. The life cycle costs also include the planning costs, abandonment, disposal, or storage costs. *Marginal or incremental cost* is the one associated with the production of one additional output unit, e.g., product, energy unit, etc., while the *marginal or incremental revenue* is the revenue resulting from the production of one additional output unit. The opportunity to use scarce resources, such as capital, to achieve monetary/financial advantage and the associated costs are the opportunity costs, e.g., an opportunity cost to building a new power plant for a company is not to build and invest their capital in 7% interest bearing securities. *Time horizon* is the time (expressed usually in years) from the project initiation to its end, including any disposal or storage of equipment and products. *Variable costs* are costs associated with the level of business activity, production or output levels, such as fuel cost, materials cost, labor cost, distribution cost, storage cost, etc. The variable costs increase monotonically with the number of units produced. *Project term* is the planning horizon over which the project cash flow is assessed, usually divided by a specific number of years. Initial cost is the one-time expense occurring at the begging of the project of investment, such as purchasing major assets for an energy project. *Annuity* represents the annual increment of the project cash flows, as opposed to the one-time quantity, such as the initial cost. Annuities can be either positive (e.g., the selling energy annual revenues) or negative (e.g., O&M annual costs). *Salvage value*, usually very small compared to the initial cost, is the one-time positive cash flow at the project planning horizon end, consisting of the assets, equipment, buildings or business sold in its end of the project actual condition. In order to gain a true perspective as to the economic value of the renewable energy systems and projects, it is necessary to compare the system technologies to conventional energy technologies on the *LCC* basis. The *LCC* method allows the total system cost calculation during a specific time period, usually the system lifespan considering not only the initial investment but also the costs incurred during the useful system life or a specific period. The *LCC* is the *present value* life cycle cost of the initial investment cost and the long-term costs related to repair, operation, maintenance, transport to the site and fuel used to run the system. *Present value* is the calculation of expenses that are realized in the future but applied in the present. An *LCC* analysis gives the total system cost, including all expenses incurred over the system lifetime. The main reasons for an *LCC* analysis are to compare different energy technologies, and to determine the most cost-effective system designs. For some renewable energy applications, there are not any other options to such systems to produce electricity because there is no power or the conventional energy methods are too expensive. For these applications, the initial cost of the system, the infrastructure to operate and maintain the system and the price people pay for the energy are the main concerns. An *LCC* analysis allows the designer to study the effect of using different equipment, components with different characteristics, performances and lifetimes. The common *LCC* relationship applicable for energy project is:

$$LCC = INV0 + \sum OM_t + \sum FL_t + \sum LRC_t - \sum SV_{sys+parts} \tag{5.5}$$

Here, $INV0$ is the initial overall installation costs, consist of the present value of the capital that will be used to pay for the equipment, system design, engineering and installation (the initial cost incurred by the user), the $\sum OM_t$ represents the sum of all yearly O&M (operation and maintenance) costs, the present value of expenses due to operation and maintenance programs (O&M costs include the salary of the operator, site access, guarantees and maintenance), the $\sum FL_t$ is the energy cost, sum of all yearly fuel costs, an expense that is the cost of fuel consumed by the conventional power or auxiliary equipment (the transport fuel cost to site must be included), the $\sum LRC_t$ is sum of all yearly replacement cost, the present value of the cost of replacement parts anticipated over the system life of the system, and the $\sum SV_{sys}+_{parts}$ is salvage value, the net worth at end of final year, typically up to 10% for the energy equipment. Future costs must be discounted because of the time value of money, so the present worth is calculated for costs for each year. The RES lifespans are assumed to be in the range of 20–30 years. Life cycle cost analysis is the best way of making acquisition decisions. On the *LCC* analysis, many renewable energy systems are economically viable. The financial evaluation can be done on a yearly basis to obtain cash flow, breakeven point and payback time, discussed later. Notice that social, environmental and reliability factors are not included here, but if they are deemed important they can be included.

> **Example 5.2:** A small hybrid power system (1 kW wind turbine, PV array and battery bank) was installed at a remote weather station for the following costs: (1) installed cost of $36,000, the loan for 10 years, with the total interest paid, minus the tax credit of $22,500, the total operation and maintenance of 2.5% of the initial cost, or $900 per year, and the total replacement cost is 3.5% of the investment. If the system salvage value is estimated at 7.5% of the initial coast, and no fuel is used for the system operation, calculate the *LCC* for the investment period.
>
> **Solution:** Applying Equation (5.5), the *LCC* for the hybrid power system is:
>
> $$LCC = 36,000 + 10 \times 900 + 0 + 2700 - 1260 = \$46,440.00$$

Capital budgeting is defined as a process in which a business determines whether projects such as building a new power plant, a wind energy development or investing in a long-term venture are worth pursuing. A *capital investment* is expenditure by an organization in equipment, land, or other assets that are used to carry out the objectives of the organization. Most of the times, a prospective project's lifetime cash inflow and outflows are assessed in order to determine whether the returns generated meet a sufficient target benchmark. Capital budgeting, an essential managerial tool, is also known as *investment appraisal*. Project proposals which scale through the preliminary screening and evaluation phase are further subjected to rigorous financial appraisal to ascertain if they would add value to the organization. This stage is also referred to as the *quantitative analysis, economic and financial appraisal, project evaluation* or *simply project analysis*. The financial appraisal of the project may predict the expected future cash flows of the project, analyze the risk associated with those cash flows, develop alternative cash flow forecasts, examine the sensitivity of the results to possible changes in the predicted cash flows, subject the cash flows to simulation and pre-pare alternative estimates of the project's net present value. For the clarification and good understanding of the economic evaluation and analysis of the renewable energy sources is necessary to define basic terms and concepts used in common practice.

The economic value is the asset or equipment value expressed through money. Different experts or economic schools explain it differently. There are two basic approaches—the *subjective* and *objective* understanding of the value. Subjective understanding of the economic value is based on individual preferences of an individual, while the objective understanding is the relationship between preferences (individual and collective) and the cost of meeting the needs. *Utility (use value)* represent the ability of an asset, equipment, installation, service or product to meet specific needs. *Non-use value* or *passive use value* is the utility of good for others (subjective economics). *Environmental (internal) value* is the result of the belief that nature has a positive value for the environment independently of human preferences and direct benefit to mankind. *Discounting* is a concept used to evaluate the present (the costs and benefits) higher than the future (costs and benefits), there is decline in the value. Discounting relays on the premise that the value of money is declining over the time, therefore the future values should be discounted relative to the present. There two important parameters related to the discounted cash flows. *Interest rate* is the investment percent return, or percent charged on the amount of money borrowed that tome horizon begging. Usually the interest is com-pounded at the end of each year, and the one year unit is referred to as the *compounding period*. The *minimum attractive rate of return (MARR)* represents the minimum interest rate required for project returns to make the project financially attractive, as set be the business, government offices or other entities and organizations that are making decisions about the project and investment. There are three major reasons for the demanding of the MARR on the foregoing present value of an amount of money invested: the inflation (reducing the future money value), the possible investment alterations (even if there are incentives for energy projects, still the investors prefer the ones with high returns), and the most important the risks associated with any investment. Notice that when the MARR equals the inflation rate, the real dollar return is zero, but the there is no actual value losses. *The nominal discount rate* is a summary rate for investment or capital that includes the inflation. *The real discount rate* is the net discount rate, a nominal rate minus the inflation rate.

In the decision-making process a short list of ideas/alternatives is identified and selected, a detailed economic evaluation for the projects in the short list of the project alternatives is per-formed and is critical for the project viability. This evaluation takes into account not only the economic aspects of the chosen alternatives and determines the profitability of the projects, but also other issues, which may not be quantified and do not affect materially the cash flow and the profitability of the project, the so-called *intangible items,* such as public good, national security, environmental issues, etc. However, the project economic analysis is treating the projects strictly as investments and the final decisions are based on the project profitability. The critical concept of the time value of money is based on the premise that one today monetary unit is worth more than the same monetary unit a year from now, the latter is worth more in two years from now, and so on. The time value of the monetary funds is intricately related to the concepts of the capital return, stipulating that a capital invested must yield more capital at the investment period end, the *interest (discount) rate, r,* the percentage of additional funds that must be earned for the lending of capital, and the current and expected future inflation, is increasing the cost of goods in the future. *Money time value* is an important parameter in all economic calculations. However there are situations, such as short lifespan investments, where the discount impact is quite limited, being reasonable to use only the value monetary amount, ignoring the discounting adjustments. When capital investments, such as energy production or conservation investments, are appraised there are inherent risks, associated with any investment and all or part of the capital may be lost. The *investment risks* are one of the justifications for the charging of an interest rate and the expectation of higher return on the invested capital, and the higher the investment risk the higher would be the expected return on the capital. Two concepts often used in the economic analysis are: the *simple payback* and the *capital recovery factor* (CRF). In simple payback, the value known as the *net present value* (NPV) is computed by summing all into and out cash flows, such as the initial costs, annuities and salvage value amounts. If the NPV is positive, then the project is economic viable. In the case of multi-year project with positive simple payback, the break-even point (BEP) is the year where the project total annuities equal the initial costs, which have been paid back at this point. CRF is a parameter that is used to measure the relationship between cash flows and investment costs, usually applied for short-term investments, up to 10 year. CRF is the ratio of the *annual capital cost* (ACC) to the NPV, estimated for a period of N years:

$$CRF = \frac{ACC}{NPV} \qquad (5.6a)$$

and

$$ACC = \text{Annuity} - \frac{NPV}{N} \qquad (5.6b)$$

Example 5.3: A utility is making an investment of \$75 million to improve the transmission capacity and reduce the losses, over a period of 10 years with an annuity of 15 million. What is the CRF in this case?

Solutions: The investment net present value is the difference between the total 8-year period annuity and \$75 million:

$$NPV = -75 + N \times Annuity = -75 + 10 \times 15 = 75 \text{ million}$$

By using Equation (5.5b) and (5.5a) the ACC and CRF are:

$$ACC = Annuity - \frac{NPV}{N} = 15 - \frac{75}{10} = 7.5 \text{ million}$$

and

$$CRF = \frac{7.5}{75} = 0.1 \text{ or } 10\%$$

This value is less than the maximum 15% CRF, as the recommended by the U.S. Electric Power Research Institute (EPRI) for power and energy industries.

Factors affecting the evaluation of natural resources are: the amount of expected future benefits from the use of resources, and the time factor. Time factor (discounting) is related to the fact that the economic analysis is based on the fact that the value falls over time. A positive discount rate expresses the rate of decline of economic indicators over time. Discounting is a normal part of the economic efficiency evaluation. Reasons for positive discount rates: preference for current benefits against future ones, and the capital productivity (the expectation that the preference of investment instead of the immediate consumption results in future higher consumption). In some cases it is appropriate to use a zero discount rate. Assumptions of discounting are using the facts that all incomes during the certain period of investment will be invested, and future value of the evaluated good is decreasing (e.g., lower quality or utility), or its amount will rise. The rule for the investment process include: the marginal productivity of capital is higher than the marginal productivity of time (the income of last unit of input does not fall below the value of time preference), and the nominal discount rate higher than inflation rate. In general, when the electricity price, produced from renewable energy sources is similar or less than the price produced from conventional energy sources, the economic considerations are favoring the development of more geothermal units, solar power plants, wind energy parks, etc. A combination of rising fossil fuel prices, incentives and a favorable regulatory environment, providing tax credits, investment guaranties and accelerated depreciation change the economic and financial circumstances for alternative energy.

Requirements for relevant information and analysis of capital budgeting decisions taken by management has paved way for a series of models to assist the organization in amassing the best of the allocated resources. The most common methods of capital budgeting techniques include: *the payback period, the net present value, the internal rate of return* (IRR) *and the real options approach.* The payback period method of financial appraisal is used to evaluate capital projects and to calculate the return per year from the start of the project until the accumulated returns are equal to the cost of the investment at which time the investment is said to have been paid back and the time taken to achieve this payback is referred to as the payback period. The payback decision rule states that acceptable projects must have less than some maximum payback period designated by management. The payback method, by definition, only takes into account project returns up to the payback period. It is therefore important to use the payback method more as a measure of project liquidity rather than project profitability. The discounted payback period method proposed as an improved measure of liquidity and project time risk over the conventional payback method and not a substitute for profitability measurement because it still ignores the returns after the payback period, stating that the proper role for the discounted payback period analysis is as a supplement to profitability measures and thus highlighting the supportive nature of the payback method, whether conventional or discounted payback period. The net present value is defined as the different between the present value of the cost inflows and the present value of the cash outflows. In other words, a project's net present value, usually computed as of the time of the initial investment is the present value of the project's cash flows from operations and disinvestment less the amount of the initial investment. NPV is used in capital budgeting to analyze the profitability of an investment or project and it is sensitive to the reliability of future cash flows that the investment or project will yield. For instance, the NPV compares the value of the dollar today to the value of that same dollar in the future taking inflation and returns into account. The internal rate of return is the discount rate often used in capital budgeting that makes the net present value of all cash flows from a certain project equal to zero. This in essence means that IRR is the rate of return that makes the sum of present value of future cash flows and the final market value of a project (or investment) equal its current market value. The higher a project's internal rate of return, the more desirable it is to undertake the project. Internal rate of return is the flip side of the NPV, where NPV is discounted value of a stream of cash flows, generated from investment. IRR computes the break-even rate of return showing the discount rate. The real options approach applies financial options theory to real investments, such as manufacturing plants, line extensions and research and development investments. This approach provides important insights about business and strategic investments which are very vital given the rapid pace of economic change. A financial option gives the owner the right, but not the obligation, to buy or sell a security at a given price, e.g., companies that make strategic investments have the right but not the obligation to exploit these opportunities in the future.

5.2.3 Critical Parameters and Indicators

A capital investment is expenditure made by an organization in equipment, land, or other assets that are used to carry out the objectives of the organization. There are several that affect decisions between various investment alternatives, and knowledge and understanding of their significance is critical for a good business decision. Fundamental to finance over long spans of time of any project or the project lifetime is the time value of money. The present value (PV) is the worth of an asset, money, or cash flows in today's dollars when rate of return is specified. The *money future value* (FV) is the asset worth or cash flow, being evaluated in today's dollars. It should make sense that *fuel costs* (FC) or *fuel savings* (FS) have values follow as the annual (or other time interval) cash flows. A *discount rate* is sometimes referred to as a *hurdle rate, interest rate, cutoff rate, benchmark,* or the *cost of capital.* Many companies and organizations have a fixed discount rate for all projects. However, if a project has a higher level of risk, a higher discount rate commensurate with that risk must be used. When cash-flow-related consequences occur in a very short duration, the income and expenditures are added and the net cash balance can be calculated. When the time span is longer, the effect of interest on the investment needs to be calculated. An economy havening low interest rates is encouraging money borrowing for investments and projects, whereas higher interest rates are encouraging money savings. Usually most of RES and DG projects are requiring higher initial investments, so the interest rates are very important here. Therefore, if the money is borrowed for such projects the interest rates are good indicators whether the projects are cost-effective. The capital borrowed to cover the partial of full initial cost of an RES or DG project or development, a fee, *the interest (I)* is charged for the *principal (P)*, the borrowed money. The interest charges are expressed as the percentage of the total amount, the principal, *the interest (discount) rate (r)*, as:

$$r \ (\text{interest}/\text{discount rate}) = \frac{P}{I} \qquad (5.7)$$

When considering projects with a useful life of several years, the time value of money has to be taken into account. There two ways to calculate the total interest charges over a project lifetime. In the *simple interest charges*, the total interest fee, I is paid at the loan end, N and is proportional with the interest rate r and the lifetime, N, while both are expressed in the same unit (i.e., one year, one month, etc.), as:

$$I = N \cdot r \cdot P \qquad (5.8)$$

The total amount paid, TP, due at the end of the loan period, includes the principal and the interest charge:

$$TP = P + I = P \cdot (1 + N \cdot r) \qquad (5.9)$$

In the second method, the *compound interest charges*, the loan lifetime (period) is divided into smaller periods, the *interest period* (usually one year) and the interest fee is charged at the end of each interest periods and accumulated from one interest period to the next. Therefore, the total payment at the end of an interest period is computed with Equation (5.8), P being replaced by total payment at the previous interest period end. If the principal is P, the total amount due at the loan period end is then expressed by:

$$TP = P \cdot (1 + r)^N \qquad (5.10)$$

Equation (5.10) states that in the second method, the amount of loan payment is increasing exponentially with N, if the interest charges follow the law of compound interest. For most of the RES, DG or energy efficiency and conservation projects the interest rates are usually constant throughout the project lifetime, otherwise is a common practice in economic analysis to use average interest rates.

> **Example 5.4:** A wind energy developer decided to invest in wind project to borrow a loan of $2,500,000. One bank offer 10-year loan with a 5% compound interest, while the second bank offer the same 10-year loan with 6.5% simple fix interest rate. Which is the most advantageous loan for the wind developer?
>
> **Solution:** If the 6.5% simple interest is paid over a 10 years period the total amount, Equation (5.9) is:
>
> $$TP_{\text{fix-rate}} = 2,500,000 \cdot (1 + 10 \cdot 0.065) = 4.125 \times 10^6 \text{ USD}$$
>
> In the case of 10-year, the loan with 5% compound interest rate the total amount, Equation (5.10) is then:
>
> $$TP = 2,500,000 \cdot (1 + 0.05)^{10} = 4.07223 \times 10^6 \text{ USD}$$
>
> The loan with 5% compound interest is more advantageous.

The process of calculating the future value of the money (future cash flow) in the present value is called present worth. The present worth PW is calculated from Equation (5.8) by replacing TP with FV, and solving for PW parameter:

$$PW = FV \cdot \left(\frac{1}{(1+i)^N} \right) \qquad (5.11)$$

where FV is the money value or the cost expected at time, after N years (future value) and i is the interest (discount) rate. If FV, N and the discount rate i, are known, then the PW can be calculated.

> **Example 5.5:** Assuming an interest rate of 7.5%, what is the present value of $100,000 will be received after 5 years from today?
>
> **Solution:** By using Equation (5.11) for the above values, the present value is:
>
> $$PW = \frac{100,000}{(1 + 0.075)^5} = 69,655.86 \text{ USD}$$

Inflation is occurring when the good and service costs increase from one period to the next one, while the interest rate is defining the cost of money, the inflation rate, r_{ifl} is expressing the cost increase of goods and services, therefore a future cost of a commodity or asset, FC is higher than its present cost, PC. The inflation rate is usually considerer constant over the energy project life, similar to the interest rate. Notice also that if the energy cost changes, especially if increases (the so-called energy cost escalation), which it is an important factor that must be taken into consideration into any energy project (development) economic and cost analyses. The future commodity or asset cost, FC estimated from the present commodity or asset cost, PC and the future value (worth), FV calculated from the present value (worth), PW, considering the combined effect of the interest rate, r, and the inflation rate, i, over the energy project (development and operation) period, N are expressed by these relationships:

$$FC = PC(1+i) \tag{5.12}$$

And, respectively for the future value:

$$FV = PW\left(1+\frac{r-i}{1+i}\right)^N \tag{5.13}$$

Example 5.6: A residential complex owner decided to invest $350,000 in a building integrated PV system. The money is borrowed through a 10-year load with an interest rate of 6.5%, and the inflation experienced by the economy is 2.75%. Determine the future value and the total cost of the investment.

Solution: Applying Equations (5.12) and (5.10) the future value and the total investment cost are:

$$FV = 350,000 \cdot \left(1+\frac{0.065-0.0275}{1+0.0275}\right)^{10} = \$350,475.83$$

and

$$TP = 350,000 \cdot (1+0.065)^{10} = \$656,998.11$$

An important parameter and indicator is the annual cash flow analysis is *future worth* (FV). Supposing that an amount A_{dep} is deposited at the end of every year for N years, and if r is the loan interest rate compounded annually, then the future worth amount FV is given by this relationship:

$$FV = A_{dep} \cdot \left[\frac{(1+r)^N - 1}{r}\right] \tag{5.14a}$$

Or for an annuity stream, and certain time horizon, the future value, FVA, of the annuity at the end of the Nth year is then given by:

$$FVA = A \cdot \left[\frac{(1+r)^N - 1}{r}\right] \tag{5.14b}$$

from investment point of view the translation of the future money to this present value, the discounted cash flow analysis is important, and for N equal annual amount (annuity) A, and a discount rate, i is calculate as:

$$V_P = A \cdot \left(\frac{(1+r)^N - 1}{r \cdot (1+r)^N} \right) \tag{5.15}$$

Equations (5.10 through 5.15) are assuming a constant (fix) annuity value, the non-constant (irregular) annuity case is considered below. If some parameters are known in these relationships the others can be computed directly. In the case of non-constant annuities, the present worth value, Equation (5.11), by treating each annuity as a single payment to be discounted from the future to the present, changes to:

$$PW = \sum_{k=1}^{N} \frac{A_k}{(1+r)^k} \tag{5.16}$$

Here A_k is the yearly predicted annuity in year, k from 1 to N (the last project year). Predicted costs varying from year to year are predicted to the net present value in a similar way. If the capital cost is subtracted from the present value of the revenue, the *net present value* (*NPV*) is obtained as:

$$NPV = V_P - C_{\text{Capital}} \tag{5.17}$$

From above equation, the *rate of return* (ROR) can be computed. The rate of return, r_{rt} is the discount rate that is making the net present value zero, and is found by solving numerically the equation:

$$0 = V_P - C_{\text{capital}} \Rightarrow C_{\text{capital}} = A \frac{1 - (1 + r_{rt})^{-N}}{r_{rt}} \tag{5.18}$$

This equation is usually solved numerically, e.g., Newton-Raphson iteration; however, an approximate value of the rate of return can be obtained by trial and error approach. The approximate method consists of finding two r_{rt} values, for which the NPV is slightly negative and slightly positive, and then by linear interpolation between these two values, the value of the rate of return is determined. It is important to keep in mind that a solution of this ROR may not exist. Once, r_{rt} value is computed for any project alternative, the actual market discount rate or the minimum acceptable rate of return is compared to the found ROR value, and if ROR is larger the project is cost effective. From the determined discount rate (ROR), making the NPV zero, the *annual energy cost*, A_{cost} can be computed by:

$$A_{\text{cost}} = \frac{C_{\text{capital}} \times r_{rt}}{1 - (1 + r_{rt})^{-N}} \tag{5.19}$$

The cost of energy unit (specific energy), C_{eng}, considering the return rate is computed by dividing A_{cost} by the annual generated energy, E_{anl}, in similar way as in Equation (5.3).

$$C_{\text{eng}} = \frac{A_{\text{cost}}}{E_{\text{anl}}} \tag{5.20}$$

When electric equipment, DG and RES installations and systems, such as PV systems, wind turbines, energy storage units, or transformers used in a wind farm, are purchased, there are several associated component costs. These include the equipment capital cost, installation cost, maintenance cost, salvage value and the annual returns. There are other factors related to income tax, depreciation, tax credit, property tax and insurance, which must be taken into account in order to evaluate the project economic viability. The maintenance cost depends on the design and the installation location. Usually any electrical and energy equipment or installations are designed for certain useful life, depending on the equipment type, design life, equipment maintenance and operating conditions. Sometimes the equipment may have zero value at the useful life end, the so-called salvage cost. Cost comparisons between different energy sources and electricity generation systems are made by the levelized cost of energy (*LCOE*). Levelized costs represent the present value of building and operating a power plant or a renewable energy system over an assumed plant or system lifetime, expressed in real terms to remove the effect of inflation. The levelized cost of energy, the levelized cost of electricity and/or the levelized energy cost are economic assessments of the average total cost to build and operate a power-generating system over its lifetime divided by the total power generated of the system during that lifetime. *LCOE* is often used as an alternative for the average price that the power generating system is receiving in a market to break even over its lifetime. It is a first-order cost competitiveness economic assessment of an electricity generating system that incorporates all costs over its lifetime accounting for the initial investment, O&M cost, fuel cost and capital cost. *LCOE* is a metric used to assess the cost of electric generation and the total plant-level impact from technology design changes, which can be used to compare electricity generation costs. There different methods to calculate *LCOE*, the ones included here are the most common found in the literature and used by practitioners. *LCOE* calculations are in framework of the annual technology baseline studies that are providing a summary of current and projected future cost and performance of primary electricity generation technology in the United States, including renewable energy technologies. For energy sources that are requiring fuels, assumptions are also made about future fuel costs. The levelized construction and operations costs are then divided by the total energy obtained to allow direct comparisons across different energy sources. In order for renewables to be cost competitive, their costs need to fall to the wholesale electricity price, or the price at which fossil-fuel power plants sell electricity to the grid. This has already occurred for some renewable energy sources, such as hydropower, small solar-thermal installations and biomass. Solar PV systems are still more expensive than other energy sources, but since PVs are often installed by individual consumers the price of PV only needs to fall to the retail power price that consumer's pay, which is greater than the wholesale price. Organizations use several different methods to determine which investment is best. There are many methods available for the evaluation of the economic efficiency of project options and alternatives, including the following: the net present value, payback period method, rate of return method, the benefit-cost analysis and return on investment. Usually, the following assumptions are made: one of the parameters in the present worth analysis is assumed, which may be either the interest rate or the useful life, the salvage value is usually taken as zero, even if the interest rate is provided, it is likely to change in the future, and when comparing products from a superior technology versus the older technology, the latter is attractive choice. In such cases, the economic advantages and the technical merits are to be weighted together.

5.2.4 Natural Resource Evaluation

For energy resources it is important to have some estimate of the resource viability. Basic methods for natural and energy resources evaluation include: the comparative method (derived from the price of other similar good); the cost method (according to the cost incurred in obtaining the subject); and the method of return (according to useful effects, which source provides). The resource evaluation consists primarily of the natural or energy source price, C_{NR} computed from an application of the present value relationship:

$$C_{NR} = \sum_{t=1}^{N} \frac{An_t}{(1+r_t)^t} \qquad (5.21)$$

Here, t is the time period, An_t is the expected value of the annuity for the period of time, r_t is expected value of interest (discount) rate for one period of time (coefficient), and N is the number of periods. Interest rate and the discount rate are considered and are quite often to be variable in time. The expected value of an annuity for a period of time is a function of several variables, such as the type and cost of production, input prices, taxes, interest rates, inflation, etc. for proper expected value estimate these factors must be known and defined. The expected value of the interest rate for a period of time is also a function of several variables, such as the money time preference, risk, inflation and other characteristics, etc. that are also needed in order to and accurate estimate and must be also defined. Frequently used assumptions are: a constant of value of the expected annuity value, An at the time (usually for long-term contracts) and a constant of value of the discount rate, r at the time are assumed, as well as because in most of the applications, there is usually very long time horizon, the infinite time series assumption, $N \to \infty$ is considered. Then a simpler relationship for natural or energy resource evaluation is used:

$$C_{NR(simple)} = \sum_{t=1}^{N} \frac{An}{(1+r)^t} = \frac{An}{r} \qquad (5.22)$$

Example 5.7: A geothermal energy source has an estimated annual annuity of $180,000 and a discount rate of 7.5%, compute the estimated source price.

Solution: From Equation (5.22) the geothermal source prices is estimated at:

$$C_{NR(simple)} = \frac{r}{i} = \frac{180,000}{0.075} = \$2,400,000$$

5.2.5 Methods for Payback Period Estimates

The payback method is commonly used for the appraisal of the capital investments and engineering projects despite its theoretical deficiencies. The payback period is the time required for the benefits of an investment to equal the investment costs. The payback method is often used when aspects such as project time risk and liquidity are the focus and where pure profit evaluation is the single criterion. In practice, the maximum acceptable payback period is often chosen as a fixed value, for instance for a certain number of years, e.g., the payback period for the domestic consumer is up to 5 years, often only 2 or 3 years. In some cases, the payback period limit value is chosen in relation to the project economic life, for example the payback period could be shorter than half the economic life of the investment. In many companies the payback period is used as a measure of attractiveness of capital budgeting investments. Most often the payback method is used as a first screening device used to sort out the obvious cases of profitable and unprofitable investments, leaving only the middle group to be scrutinized by means of more advanced and more time-consuming calculation methods based on discounted cash flows (DCF), such as the internal rate of return (IRR) and net present value (NPV) methods. However, it should be noted that payback method can be developed to handle cases with varying cash flows although some of its simplicity is lost in the process. Due to the fact that the decision situations in the evaluation of capital budgeting investment typically are uncertain concerning the time pattern and the duration of cash flows, the use of the simple and more robust payback method can be justified even if there will be time for more advanced analyses or methods. The payback period goes on decreasing for high capacity system used by commercial consumers if electricity is replaced. The payback period method (PBP) of capital budgeting calculates the time it takes to recover the initial investment cost. There are two approaches, the short-cut (simple) payback method and the unequal cash flow method, both are based the calculations on the annual net cash flows (cash outflows minus cash inflows). As is true with NPV and IRR methods, each year may have different cash flow amounts. In short-cut method, the amounts of annual operating cash flows expected from a potential capital asset acquisition are equal each year, and the short-cut calculation is used to determine the payback period. The simple payback period (SPP) is calculated as:

$$\text{SPP} = \frac{\text{Initial Investment (Cost)}}{\text{Net Benefits per Year}} \tag{5.23}$$

Example 5.8: To improve the power factor a capacitor back is installed at a 100 kW wind turbine. The cost of the unit is \$20,000 and the interest rate is 7.5%, the combined federal and state tax credits are 40%, and the incremental tax rate is 45%. The system-related costs are: the loss factor 0.4, maximum reactive power demand 80 kVAR, maximum demand cost of kW, kVAR, kVA is \$3.80/kW/month, the Cost of released transformer kVA is \$12 per kVA/year, cost of reactive energy is equal to \$0.0025/kVARh, the combined feeder and transformer resistance per phase is 0.040 Ω, the current before installation 55 A, and after installation 80 A. Assuming a power factor improving from 0.65 to 0.95 due to the power factor correction unit installation calculate the simple payback period.

Solution: Savings due to the reduction in the reactive power penalty per year is:

$$C_{PFP} = 3.80 \times \left(100 \frac{\sqrt{1-0.65^2}}{0.65} - 100 \frac{\sqrt{1-0.95^2}}{0.95} \right) \times 12 = \$3832.4$$

The saving due to the reduction in kVA demand transformer is:

$$C_{kVA} = \left(\frac{100}{.65} - \frac{100}{.95} \right) \times 12 = 582.9959 \approx \$583.0$$

Savings due to the transmission line loss reduction is:

$$C_{loss} = 3 \times 0.06 \times 0.040 \times \left(80^2 - 55^2 \right) \times 8760 \times 0.4 \times 10^{-3} = \$85.1$$

Savings due the reduction of the reactive energy cost is:

$$C_{Reactive} = 0.0025 \times \left(100 \frac{\sqrt{1-0.65^2}}{0.65} - 100 \frac{\sqrt{1-0.95^2}}{0.95} \right) \times 8760 = \$1840.6$$

The annual benefit is \$6341.20 and the simple payback period is then:

$$SPP = \frac{20,000}{6341.2} = 3.15 \text{ years}$$

The payback period indicates how long it takes to recover the investment used to acquire an asset or for an equipment installation and purchasing. The payback period is equal to the number of full years plus the final investment recovery year fraction. The simple payback period method ignores the money time value. One of the ways to compare mutually exclusive economic aspects is to use of the present time using the present worth method. The present worth of annual benefits is obtained by multiplying the benefits with the present worth factor defined for a discount (interest) rate, r and N years, as:

$$PVF = \frac{1}{(1+r)^N} \tag{5.24}$$

Then the cumulative value of the benefits, for AB, the benefits per year is given by:

$$B_{\text{cum}} = AB\left(\frac{(1+r)^N - 1}{r \cdot (1+r)^N}\right) \tag{5.25}$$

In the payback evaluation method, the period Y_{proj} (usually expressed in years) required recovering the cost of the project, the initial investment, $CF0$ is determined by solving the following equation (Equation 5.25), expressed in the interest (discount) rate, i and the average annual net savings, CF_k, as:

$$CF0 = \sum_{k=1}^{Y_{\text{proj}}} \frac{CF_k}{(1+r)^k} \tag{5.26}$$

If the payback period Y_{proj} (often called the *discount payback period, DPB*) is less than the project lifetime, N, then the project is economically viable. In large majority of applications, the time value of money is neglected, and Y_{proj} is SPP, as discussed before, being the solution of the simplified Equation (5.26), the following equation:

$$CF0 = \sum_{k=1}^{Y_{\text{proj}}} CF_k \tag{5.27}$$

In the case of where the annual net savings are constant, A_{SV}, the simple payback period is easily calculated as the ratio of initial investment and the annual saving:

$$Y_{\text{proj}}(SPP) = \frac{CF0}{A_{SV}} \tag{5.28}$$

The SPP values are shorter than the DPB values because the undiscounted net savings are larger than their discounted counterparts. Therefore, acceptable SPP values are usually significantly shorter than the project lifetimes. Usually, the investments and the economic returns are expressed on time-varying basis. In energy projects the payback period calculations, using the present worth analysis involves the present worth factor evaluation, annual benefits and the present worth of the cumulative benefits. In order to perform the economic analysis, the income tax effects must be included, and the available tax credits, if any for the project. Businesses can deduct a percentage of the new equipment costs as a tax credit, usually after the tax return. At the same time the basis for the depreciation remains the full cost of the equipment. The main components taking into account the tax effects are cash flows before taxes, depreciation, change in taxable income, equal to the cash flow before taxes, less the depreciation, income taxes (taxable income times the incremental income tax) and after tax cash flows, equal to before tax cash flows, minus the income tax. In general, the prices and cost of services change with time and hence the inflationary trend should be included in the payback period calculations. The inflation adjustments must be included into the payback period analysis. If the annual inflation rate is i, the actual cash is computed by cash flow before taxes multiplying with $(1+i)^N$, the depreciation and the income taxes are computed as before, while the after cash flow is also adjusted by the inflation rate multiplying with $(1+i)^N$.

> **Example 5.9:** In DG system an old less efficient (60%) boiler is about to be replaced by a more efficient (85%), having the same annual O&M costs as the older one, $150. The equipment cost is $20,000 and the discount rate is 5%, after adding the tax credit per year and its lifetime is 10 years. The boiler consumes 9000 gallons of fuel at $1.30 per gallon.
>
> **Solution:** The annual cash flow is computed from the net savings of using the more efficient equipment:
>
> $$A = C_k (k = 1...10) = 9000 \times \left(1 - \frac{0.6}{0.85}\right) \times 1.30 = \$3573.53$$

Assuming a constant annual savings, $CF_1 = ... = CH_{10} = A_{SV}$, $d = 0.05$ and $N = 10$, and then solving numerically Equation (5.26), the DPB is 4.68 years, shorter than the project lifetime, 10 years, therefore replacing the old equipment is cost effective. The cumulative benefits equation is given here:

$$20,000 = \sum_{k=1}^{Y} \frac{3573.53}{(1+0.05)^k} = 3573.53 \frac{1-(1+1.05)^{Y+1}}{1-1+0.05}$$

When the dollar amount of operating cash flows is not expected to be the same amount each year, a longer interval is expected for the recovery, and is estimated by the so-called *unequal cash flow method*. In essence, by beginning with the acquisition cost, the amount to be recovered, and then subtracting the expected cash flows for each year until the point in time at which the cash is recovered. The calculation begins by subtracting the first year cash inflows from the initial investment, and if the remaining cash flows to recover are greater than or equal to the cash flows of the second year, the cash flows expected for the second year it is subtracted, and the process continues for each subsequent year until the remaining cash flows not yet recovered are less than the cash flows expected for the next year, and the point during the next year that the remaining cash is recovered. The payback period indicates how long it takes to recover the cash investment used to acquire the asset. It is expressed usually in years with two decimals, such as 4.25 years. Usually, shorter payback periods are more attractive than longer ones. If the cash is expected to be recovered in a time period shorter than the investment lifetime, it is tentatively deemed acceptable. However, other capital budgeting methods must always be used in conjunction with payback period method, because even when it appears to be an acceptable investment, it may not be acceptable under a method that considers the time value of money. If the shortcut method is used to calculate the SPP, the results may indicate a payback period greater than the useful asset life. It is not possible to have a useful life greater than the payback period because the *lifetime end* indicates the asset is no longer used in the production, so it no longer brings in economic resources. As such, when the numerical result using the shortcut method appears to have a payback period exceeding the useful life, *the interpretation is the investment is never recovered*. The payback period method has some faults that create limitations on its usage. First, it does not consider the total stream of cash flows. The second drawback of the payback period method is that it does not consider the time vale of money in its calculations. The payback period method also ignores the timing of the cash flows and evaluates both investments as equal options.

5.2.6 Rate of Return Analysis, Benefit-Cost Ratio Analysis and Net Benefits Method

The rate of return is defined as the interest rate paid on the unpaid balance such that of the loan is exactly repaid within the schedule of payments. Such economic analysis is useful when the economic activity is very slow and the interest rates are very low (e.g., years 2002, 2003, or the interval 2008–2012). If the interest rates are around 2%, or even lower the effect of compounding interests is insignificant. To calculate the rate of return (*RRet*) on an investment is given by the ratio:

$$RRet = \frac{\text{Present Worth of Benefits}}{\text{Present Worth of Costs}} = 1 \qquad (5.29)$$

In Equation (5.29) the only unknown is the interest rate, *i*. A second simplified approach to capital budgeting is the accounting rate of return method, very similar to the rate of return analysis, but can be applied regardless range of the interest rates or economic activity. It is considered to be *simplified* because it is not using time value of money in evaluating capital investments. This capital budgeting method uses net income, not cash flows. The accounting rate of return (ARR) method calculates the return generated from the average net income expected for each of the years the proposed capital investment is expected to be used in operations. It is like the rate of return concept; however, this return is based on a single proposed asset acquisition, while the rate of return in financial accounting is based on the return generated by a company, project or installation total assets. The calculation of accounting rate of return is calculated as:

$$ARR = \frac{\text{Average Net Income}}{\text{Average Investment}} \qquad (5.30)$$

Average net income is calculated by dividing to the investment number of years, in which is expected to generate economic resources, the total net income over the same years. The average investment is based on the book (accounting) value of the potential capital budgeting acquisition. The beginning book (accounting) value and the ending book (accounting) value are averaged to obtain the average investment. Beginning book value is the book value at the beginning of year 1 and ending book value is the book value at the end of the proposed investment lifespan. The accounting value is the cost of a long-term asset minus the accumulated depreciation. At the life-end of an asset, the asset cost minus accumulated depreciation equals the salvage value. When the asset is sold for the salvage value amount, there is no gain or loss on the sale. The ARR is expressed as a percentage return with two decimals displayed, such as 6.93%. This amount tells you that the company is expected to earn about 7 cents of profit out each dollar that is tied up in the investment. If the ARR is equal to or greater than the required rate of return, the project is acceptable. If the investment is expected to generate a return that is less than the desired rate of return, it should be rejected. When comparing investments, the higher the ARR, the more attractive the investment. ARR limitation is that the ARR ignores the time value of money in its computations. By doing this, it views amounts generated in the first year to be equal to the amounts generated in the last year on a dollar per dollar basis. In essence, it ignores the timing of the cash flows within the useful life.

> **Example 5.10:** An investment of $75,600 was made to install at a remote farm a 1500 liter per day solar water heating system. The expected project lifetime is 15 years. It was estimated that the cost savings due to kVA reduction and loss reduction is $15,250 per year. Calculate the rate of return on this investment.
>
> **Solution:** The input data in Equation (5.25) are B_{cum} = 75,600, N = 15 years, and AB = 15,250, solving for the discount rate, r, by a numerical method of this nonlinear equation:
>
> $$75,600 = 15,250 \left[\frac{(1+r)^{15} - 1}{r \cdot (1+r)^{15}} \right]$$

The estimated interest (discount) rate is approximately, 6.0%.

The rate of return analysis is one of the most used analysis methods in the engineering projects. The rate of return gives a measure of the project desirability in terms that are easily understood. However, the rate of return problem may not be solved with a positive interest rate due to the cash flow nature. Therefore, there is recommended to use it carefully. In energy projects, besides of the NPV and MARR, it is very useful to know the interest rate in which the year the investment (project) breaks even at the end of its lifetime, and this information is used to evaluate the project attractiveness. This method is known as the *internal rate of return* (IRR) or the *discounted cash flow return on investment method*, being quite similar to the ARR method. For a project, regardless the annuity type (constant or non-constant) is useful to compute the IRR in a tabular format to calculate the project NPV as a function of the interest rate that is giving NPV equal to zero. Notice that there are software packages with built-in IRR functions.

Benefit-cost ratio analysis states that at a given minimum attractive rate of return, an alternative can be acceptable if the benefit-cost ratio (BCR), as defined below, is greater than 1.

$$BCR = \frac{\text{Present Worth of Benefits}}{\text{Present Worth of Costs}} > 1 \qquad (5.31)$$

This analysis can be used for all the three types of economic problems namely, fixed input, fixed output and where neither input nor output is fixed. In this case, obtain the benefit-cost ratio. If it is greater than 1.0, then the project is viable. The benefit-cost ratio or the savings-to-investment (SIR) ratio provides a measure of the project net benefits (savings), relative to its net cost. The net values of the benefits (B_k) and the costs (C_k) are compared to a base case and the present value of all cash flows is used into BCR analyses. Therefore, the project (investment) benefit-cost ratio, lasting N years, is computed as follows:

$$BCR = \frac{\sum_{k=1}^{N} \frac{B_k}{(1+r)^k}}{\sum_{k=1}^{N} \frac{C_k}{(1+r)^k}} \qquad (5.32)$$

Example 5.11: An investment of $930,000 was made to install a 1 MW wind energy conversion system.

The prevailing interest rate is 6.0%. The projected life duration is 25 years according to the manufacturer. The lifespan is about to be de-rated due to a future expansion of protected wildlife area. Lifetimes of 20 or 15 years are under consideration. Calculate the benefit-cost ratio for each of the life durations. The estimated annual benefits due to this project are $95,800.

Solution: Applying Equation (5.32), and by using Equation (5.25) to estimate the present worth benefits for the 25, 20 and 15 years lifespans yields to:

$$BCR_{15-year} = \frac{95,800\left(\dfrac{(1+0.06)^{15}-1}{0.06\cdot(1+0.06)^{15}}\right)}{930,000} = 1.0005 > 1$$

$$BCR_{20-year} = \frac{95,800\left(\dfrac{(1+0.06)^{20}-1}{0.06\cdot(1+0.06)^{20}}\right)}{930,000} = 1.181 > 1$$

$$BCR_{25-year} = \frac{95,800\left(\dfrac{(1+0.06)^{25}-1}{0.06\cdot(1+0.06)^{25}}\right)}{930,000} = 1.316 > 1$$

All the project lifespans 15, 20 and 25 years provide benefit-cost ratios greater than 1.

In addition the general benefit-cost ratio, there are two alternatives often used in the energy sectors, the so-called, the *conventional BCR* and *modified BCR* formulas, depending on the ways that the incurred O&M costs are considered by the operators. The two BCR relationships are:

$$BCR_{conv} = \frac{\text{Total Benefits}}{\text{Initial Cost (INV}_0) + \text{O\&M Costs}} \tag{5.33a}$$

and

$$BCR_{modified} = \frac{\text{Total Benefits} - \text{O\&M Costs}}{\text{Initial Cost (INV}_0)} \tag{5.33b}$$

Note that there is no computational advantage by using either BCR formulas, the preference is matter of choice or traditions.

Net present value or net benefits method computes the difference between the benefits and costs, while all amounts are discounted for their time value, and if the costs exceed the benefits, the net losses are the result. The NPV method, the *net present worth* or *net savings method* is used for evaluating a cost reducing investment, the cost savings are the benefits, and it is often called the net savings (NS) method. Following is a formula for finding the NPV from an investment, such as an investment in energy efficiency or renewable energy systems:

$$NPV_{A_1 \div A_2} = \sum_{k=1}^{N} \frac{B_k - C_k}{(1+r)^k} \tag{5.34}$$

where $NPV_{A_1 \div A_2}$ (NB – net benefits) is the present value of the net benefits (savings) over the present value costs for the alternative A_1 as compared with alternative A_2, B_k is benefits in year k, which may be defined to include energy savings, C_k is costs in year k associated with alternative A_1 as compared with a mutually exclusive alternative A_2, and r is the discount rate. The NPV method is particularly suitable for decisions made on the basis of long-run profitability. The NPV (NB) method is also useful for deciding whether to make a given investment and for designing and sizing systems. It is not very useful for comparing investments that provide different services.

5.3 Life Cycle Cost Analysis of Renewable Energy

Life cycle cost analysis (*LCCA*) of any system is one of the most important analyses for the project success and viability, giving a clear idea and an informed decision to accept or not the project (system) for implementation. *LCCA* is a methodology used first time by the Department of Defense of United States, it's an economic calculation of all costs propagated during the life span of any technical system. For RES systems, *LCCA* is a good methodology, showing the cost-effectiveness of using RES or DG as an alternative energy source. Alongside with the growth of renewable energy (RE) market, the need for an accurate and precise evaluation for an economic feasibility of these technologies is a pressing issue. Accordingly, all costs regarding the project should be taken into consideration from the conceptual to disposal phases, or what it was called life cycle cost analysis (*LCCA*). The concept of *LCCA* was firstly introduced in 1970s by the U.S. Department of Defense. Since then the *LCCA* concepts and approaches were widely adopted in a wide range of industrial sectors including energy, construction, manufacturing, transportation and healthcare. The cost of any RES systems, such as solar PV systems or wind energy conversion systems, was initially measured by cost of power ($/W) which is lacking several aspects (e.g., financial policies, system lifetime and equipment performance). The *levelized cost of electricity* (*LCOE*) is a more accurate energy cost calculation, being well-accepted and widely used technique and it has been adapted by researchers and agencies. *LCOE* is commonly used within energy contracts to determine a fair energy price and to set the contract price and the price schedule. The cost of energy for each unit of energy determined in the energy purchase agreements, negotiated usually around the standard *LCOE* accounting for the possible risks that could raise the actual *LCOE*. Normally, *LCOE* models include all the costs associated with a project, such as the capital costs, operational and maintenance costs over the project lifetime, the energy produced and the weighted average cost of capital (WACC). The *LCOE* is calculated as the ratio between the life cycle cost (*LCC*) of the energy generating system, e.g., an RES or a DG system over the entire life of energy production (*LCE*), the power plant lifespan, expressed as:

$$LCOE = \frac{LCC}{LCE} \tag{5.35}$$

Example 5.12: Calculate the levelized cost of energy, for the hybrid power system of Example 5.2, assuming the hybrid power system is generating 63,000 kWh per year and has a parasite energy consumption (the energy used by the system itself the operate) of 5% of the generated energy.

Solution: The calculated *LCC* for the system of Example 5.2 is $46,440.00 per 10-year period or $4644.00 per year. The useful annual energy generated by the system is:

$$LCE = 63,000 - 0.05 \times 63,000 = 59,850 \text{ kWh/year}$$

The system *LCOE*, Equation (5.35) is then:

$$LCOE = \frac{LCC}{LCE} = \frac{4644}{59,850} = 0.07759398 \approx 0.078 \ \$/kWh$$

However, the definition of *LCOE* is the cost, assigned to every unit of energy produced by the system over the analysis period, equaling the total life-cycle cost (*TLCC*) when discounted back to the base year, where a discrete compounding is assumed. *TLCC* is the *INV0* in the initial investment, and *PV-OM* is the present value of the O&M costs. In practice, the *TLCC* and *PV-OM* parameters are given by the following relationships:

$$\sum_{k=1}^{N} \frac{EGen_k \times LCOE}{(1+i)^k} = TLCC, \quad \text{and } TLCC = INV0 + PV\text{-}OM = \sum_{k=0}^{N} \frac{COE_k}{(1+i)^k} \tag{5.36a}$$

$$PV - OM = \sum_{k=1}^{N} \frac{OM_k}{(1+i)^k} \tag{5.36b}$$

Since *LCOE* is by definition constant once calculated, it can be factored out of the summation and the model is rewritten in Equation (5.36a) as:

$$LCOE = \sum_{t=1}^{N} \frac{TLCC}{\frac{EGen_t}{(1+i)^t}} = \frac{\displaystyle\sum_{t=0}^{N} \frac{COE_t}{(1+i)^t}}{\displaystyle\sum_{t=1}^{N} \frac{EGen_t}{(1+i)^t}} \tag{5.37}$$

Here, $EGen_k$ is the amount of energy produced in year *k*, *i* is the discount rate, and *N* is the number of years over which the *LCOE* is calculated OM_k is the O&M cost in year, *k*, and COE_k is the cost of generate energy or electricity and each parameter is given in the *kth* year. Based on the derivation of *LCOE*, the *LCOE* model must incorporate all financial (cost) parameters that contribute to the *TLCC*. Notice that in the equations above contrary to the appearance, the costs and not the energy are discounted. In RES applications, the *LCOE* is calculated based on the expected cash flows for O&M and capital expenditures. In such applications it is also important to include, besides the cash flows, used to determine the actual money spent and costs involved in an RES project, the penalties, production losses and tax credits in the *LCOE* model and calculations.

Example 5.13: A nuclear power plant has an initial cost of $1.25 billion and an expected lifetime of 36 years, is producing 2.85 billion of kWh per year, with a capital cost of 6.7% per year at a total fuel and O&M costs $35.0 million per year. If the annual return to the investors is 9.5% of the O&M and capital repayment costs, calculate the *LCOE*.

Solution: The annualized capital cost, estimated from Equation (5.15), the annual capital cost and the total annual cost are, respectively:

$$\frac{A}{P} = \frac{0.067 \times (1+0.067)^{36}}{(1+0.067)^{36} - 1} = 0.074184$$

$$ACC = 0.074184 \times 1.25 \times 10^9 = \$92.73 \text{ million}$$

and

$$TACC = 45.0 + 92.73 = \$137.73 \text{ million}$$

Then the investor return is \$13.084 million and levelized cost of energy is:

$$LCOE = \frac{(137.73 + 13.084) \times 10^6}{3.60 \times 10^9} = 0.040217 \text{ \$/kWh} \approx 4.022 \text{ cent/kWh}$$

When investing in an energy system or in any other engineering project, the money (capital) valued today is known, as well as from the time value of money that future costs or savings are discounted over any time horizon in terms of *present worth* (PW). The *net present worth* (NPW) of a system over the period of evaluation is called the *life cycle savings* (LCS). The LCS is the difference between the life cycle costs of a conventional fuel-only energy system and the life cycle costs of an RES or DG system with auxiliary fuel costs. We can evaluate the LCS using various mathematical methods. In the discrete methods, first calculate the annualized life cycle costs, or average yearly cash flow summed from the contributing costs and savings (in complement, it can be evaluate the annualized life cycle savings), then the time value of money is applied to place the annualized cash flows in terms of present worth (PW). Finally, the sum of the annualized cash flows in present worth is computed to arrive at the LCS or NPW. Using a continuous method of analysis, the net present worth of each cost and savings contribution for the full period of evaluation is obtained, and then directly the sum of the contributing factors to arrive at the LCS. The *LCOE* relationship discussed here is the standard method that includes four basic inputs: the capital expenditures, the operational expenditures, the annual energy production (AEP) and the fixed charge rate (FCR), a coefficient that captures the average annual carrying charges including return on installed capital, depreciation and taxes. However, the *LCOE* is not traditionally defined as a measure of all societal costs and benefits associated with power generation resources. The most common *LCOE* equation of the standard energy economics methodology includes four basic inputs: *capital expenditures, operational expenditures, annual energy production (AEP)* and the *fixed charge rate* (FCR, a coefficient that captures the average annual carrying charges including return on installed capital, depreciation and taxes). Usually the *LCOE* indicator is defined, in the electricity production field, as the produced electrical energy price in present value (\$/kWh), taking into account the power plant economical lifespan and the costs incurred into the plant construction, O&M, and the fuel(s), expressed as the generation costs over the construction and production periods, by this relationship:

$$LCOE = \frac{\left(\displaystyle\sum_{k=-N_{\text{cnstr}}}^{k=-1} \frac{INV_k}{(1+i)^k}\right)_{\text{Cnstr}} + \left(\displaystyle\sum_{k=0}^{k=N_{\text{Gen}}-1} \frac{FC_k + OM_k + DC_k + TX_k + RL_k}{(1+i)^k}\right)_{\text{Prodct}}}{\left(\displaystyle\sum_{k=0}^{k=N_{\text{Gen}}-1} \frac{EGen_k}{(1+i)^k}\right)_{\text{Prodct}}} \qquad (5.38)$$

Here, *LCOE* is expressed in (cent/kWh); INV_k is the investment made in the year, k in USD; OM_k is the O&M cost in year, k in USD; DC_k is the depreciation credit in year, k in USD; TX_k is the tax levy in year k, in USD; RL_k are the royalties paid in year, k in USD; FC_k is the fuel cost in year, k in USD; $EGen_k$ is the generated electrical energy in year, k in kWh; i is the discount (interest) rate; N_{cnstr} and N_{Gen} are the duration of the construction period and the duration of the generation period (the plant lifespan), both expressed in years.

LCOE is estimated over the lifetime period of the energy generating technology, about 20 years for wind energy and about 30 years for PV systems. The discount rates selected in Equation (5.38), as well as in previous chapter equations, depend on the costs and the available capital sources, considering a balance between the equity and the financing debt, and the risk estimates. In the energy economics practice, it is advisable and is recommended to include the inflation effects and to evaluate and analyze future expenditures, incomes and investments, in consequently, the discount rate is adjusted to the actual (real) discount rate, i_{real}, defined as the sum of the discount rate, i and the inflation rate, r_{ifl}, as:

$$i_{\text{real}} = i - r_{\text{ifl}} \tag{5.39}$$

The project or development net present value is defined as the value of all payments, discounted back to the project or development start (begging), as defined previously. If the net present value is positive, the project has a rate of return larger than the real interest rate, while if NPV is negative the project has a lower rate of return. The net present value is computed by adding the annual terms, computed by dividing, P_k, the payment in the year, k to $(1 + i_{real})^k$, and the NPV is then expressed as:

$$NPV = \sum_{k=1}^{N} \frac{P_k}{\left(1 + i_{\text{real}}\right)^k} \tag{5.40}$$

5.4 Economic Analysis of Major Renewable Energy Systems

The main problem of calculating the costs consists of the fact that there is not one universal approach to calculate the costs. Between the renewable energy sources exist relatively large differences, e.g., specific application conditions and characteristics of the land, land size, price of fuel and labor costs, distance to the grid, regulations and permitting, etc. (Short, Packey and Holt, 1995, Kaltshmitt, Streicher and Weise. 2007). The problem of quantify the real costs residing in the fact that the costs are quite possible to accurately quantify, but question is which items to include among them, for example the problems connected with interface, the problems with information technology, etc. From financial point of view, the potential investors need to calculate these basic costs, such as an investment to a biogas station: depreciation of technology and buildings, insurance; consumption of electricity and heat by biogas station; laboratory testing, certificates; purchasing (other relevant costs) of biomass; stuff costs; water costs; and costs connected with waste disposal. There are several types of risks associated with any RES and DG project. In order to construct and operate a power plant, developers must obtain several permits. The total time required to obtain these is referred to as administrative lead time, the longer the higher risk and chance of project failure incurred. Additionally, for offshore wind and marine renewable energy developments, factors that increase lead time are the lack of experience and of communication with other sea users. Administrative risks are defined as investment risks related to approval needed from the authorities. The infrastructure required to generate power from renewable sources is capital intensive.

For renewable energy, almost all investments take place in the first stage of development, requiring the availability of capital such as equity, but also public financing support such as grants and soft loans enabling investments. If this is not available, this can lead to capital scarcity. Main reasons for capital scarcity are under-developed and unhealthy local financial sector or financial distresses. Furthermore, limited experience with renewable energy projects combined with tighter bank regulations could result in inability of developers to finance their projects. Risks that arise from the scarcity of available capital are called financing risks. Technical and management risks refer to the availability of local knowledge and experience and to the maturity of the used technology. Uncertainties arise due to the lack of adequate resource assessment for future potential or the use of new technologies. The probability that a loss incurs due to insufficient expertise, inability to operate, inadequate maintenance of the plants, lack of suitable industrial presence and limitation of infrastructure are parameters that are included in technical and management risks. To become operational, the RES projects should be connected to the electricity grid. This process includes the procedure to grant grid access, connection, operation and curtailment. The convenience of connecting is influenced by different factors, such as the capacity of the current grid, the possibilities for expansion, planned reinforcements and whether the connection regime allows for RES priority. If this is all well-regulated, new RES projects can be connected to the grid at low risk. However, in the case that the conditions are less convenient and grid connection lead times are long and the connection procedure is unclear, grid access risks can seriously affect the project. Often, these risks are due to an inadequate grid infrastructure for RES, suboptimal grid operation, lack of experience of the operator and the legal relationship between grid operator and plant operator. Support mechanisms are needed for renewable sources to be competitive, as there is still a cost gap between renewable and conventional energy technologies. Uncertainties arise when the policy design does not account for all revenue risks, such as RES yield, demand and price fluctuations. Market design and regulatory risks refer to the uncertainty regarding governmental energy strategy, power market deregulation and liberalization.

The benefits can be divided into direct type and indirect type. Direct benefits are savings associated with the purchase of primary energy raw materials and to create of new less risky portfolio. Indirect benefits are not easy to define and quantify, and sometimes even questionable. This is a space for discussions, lobbying, policy, as well as changes in the innovation potential and knowledge in technology, computer technology, high-tech industries, electronics, energy conservation, new areas for scientific research (natural, social and economic sciences), opportunities for new jobs, creation of a new more progressive macroeconomic environment related to new technologies, benefits related to environmental technology perceptions with direct impact, especially on younger generations. However, all the indirect benefits have one issue in common it is very hard to quantify them. Major problems connected with investing that must be known before realization of investment include: (a) technology, (b) legislation, (c) economic conditions, transport and installation costs, (c) weather and climate conditions, (d) property rights, (e) environmental protection issues, (f) biodiversity and pant production for food industry, (g) the availability of needed skilled engineers, technicians and workers or (h) if there are required laboratory testing of the energy sources characteristics. There also need to be taken into considerations if there are needs for regular specialized service of the used technology, legal conditions (e.g., legal conditions are different in different countries), to follow local and regional legal regulations for support the RES and DG projects, for example state-purchasing price for heat and electricity produces from renewable resources, the price is guaranteed for the same period, state aid available also for producers which does not deliver the energy to public energy network, mandatory purchasing by energy suppliers, priority to connection to distribution network, a financial support, etc. Possibilities for co-financing the RES and DG investments include: general government grants and aids and structural funds of state and/or federal agencies.

Renewable energy conversion systems, such as wind turbines, PV and solar-thermal power plants, can produce electricity for consumption on or near the site, to sell to a utility, or both. The higher the selling price, the more economically feasible the project becomes. In general, where there are systems of wind turbines or PV arrays, the owner are using part of the generated energy and sell the excess to the utility. The electricity used on site is displacing the electricity at the retail rate. For those regions that have net energy billing (in general, the size is limited to the small wind turbines or PV units), even the energy fed back to the utility is worth the retail rate. If larger energy amounts are produced than used during the billing period, then that energy is sold for the avoided costs. For the locations where the retail rates are higher than the avoided cost paid for excess energy fed back to the utility, economic feasibility improves with increased on-site electricity consumption. The price paid by the utility is either negotiated with the utility, set by law, or decided by a public regulatory agency. The costs of operation and maintenance represent the off time and the costs of the parts and repairs. Until system reliability and durability are known for long time periods, the costs of repairs are quite difficult to estimate. Estimates are often made on costs of repairing the most probable failures. Insurance costs may be complicated by companies that are uncertain about the risks involved in a comparatively new technology. Inflation has its main impact on the expenses incurred over the system lifetime. Overall renewable energy technologies tend to substitute fuels, leading the *fuel savings* (*FS*), while RES savings are calculated sum of *FS* less the fixed and variable costs for the intended renewable energy conversion system (C_{RES}). The *renewable energy savings* are further evaluated over the tie horizon (evaluation period) that include any load period (life) tied to the RES or DG system and large part of the RES or DG lifespan. Such approaches were initially used to evaluate domestic solar thermal systems and extended later to PV systems.

The factors, concepts and economic methods discussed in the previous chapter sections are applied with a few adjustments to the project evaluation and economic viability in the case of DG and RES projects. For example the economic analysis in the case of renewable energy projects does not include the full spectrum of drivers that affect renewable energy prices. For example, not all policy incentives or factors from underlying economic conditions (such as an economic recession), the cost of building interstate transmission, or potential integration costs are included into all economic analyses. These important variables can significantly impact renewable energy generated electricity costs by reducing total costs, adding expenditures, delaying projects, or halting projects altogether. *LCOE* is highly sensitive to even small changes in the input variables and assumptions, so careful assessment and validation of assumptions used for different technologies when comparing the *LCOE* are important. The cost of energy (value of the energy produced by the RES or DG unit) gives a levelized value over the system lifetime (e.g., 20–25 years for wind energy and PV systems). The cost of energy (COE) is primarily driven by the installed cost and the annual energy production. The COE is one measure of economic feasibility, and is compared to the price of electricity from other energy sources (primarily the utility company) or the price for which RES generated electrical energy can be sold. A simple COE relationship that is often used for wind energy and solar energy technologies, easily adaptable to any other RES and DG systems. The relationship to estimate the system COE is determine by the capital expenditure (CapExp), or the initial cost, usually in $/kWp, the r_{FCR} is the fixed charge rate, expressed in %, the operational expenditures (OprExp, the operation and maintenance costs), expressed in $/kW/year, the annual energy production, AEP, expressed in kWh/year and levelized replacement costs (major repairs), *LRC*, expressed in $/year, as:

$$COE = \frac{CapExp \times r_{FCR} + LRC + OprExp}{AEP} \qquad (5.41)$$

Example 5.14: A 2.0 kW wind turbine is installed at a ranch, the initial cost is $3200.00, the fixed discount rate is 7.5%, the expected annual generated electricity is 6000 kWh/year, the averaged LRC is $150/year and the annual operational expenditure is $0.008/kWh. Determine the turbine COE.

Solution: By using Equation (5.41), the COE is:

$$COE = \frac{3200 \times 0.075 + 80}{4500} + 0.008 = 0.079111 \approx 0.08 \; \$/kWh \text{ or } 8 \text{ cents}/kWh$$

Levelized replacement costs are distributing the major costs for major replacements and repairs over the system lifespan. It can be calculated from the replacement year and replacement costs, and their present values. Once the COE is determined, major assumptions can be made in the *LCOE* calculation. These are the discount and inflation rates, average system cost, financing method and incentives, average system lifetime and degradation of energy generation over the lifetime. For these reasons, the *LCOE* of renewable energy technologies varies by technology, country and project, based on the renewable energy resource, capital and operating costs and the technology efficiency and performances. The analysis is often based on a simple discounted cash flow (DCF) analysis. This method of calculating the cost of renewable energy technologies is based on discounting financial flows (annual, quarterly or monthly) to a common basis, taking into consideration the time value of money. Given the capital intensive nature of most renewable power generation technologies and the fact that fuel costs are low, or often zero, the weighted average cost of capital (WACC), also referred to as the discount rate in this report, used to evaluate the project has a critical impact on the *LCOE*. The standard *LCOE* equations can be simplified for each technology, which for renewable energy systems are resulting into the following equation used to calculate *LCOE*, with specific adaptations for wind, solar, geothermal, biomass, marine, hydropower, etc.

$$LCOE = 1000 \times \frac{CapExp \times r_{FCR} + OprExp_{\text{total}}}{AEP_{\text{net}}} \tag{5.42}$$

In above equation, the *LCOE* is the levelized energy cost is expressed, usually in $/MWh, the $OprExp_{tot}$ is the total expenditures, including the LRC, expressed in $/kWh/year, and the AEP_{net} is the net annual energy production in MWh/MW/year. The definitions and relationships for these parameters are:

$$r_{FCR} = \frac{i \cdot (1+i)^N}{(1+i)^N - 1} \cdot \frac{1 - (T_{\text{eff}} \times PV_{\text{dep}})}{1 - T_{\text{eff}}} \tag{5.43a}$$

$$AEP_{\text{net}} = P_{\text{sys}}(MW) \times 8760 \times CF_{\text{net}} \tag{5.43b}$$

and

$$OprExp = LLC + (O\&M)_{\text{cost}} \tag{5.43c}$$

where *r* is the discount rate weighted average cost of capital) (%), *N* is the economic operational life (yr.) or the investment lifespan, T_{eff} is the effective tax rate (%), PV_{dep} is present value of the depreciation (%), P_{sys} is the system installed power, in MW, CF_{net} is the net energy system capacity factor (%), $(O\&M)_{cost}$ is the levelized pre-tax operation and maintenance costs ($/kW/yr.), and *LLC* is annual levelized land lease cost ($/kW/yr.). The first three basic parameters into the *LCOE* equation, the capital expenditures (*CapExp*), operational expenditures (*OprExp*) and annual energy production (AEP_{net}), are enabling this equation to capture the system-level impacts from design changes (e.g., new technologies, larger systems, etc.). The fourth basic input parameter, the fixed charge rate (r_{FCR}), represents the amount of revenue required to pay the carrying charges, as applied to the *CapExp* on that investment during the expected project lifespan on an annual basis. Usually, the lifespan of wind energy, geothermal or solar energy projects is from 20 years up to 30 years, or even higher.

5.4.1 Wind Energy Economics

The most critical factors in determining whether it is financially worthwhile to install wind turbines are the installation initial costs and the annual energy production. In determining economic feasibility, wind energy must compete with the energy available from competing technologies. If the wind energy system produces electrical energy for the grid, the price for which the electrical energy is sold is also a critical factor. The main driver of the wind energy expansions is the declining in the costs of components, installation, operation and maintenance. Over the last two decades, there is steadily decline in the costs of the wind energy with a typical cost of onshore (land) wind farm of $1000/kW and about $1500/kW for offshore wind farms. However, the corresponding generated electricity costs vary due to factors, such as wind velocity variations, locations, tax credits and tax policy, country or state specific institutional frameworks, transmission, operation and maintenance costs. The wind energy growth is mainly driven by increasing energy demands, limited supply and costs of the fossil fuels, tax incentives, declining wind energy system costs and wind turbine technological advances, such as lighter blades, taller towers, or smart control, protection and management of wind energy conversion systems. At lower fossil fuels' price the wind energy, like most of renewable energy sources is not competitive and vice versa in the case of higher fossil fuels' prices. However, the wind energy generation capacity, since 2012 is about 5% of the U.S. installed generation capacity. Developers of wind energy installations are looking for an investment span of about 25 years, so any large fluctuations in energy cost may significantly affect the wind energy developments. Notice that one the major issue in any wind energy project is the transmission, while many of the best wind development areas and locations are often located in regions with little power demands and far from transmission lines. For example, one of the highest wind energy exists in the remote Aleutian Islands of Alaska, however with no transmission capabilities. Among the options to mitigate such issues is energy storage, using the generated electricity to convert water, where water is abundant in the proximity of wind farms to hydrogen and transfer the hydrogen to various options. The wind turbine sizes for residences, farms, ranches and rural applications are strongly dependent on the amount and price of electricity from the grid, if net metering is available and local infrastructure. The consumed energy (kWh) can be obtained from the monthly electric bills or directly from the local utility to obtain the monthly average use. To maximize the return on the wind system, most of the energy should be used on site, because that energy is worth the retail rate. However, net energy billing allows for larger-sized systems, as the system can be sized for producing all the energy needed on site within the billing period. A wind turbine is economically feasible if its overall earnings exceed its overall costs within a time period up to the lifetime of the system. The relatively large initial cost means that this period can last for several years, and even in some cases earnings may never exceed the costs. Of course, shorter payback periods are preferable to longer ones, and a payback period of 5–7 years is usually acceptable.

An important characteristic of the wind power generation, compared to other generation options, is the lack of point price of electricity, due to the wind variability and intermittent nature, rather having for wind energy a price interval or range, depending on the wind velocity. The costs of electricity generated by the wind energy conversion systems must include the following components: capital equipment economic depreciation, borrowed money (capital) interests, O&M costs, local, state and federal taxes, tax credit and governmental incentives (if any), royalties paid to the land owners, cost of the electricity used in standby mode, cost of energy storage units (when are used), and zero cost of wind energy as fuel. Approximately 75% of the total cost of energy for a wind turbine is related to upfront costs such as the cost of the turbine, foundation, electrical equipment, grid connection equipment, energy storage and so on, while the fluctuating fuel costs have no impact on power generation costs. Thus a wind turbine is capital-intensive compared to conventional fossil fuel fired technologies such as a natural gas power plant, where as much as 40%–70% of costs are related to fuel and O&M. In the wind energy conversion, no fuel is used so the fuel cost is zero, while the wind turbines are usually assembled into manufacturing facility and directly delivered to the wind park (farm) resulting in a short construction and installation time and costs. The standard *LCOE* equations are simplified for wind energy, resulting in the following equation:

$$LCOE = \frac{\displaystyle\sum_{k=1}^{N_{Gen}} \frac{INV_k + OM_k - PTC_k - DC_k + TX_k + RL_k}{(1+i)^k}}{\displaystyle\sum_{k=1}^{N_{Gen}} \frac{EGen_k}{(1+i)^k}} \tag{5.44}$$

Here, PTC_t is the production tax credit, TX_t is the levy tax, and CF is the wind turbine capacity factor, sometimes also called the intermittence factor, all for year t. *LCOE* is estimated for the wind energy conversion system lifespan, usually 20 years or 25 years for larger wind parks. An alternative and simpler *LCOE* relationship is:

$$LCOE = \frac{CRF}{EGen}(INV0 + OM) \tag{5.45}$$

Here, *CRF* is the capital recovery factor, as defined in previous sections, and the *OM* is the total operational and maintenance costs. Wind turbine prices vary due its size, power, technical characteristics, generator, drive-train, tower height and type, costs of the transportation and foundations, switchgear, power electronics units, or connection to the grid. Usually the price of the wind turbines is the one form the manufacturer price lists, often expressed in $/m² of the rotor swept area. One distinct wind energy generation characteristics is that the productivity and costs depend on the electricity price and not vice versa as the case in many other energy systems. Moreover, the high electricity generation costs in the wind energy are not related to the level of the installation costs, but rather by the wind energy resource quality. The operation and maintenance costs are estimated as either a fixed amount or a percentage of the wind turbine cost. In wind energy the price per unit of rotor area is used rather than the $/kW, because the rotor area is critical for energy production not the generator power. Another feature of the wind energy is that the royalties (land rents) depend of the project productivity, and are sometimes ranked as the cost of wind energy. The three indicators that are used in the wind energy economics are the equipment cost, the total installed project cost, including fixed financing costs, and the levelized cost of electricity, *LCOE*. Note that the capital costs of the offshore wind energy projects at the present time are relatively quite high, even compared to the onshore wind projects. The main reason is that the offshore wind turbine support structures are considerably more expensive than in the onshore installations. Installation cost of the turbines and support structures is also high due to the need for specialized and expensive ships. The installation of the submarine power cables is also expensive compared to installing land power lines on land. On the positive side, the energy production per square meter of swept area from offshore turbines is generally considerably higher offshore than on land, due to the excellent wind resources available offshore. Operation and maintenance, however, is usually much more expensive offshore than it is on land, primarily to the difficulties of access. When considered on a per MWh basis, O&M may be one of the most significant contributors to the overall cost of energy. The average cost ($/MWh) to produce electricity from an offshore wind energy units, operating at about 0.4 capacity factor are in the range of $100–$150 per MWh. An example of the *LCOE* of an onshore wind energy unit is presented here.

Example 5.15: A 1.2 MW wind turbine having a rotor diameter of 90 m, a capacity factor of 0.25 and lifespan of 20 years is installed in a new wind park. The overall installation cost is $2250 per kW, including grid connection and foundation. The cost of wind turbine is 60% of the installation cost, and O&M cost is 1.5% per year of the turbine cost. The tax credit per year is 7.5% of the overall installation cost, while a fixed combined tax, depreciation and royalty per year. Calculate the *LCOE* for this wind energy project.

Solution: The total (overall) wind turbine installation cost is $2.70 million, while the wind turbine cost is 1.62 million. The O&M cost is then:

$$OM_{20\text{-year}} = 0.015 \times 1.62 \times 10^6 \times 20 = \$486,000$$

The total expenditure cost (overall installation and O&M cost) is $3.186 million, and the tax credit is $11,947.5 per year. The energy produced in one year is then:

$$GenE_{yr} = 0.25 \times 1.2 \times 10^6 \times 8760 = 2.628 \text{ GWh}$$

Plugging the input data into Equation (5.43), and by using a MATLAB script, the *LCOE* value is equal to 0.035 $/kWh.

5.4.2 Solar Energy System Cost and Economic Analysis

In the case of solar energy technology and projects, it is necessary to determine its economic viability so that the users of the technology may know the importance and project viability. Two crucial parameters are in solar energy conversion system analysis, the *solar gains* and *loads*. The solar gains are defined as useful system converted energy, varying over the days, week, season and year. The useful energy is the converted energy less system losses. The loads are representing the energy needs and also are varying during the day, week, season and year. The total load, in the case of solar energy conversion system, can be separated into the fraction supplied by the solar energy conversion system, the so-called solar fraction and the rest supplied by other energy sources or by grid. The solar fraction is used as the proportionality factor in the solar energy project financing to estimate the optimal system size at a given site or location. A smaller solar fraction is entailed higher number of units and BOS components, meaning higher initial costs, according unit cost. The cost solar energy conversion systems, like any energy system is of the in discrete increments, because for example PVs are purchased in one PV panel increments, not half of module or panel, similar like in the case of electrical generators of transformers.

Maximizing the annual energy is usually the optimal system design choice and goal, but not for all solar energy applications. The reasons for this are different for the PV off-grid versus on-grid systems. For the off-grid PV systems, it is very common to design systems that are designed to maximize the winter energy output at the expense of the annual energy output, because there is usually a surplus of energy produced in the summer and it often more important to minimize either outage time or generator run time. For on-grid PV systems, the structure of system incentives, i.e., subsidies and tax credits, can have a strong influence on system design, as can the structure of utility rates. Feed-in-tariffs promote designs that maximize power yield. One-time upfront rebates and tax credits have an increased emphasis on the system capacity. Many utilities offer rates, varying by time, a trend being emphasized in the emerging smart-grid concept. Commercial rates for larger customers include demand charge components, usually much smaller than the electricity bill energy-based component, which for some customers is significant enough to influence the size of the PV system. Finally, residential utility customers often pay increasing rates for progressive blocks or energy use. Each of these considerations is influencing the system design. When performance-based benefits such as feed-in rates, power purchase energy sales, avoided energy costs and renewable energy credits (i.e., green tags) are factored in, more comprehensive and traditional figures of economic merit such as net present value, benefit-cost ratio, payback and internal rate of return are commonly used, as discussed before. However, one of the strongest influences on economics and design is the type of incentive. These are broadly termed as either capacity-based, as in upfront rebates, or energy-based, as in feed-in tariffs. Inasmuch as PV installations and solar thermal energy systems are extremely capital intensive, the estimated cost of the electricity or thermal energy generated by them is largely determined by the interest rate and installation service suppositions on which such projections are based. Notice also, that standalone PV systems are significantly more cost intensive than grid-connected systems; the former ones are quite often the most economical solutions in the remote areas or for the sites that are at a considerable distance from the electric grid, making the grid connection very expensive. However, the PV electricity from stand-alone installations is far cheaper than power obtained from a battery bank and less expensive than using a diesel generator. Thus for sites in developing countries or remote mountainous areas, stand-alone installations are the best solution in terms not only of their cost, but also from an ecological and logistics standpoint, as they are emissions-free and require no grid energy.

LCCA of solar energy technology and developments are primarily depending on some critical parameters, such as the initial investment (present value or first cost) for the construction of solar energy installation (*P*), the annual operating and maintenance (*O&M*) costs, the annual energy output in term of either thermal energy or exergy, the interest (discount) rate (*r*), tax credit and incentives, impact on the environment due to the pollutant emission by embodied energy, and the overhauling cost of renewable energy system, if any, during life of the system, energy payback time and lifetime and salvage value of the system. For example, the cost of PV generated electricity is consistently declining, while the cost of conventional generated electricity has an overall trend to increase.

Advances in solar PV cell technology, conversion efficiency and system installation have allowed utility-scale PV, as well as low power residential PV installations to achieve cost structures that are competitive with other electricity sources. The breakeven cost for PV technology is defined as the point where the cost of PV-generated electricity equals the cost of electricity purchased from the grid, being also referred to as the grid parity. Grid parity is considered when the *LCOE* of solar PV is comparable with grid electricity prices of conventional technologies and is the industry target for cost-effectiveness. The key parameters that govern the cost of PV power are the capital costs and the discount rate. Other costs are the variable costs, including operations and maintenance, while the capital cost is the most significant and provides the largest opportunity for cost reduction. The capital costs themselves fall into one of two categories: PV modules and the balance of system (BOS). The BOS costs include structural system costs (structural installation, racks, site preparation and other attachments) and electrical system costs (the inverter, wiring and transformer and electrical installation costs). Usually, the breakdowns of the capital costs for a ground-mounted system as suggested are 40% for PV module costs and 60% for the BOS costs.

Utility scale PV systems are usually built directly on the ground, with fewer systems installed on large buildings, and typically have a power range between 1 MW and 50 MW. The average cost for PV systems has declined to $2.80 per direct current watts (W_{DC}) for residential systems, to $1.85 W_{DC} for commercial, $1.03 W_{DC} for fixed-tilt utility-scale systems, and $1.11 Wdc for one-axis tracking utility-scale systems, or even lower for all types and is keeping declining. The initial investment in a PV system is the total cost of the project plus the cost of construction financing. The capital cost is driven by: area-related costs which scale with the physical size of the PV system namely the panel, mounting system, and site preparation, field wiring and system protection, grid interconnection costs which scale with the peak power capacity of the system including electrical infrastructure such as inverters, switchgear, transformers, interconnection relays and transmission upgrades, and project-related costs such as general overhead, sales and marketing, and site design which are generally fixed for similarly sized projects. The depreciation tax benefit is the present value of the depreciation tax benefit over the financed life of the project asset. Public policy which enables accelerated depreciation directly benefits the system *LCOE* because faster depreciation translates to faster recognition of the depreciation benefit. In the *LCOE* calculation the present value of the annual system operating and maintenance costs is added to the total life cycle cost. These costs include inverter maintenance, panel cleaning, site monitoring, insurance, land leases, financial reporting, general overhead and field repairs, among other items. The present value of the end of life asset value is deducted from the total life cycle cost in the *LCOE* calculation. Silicon solar PV panels carry performance warranties for 25 years and have a useful life that is significantly longer. Therefore if a project is financed for a 10-year or 15-year period, the project residual value can be significant. The value of the electricity produced over the total life cycle of the PV system is calculated by determining the annual production over the lifetime of the electricity production which is then discounted based on a derived interest (discount) rate, as expressed by:

$$EGen = \sum_{k=1}^{N} \frac{\dfrac{\text{Initial kWh}}{\text{kWp}} \times (1-SDR)^k}{(1+i)^k} \tag{5.46}$$

Here, *EGen* is the total energy generated by the PV system over the lifetime (*N*, years), and *SDR* (%) is the PV system annual degradation rate. The first-year energy production of the system is expressed in kilowatt hours generated per rated kilowatt peak of capacity per year (kWh/kWp). The kWh/kWp is a function of the amount of sunshine the project site receives in a year, the system is mounting and orientation (i.e., flat, fixed tilt, tracking, etc.), the spacing between PV panels as expressed in terms of system ground coverage ratio, the energy harvest of the PV panel (i.e., performance sensitivity to higher temperatures, or sensitivity to shadowing, low or diffuse light, etc.), system losses from soiling, transformers, inverters and wiring inefficiencies and system availability is largely driven by inverter downtime. To calculate the quantity of energy produced in future years, a system degradation rate is applied to initial system performance to reflect the wear of system components. The system degradation (largely a function of PV panel type and manufacturing quality) and its predictability is an important factor in life cycle costs as it determines the probable level of future cash flows. Finally, the system's financing term determines the duration of cash flows and impact the assessment of the system residual value. A PV power plant capacity factor is a function of the insolation at its location, the PV panel performances (primarily the temperature performance), the PV panel orientation, system electrical efficiencies and the availability of the power plant to produce power.

Example 5.16: A 2 kWp residential PV costs $6500.00 has 25 years lifetime ($N = 25$) specified by the manufacturer, has an SDR of 1% per year. After 15 years the inverter, estimated at $1950 need to be replaced. If the discount rate is $i = 0$, and assume that in first year the PV system generates 850 kWh/kW, calculate the system *LCOE*.

Solution: The total system cost is $8450.00 = $6500 + $1950, the annual cost is:

$$C_{year} = \frac{8450}{25} = \$353.8 \text{ (per year)}$$

The annual amount of electricity generated by the PV system over its lifetime, Equation (5.39) and geometric series formula is:

$$E_{year} = \frac{EGen}{25} = \frac{E_a \cdot (1-SDR) \times \left[1 - (1-SDR)^{24+1}\right]}{25 \cdot \left[1 - (1-SDR)\right]} = \frac{1700 \times .99 \times \left[1 - (1-0.01)^{24+1}\right]}{25 \times 0.01}$$

$$= 1357.9433 \approx 1358 \text{ kWh/year}$$

Then, the PV system *LCOE*, Equation (5.33) is:

$$LCOE = \frac{\$353.8}{1358 \text{ kWh}} = 0.26 \text{ \$/kWh}$$

The levelized electricity cost is higher than the conventional electricity cost; however, the O&M costs are minimal, being extra revenue for the residence owner. In regions with higher annual irradiance (solar radiation) the specific electricity cost is lower making PV systems more attractive. The costs of domestic solar water heating systems are split in one-third for the solar collector, one-third for the hot water storage tank and accessories, and the remaining one-third for the installation. Costs of such systems have also a decaling trend, however less than the costs of the PV panels. Public subsidies are reducing the costs if they are available. Most of the solar heat production costs are competing only with the electric water heating systems, while to compete with natural gas heating systems, the public subsidies are required. However, in regions with higher solar irradiations, low labor and investment costs, solar heating systems can compete with any fossil fuel based heating systems.

Example 5.17: A 300-l complete forced flat-plate solar collector has an initial cost of $7,200, flat cost. Assuming a solar fraction 50% ad an annual heat demand of 6000 kWh, the O&M costs per year of $68.50, and the conventional heating system efficiency is 82%, and 20 years lifetime, determine the COE for this solar water heating system.

Solution: The total energy r replaces by the solar water heating systems is 50% of the heat demand or 3000 kWh_{therm}, and the annual substituted electricity is: 3000/0.82 is 3658.5 kWh_{el}. The total annual system (installation and O&M) cost is then:

$$C_{year} = \frac{7,200.0 + 20 \times 75}{20} = \frac{8570.0}{20} = \$428.50$$

And the COE for this system is:

$$COE = \frac{C_{year}}{E_{year}} = \frac{428.50}{3658.5} = \$0.117 / kWh_{therm}$$

In order to evaluate the costs generated by the solar thermal heat usage, a clear and detailed description of the all capital investments, the O&M costs, tax credits, inflation rate, incentives, subsidies, panning and permitting or other legal costs cost is needed. Parabolic trough solar thermal or Dish-Stirling power plant have an average lifespan of 20–25 years, nominal capacity 5 MW and 10 MW, respectively. Specific investment costs are $9000 per kW and about $6500 per kW, respectively and overall operation accounting for 1.5% of the total capital cost. In addition a cost of about $8–$10 per m² is usually required for the collector maintenance and operation in the case of through solar thermal power plant. The cost of the electricity is about $0.18 per kWh for the through solar thermal power plant, and about $0.25 per kWh for the second thermal power plant. Installed in the high solar radiation areas an average of 2400 operation hours are usually for solar thermal power plants.

5.4.3 Bioenergy Cost Analysis

The aims of a bioenergy supply chain is to satisfy likely varying useful energy demands and to provide necessary conversion and primary material supplies with the required quality and quantity. Usually supply chains consist of a life cycle sections of biomass production or biodegradable material preparation phase, conversion, utilization and disposal. The deciding factors are economical, technical and administrative framework conditions, all having significant effects on putting a supply chain in practice. The framework conditions are depending on the supply side (bioenergy production) and the demand side (final energy provisions) on the other hand. There several options to use bioenergy for electricity and/or heat generation, ranging from biomass, biowaste and biogas to those producing biofuels for transportation. Bio-power is the use of biomass to generate electricity. There are four major types of systems: direct-fired, co-firing, gasification and small modular. In direct-fired combustion, the biomass is the second most utilized renewable power generation source in the United States. The direct-fired power plants are similar to the fossil fuel power plants. Around 7000 MW of power is produced by the biomass.

Co-firing involves replacing a portion of the coal with biomass at an existing power plant boiler, which represents one of the least costly renewable energy options. Gasification represents a thermochemical process that converts solid biomass raw materials into a clean biogas (methane) form. The biogas can be used for a wide range of energy production devices. For example, they are fuel cells and turbines; it also can be used as cooking gas.

Small modular systems are bio-power systems that have a rated power capacity up to or about 5 MW, and potentially such systems can provide power at the village level to serve many of the people and their industrial. However, there is fuel costs associated with bioenergy-based electricity generation, with the exception of the biowaste and biogas produced from waste, where there are negative fuel costs, resulting from the payments of the waste disposals.

Installation and equipment costs for the electricity generation based on most of the bioenergy options are similar to the costs that are found in the conventional power generation units. The installed costs are in the range of $1000–$2000 per kW, while the capital costs for biomass generation are ranging from about $1,000 for co-firing installation up to $7000 per kW for combined heat and power generation units. *LCOEs* are ranging from $0.05 up to $0.25 per kW, and the capacity factors are 50%–60%. The capital costs for bioenergy station are lower in the developing countries with values in the range of $600–$1500 per installed kW. However, the production costs (*LCOE*) of electricity from biogas and/or biomass are usually in the range of $0.08 per kWh to $0.12 per kWh, or often higher, being generally higher compared to the electricity production costs from fossil fuels in most of the conational power plants (i.e., grid-connected electricity generation in large-scale power generation units). For biofuels there are costs associated for the crops, biomass transportation and conversion and processing facility operation. However, for example biodiesel and petro-diesel have comparable fuel economy when used in diesel engines, from 20% up to 40% greater than the fuel economy achievable in gasoline-powered vehicles. Costs for conventional biofuels and biomass depend primarily on the prices of agricultural crops, which are relatively volatile and dependent on factors like climatic conditions, agricultural policies, intensification and mechanization of production. Biomass supply depends largely on region-specific conditions the cost for the provision of biomass to the plants can be from 10% to 60% of total biomass supply (i.e., production and provision) costs, depending manly on the plant location and the infrastructure.

5.4.4 Geothermal Energy Economics

The economics of geothermal utilization depend on many factors, each of which is specific to each application type. Ground source heat pump installations, for example, face many market challenges that differ significantly from those that affect direct use applications or power generation. These differences reflect differences in the scale of the project, the customer base, the state of the technology, the competitive milieu and the policy and regulatory environment. The economic viability of a specific geothermal power generating facility is strongly dependent on local market conditions, the engineering approach being used at the site and the resource it is utilizing. The geothermal energy capacity factors, which are an indication of how closely an installed conversion technology comes to producing the amount of power an installed generating facility is rated to produce, are in the highest bracket with values of 0.90 or even higher. Other important factors are that geothermal power production requires no fuel cycle, and the base-load nature of geothermal energy, since geothermal energy systems are not affected by intermittency imposed by externalities such as weather, diurnal cycles, or other factors, it can function indefinitely without interruption. Additionally, geothermal power plants operate under relative modest temperature (<350°C) and pressure conditions. These modest physical conditions that geothermal power plants experience are resulting in lower stresses on the facility materials, allowing longer lifetimes and less disruptive maintenance efforts and costs.

A geothermal project consists of two elements very different in nature and risks: a thermal energy to electricity conversion system, and the heat supply, i.e., the geothermal energy resource with attributes similar to oil or natural gas fields. The geothermal project risks are associated primarily to the geothermal resource and the investment is mostly up front. However, the power is supplied at predictable cost unaffected by price fluctuations. A geothermal project development typically proceeds in two parallel paths: technical operations and commercial and legal procedures. Geothermal projects that require drilling of wells, whether for power production or other applications, follow a development timeline that usually consists of several steps. The only exception is installations of ground source heat pump systems, which usually are developed in the same way as water well projects. The sequence of steps is generally as follows: obtain rights to use the resource and the permits that allow the exploration, conduct an exploration and resource assessment program, drill exploration wells to refine resource assessment, then drill production wells and complete a feasibility study, undertake construction of facility, and finally begin operation and energy production. Conducting a resource assessment program during the early stages of a project satisfies two requirements. First one consists of the development of a preliminary database, supporting rigorous analysis of the size and characteristics of the resource being considered. Such information will provide a preliminary indication of whether the resource is sufficiently large to support the intended use. It also provides a basis for identifying potential complications in the project that could increase cost and risk. A rigorous, scientifically based resource assessment is also a prerequisite for obtaining project funding. A drilling program designed on the basis of a good preliminary resource assessment will target locations that are likely to have high geothermal gradients. If the resource assessment has concluded that there is an accessible resource suitable to meet the needs of the project, the next step is to drill a production well into the reservoir in order to establish that flow rates, temperatures and other operational conditions are suitable. Drilling production wells often require satisfying additional permitting and regulatory requirements. Power costs of geothermal projects consist of three main components: capital cost component (including cost of money), operations and maintenance (O&M) cost component (not counting debt service, which is included under the capital cost component) and make-up well drilling cost component. The factors that determine the geothermal power costs can be grouped into four categories: economy of scale, well productivity characteristics, development and operational options, and macroeconomic climate. In general, economy of scale allows both unit capital cost (in U.S. dollars per kilowatts installed) and unit O&M cost (in ¢/kWh) to decline with increasing installed capacity. The unit capital costs are estimated to vary from $1500 per kW to $3000 per kW depending on project size, location and other project-specific criteria. For the smaller project sizes, about 5 MW, a reasonable average unit capital cost of $2500 per kW and for the larger power stations, sizes of 150 MWp, the average cost of $1800 per kW is reasonable for preliminary estimates.

Costs of geothermal power stations are highly dependent on the geothermal resources, with large cost variations during the drilling phases. The overall U.S. costs are in the range of $15,000–$6000/kW, which are leading to generated electricity prices in the range of $40–$80/MWh with variations between various generation technologies, countries and locations. The costs per kW of installation are lower in the regions with high geothermal potential, such as New Zealand and Philippines. However, the overall costs and price tendencies are to decline and costs range are different for various studies. Typically, the costs of the power station, surface facilities and power transmission account for about 50% of the total geothermal station costs. The costs of the resource assessments are much higher for geothermal projects than other renewable energy systems, such as in the case of solar-thermal or PV projects. The O&M costs associated with geothermal power generation are lower than conventional power units and even compared with other RES technologies. The levelized cost for geothermal include the higher initial investment of the geothermal power station (large part of the total costs), to O&M costs, and zero fuel costs. Among renewable energy technologies the *LCOE* are in the range of $40/MWh–$50/MW one of the lowest fi not the lowest. In summary, the capital costs in the geothermal energy projects includes: the exploration cost, the power plant cost, gathering and injection system costs, and the cost of capital. Annual O&M cost includes personnel cost, general and administrative cost, insurance cost, supplies/consumables/engineering and laboratory services cost, well field maintenance cost, generator and turbine maintenance cost, and other equipment and maintenance cost.

5.4.5 Economic Analysis of Small Hydropower

Hydroelectric power stations, large or small are characterized by several design types, while the optimum design is primarily determined by the local conditions. The plant costs are mainly the expenditures for the plant structure components (e.g., dam, power house, gates, water intake, etc.), about 50% of the total installation costs, the electrical and mechanical components (e.g., generators, transformers, turbines, check valves, control units, etc.), about 30%–40% of the overall costs, and other project related expenses (e.g., land acquisition, permitting, planning, etc.). The lifespan of a hydropower plant is from 60 to 100 years for the structure and about 40 to 50 years for the equipment. The cost per installed kW tends to be lower for large station, $1200 to $3500 than for medium- and small-size station, ranging from $1500 up to $8000 or even higher. The electricity costs are lower for larger stations with a range from $0.06 up to $0.1 per kWh, with tendency to decrease over the years. The annul operation and maintenance costs are from 1% up to 4% of the overall investment. An important step in the financial review is the preparation of estimates for year-by-year project cash flows that account for project costs and benefits. Small hydropower project costs are cash outflows and include the investment costs incurred during construction and the investment and operating costs incurred during a 20 to 30 year operating period. Non-cash environmental costs such as loss of fishing or the scenic value of waterfalls that do not affect the projects financial performance are excluded from the financial analysis, but should be included in any economic analysis. Project benefits are the expected year-by-year cash inflows plus a final year residual value of the scheme. For small hydroelectric generation projects the data for assessing financial and economic performance comprises a relatively small number of variables. The data is normally developed during the project planning process and can be entered into a project financial model and evaluated.

Example 5.18: Find the levelized energy cost, *LCOE* for small hydropower station installed on a waterfall that has an average 6.0 m³/s flow rate (Q), a net head (H) of 10 m, efficiency 0.85 (η), and a capacity factor of 0.72. The system capital cost is $3000/kW, the total O&M costs, including replacement costs are 8.5% of the total initial investment, the permitting and assessment costs where 20% the initial installation cost, and the salvage value 3.5%. The small hydropower station lifetime is estimated at about 40 years. The investment was through a governmental grant, so no interest is paid.

Solution: The power of the hydropower station is computed, using relation of Chapter 8 (Volume 1), and the installation and equipment costs, the total investment (including permitting and assessment), the total LCC are:

$$P = 9.805 \times \eta \cdot Q \cdot H = 9.805 \times 0.85 \times 6 \times 10 = 500.055 \text{ kW}$$

$$C_{\text{inst}} = 3000 \times P = 3000 \times 500 = \$1,500,165.00$$

$$INV0 = C_{\text{inst}} \cdot (1+0.2) = 1,500,165 \times 1.2 = \$1,800,198.00$$

and

$$LCC_{40-\text{year}} = 1,800,198.0 \cdot (1+0.085-0.035) = \$1,890,208.00$$

The annual generate electricity, the annual *LCC* and the *LCOE* are then:

$$LCE_{\text{year}} = 0.60 \times 8760 \times 500.055 = 2,628,289.0 \text{ kWh/year}$$

$$LCC = \frac{LCC_{40-\text{year}}}{40} = \frac{1,890,208.0}{40} = \$47,255.2$$

And, finally the *LCOE* is:

$$LCOE = \frac{LCC}{LCE_{\text{year}}} = \frac{47,255.2}{2,628,289.0} = 0.01779 \approx \$0.018 / \text{kWh}$$

5.5 Summary

Even future energy systems are hard to foresee, there is no doubts about the global energy consumption growth and the necessity for more decentralized and dispersed energy sources and systems. The potential is without any doubts very large, but how quickly their contributions meet the energy demands are critically dependent on the technological advances, economic viability or the strength of government support to stimulate research and to make renewables energy cost competitive with conventional energy sources. Government support for renewable energy projects can be justified by the long-term economic, energy security and environmental benefits they can bring, though it is essential that support mechanisms are cost effective. The future energy systems are the results of significant technological innovations, as well as various social, economic and environmental changes. Important limitations to expanding the renewable energy potential and usage are the economics and the geographic energy resource distribution. Economic analysis is an important aspect and tool of the decision-making process, while the economic efficiency is a major factor in the planning of RES and DG projects. The fundamentals of present worth and future worth analyses are very important into the planning and analysis of any engineering project and in the choice of project alternative. Fundamental to finance and invest over long timespans is the time value of money. The asset, money or cash flow worth present value or their future values are important economic parameters. Usually the solution with the best economic benefits is the choice, while the technical and/or ecological aspects are of secondary importance. The economic analysis aims is to find one system out of several alternatives or possible solutions that provides the desired energy form at the lowest cost possible. The cost analysis methods, such as payback period methods, cost-benefit analysis, life cost analysis, or cost components involved in any RES and DG project are presented and discussed I details. The effects of taxes, depreciation, tax credits and inflation on economic analysis are also discussed. The payback period approach is analyzed for the simple case, taking into account the present worth value of money, taxes, depreciation, tax credits and inflation. The other important approaches suitable for the analysis of the power factor correction project are benefit-cost analysis and rate of return analysis. It should be noted that there are other economic analysis techniques available for the economic analysis of the RES and DG projects. Interested readers are directed to the references included at the end of this chapter or elsewhere in the literature. In this chapter the economic analysis, life cycle cost analysis and other economic analysis methods and practices for renewable energy systems were used to analyze the most common RES systems. It is important to remember, that cost is one of the most important elements and decision criteria in all the DG or RES project evaluations and designs. However, the cost alone is insufficient for a complete analysis and informed decision of the project, reliability, and power quality are also essential elements of the decision together with the selected RES or DG type.

Questions and Problems

1. Briefly describe the *LCCA* method.
2. Define the interest (discount) rate and real interest rate.
3. Define the capital-recovery factor (CRF) and the compound interest factor (CIF).
4. What are the different approaches and methods available for the economic analysis of engineering projects, such as RES and DG projects and developments?
5. What is the basis for the selection of a suitable and appropriate approach or methods for a specific project or development?
6. What are the most important factors that determine the cost of energy (COE)?
7. Briefly describe the life cycle costs (*LCC*) for a renewable energy system and the main factors that are affecting the *LCC*.
8. Which of the four methods (payback period, rate of return, benefit-cost and break-even) of analyses is suitable for a stand-alone PV project and for a residential wind energy conversion? Why?
9. What are the limitations of the payback period method?
10. What are the projected costs of geothermal power found in the literature search, compared to power generated from fossil fuels?
11. Briefly describe the cost-benefit and internal return rate analysis methods.
12. Briefly describe the key steps that are usually required to successfully develop a geothermal project.
13. A wind energy company is considering adding 8 new wind turbines to its wind farm, each turbine, plus the installation cost is 1,350,000 USD. A bank is offering a loan covering the investment for 15 years with a fixed rate of 7%. How much is the loan total amount paid at the of the loan period?

14. Calculate the future worth, FW, for a deposit of $60,000. The number of investment years is 6, and the interest rate is 4.5%.

15. If an amount of $7500 annual deposit is made for 5 years, what is the future worth value of those deposits? The interest rate is 3.85%.

16. The investment in solar thermal equipment installed at a farm is $50,000. Calculate the double declining depreciation schedule. Assume 10 years for the calculation.

17. If a cash flow consists of an annuity of $175,000 over an 8-year period at 6% interest rate, calculate its future value at the end of this period.

18. An investor decided to invest $1.5 million, borrowed with 7.5% interest, in a geophysical district heating system, having a lifespan of 20 years. Determine the cost of the money. Assuming the O&M cost per year of 8.5% of the initial investment and 3.5% of the initial investment, in governmental tax credit per year and electricity saving per year of 9.575 million of kWh at an average cost of 4 cents per kWh, calculate the annual revenue of the development.

19. Estimate the cost of energy for a 2.5 kW hybrid PV-wind system, assuming the fixed cost rate 8%, and overall system LRC to be 5% of the initial cost. In order to calculate COE, you must make reasonable assumptions on the initial cost, the annual generated energy by this system and the annual O&M costs.

20. An energy corporation is planning to invest $24 million to improve the efficiency of its power plants over an 8-year period, which is projected to save $850,000 per year. The investment has a salvage value of $1.35 million and governmental subsidy of $200,000 per year. If the plant has a MARR of 12%, is the investment financial viable?

21. A residential PV system costing $11,500 has a operating period of 25 years is located at mid-latitude site. Assuming that the system generated 820 kWh/year has a maintenance cost of $100 at every 5 years and after 12 years the inverter is replaced with inverter total cost $1950. Calculate the leveled electricity cost, assuming no return on capital.

22. A small size geothermal power station is expected to generate a net income of $500,000 in each of the next 25 years. What is the present value of this cash flow if the discount rate is 6.0%?

23. An isolated rural community is investing in a small energy project, using wind and geophysics energy projects with an initial cost of $4.5 million. Assuming an annuity of $300,000 for a 25 years interval, a salvage value at the end of $300,000 and a non-reimbursable federal grant of $1.35 million, calculate the net worth of this project using simple payback method. If an MRR of 4.85% is assumed what is the project present net worth by using this approach?

24. Calculate the benefit-cost ratio for the project (investment) of the previous problem.

25. The cost of 300-liter forced solar thermal collector is $7250 and the annual O&M is $45.0. Assuming a lifetime of 20 years and annual electricity saving of 2450 kWh, compute the specific energy cost.

26. A 200 MW wind park, consisting of 100 wind turbine, 2-MW each is installed in the class 5 wind regime, the installation cost is $1800/kW, fixed cost rate is 8%, the turbine capacity factor is 0.28, and the LRC is estimated at $120,000/year per wind turbine. Calculate the wind park cost of energy.

27. A wind power park has an initial cost of $85.25 million and an expected lifetime of 25 years, is producing 2.85 million of kWh per year, with a capital cost of 7.5% per year at a total O&M costs $1.5 million per year. If the annual return to the investors is 8.5% of the O&M and capital repayment costs, calculate the *LCOE*. Compare this cost with the specific energy cost in your state.

28. Consider the installation of a residential solar PV system, during the tax year 2015. The cost of the solar PV panels and installation cost is $2000 per kW. The rating of the panel is 45 kW. Assume a tax credit of 20%. The expected benefits due to the sale of electricity per year are $6300. Calculate: (a) the simple payback period; (b) the payback period taking into account the money present worth value; (c) the payback period using the PWV method along with the tax credits; and (d) the payback period for (e) taking into account the inflation. Consider the accelerated depreciation scheme of 5 years with rates at 20%, 30%, 25%, 15% and 7.5%.

29. Find the *LCOE* for a 10 kW PV electric system if the capital cost per kW is $2850.00, the interest rate is 6%, the system lifespan is 25 years, the capacity factor is 0.18, and the annual O&M costs are 1.5% of the net installation cost.

30. Calculate the present value of 2.5 MW wind turbine over a lifetime period of 25 years. The capital cost is $1500.0/kW and the discount rate is 6%. The cost of electricity charged by the operator is 4.15 cents per kWh, and turbine capacity factor is 0.265. Calculate the specific energy unit cost.

31. Compute for the investment of Example 5.3: (a) the payback period taking into account the present worth value of money along with the tax credits, and (b) the payback period taking into account the inflation, assuming a 1.5% inflation rate. Consider the accelerated depreciation scheme of 5 years with rates at 20%, 30%, 25%, 15% and 10%.

32. A project is expected to cost $25,000 to implement. Equipment costs make up $12,500, installation is estimated at $6300 and training costs are estimated at $2200. The project proposes to eliminate 140 gallons of hazardous waste that is usually managed in small containers. Since the area no longer needs a satellite accumulation area, approximately $2,100 of savings is estimated from reduced oversight and compliance requirements. Additionally, the employees are no longer required to attend hazardous waste training anymore, which is estimated to save about $450 annually. What is the payback period?

33. A 50 MW wind park has installation costs of $1500/kW, the OM costs of $0.01 per generated kWh, a parasitic energy of 0.5% of the annual generated energy, a tax credit of $0.04 per sold kWh, a discount rate of 0.85, and a neglected inflation rate. Calculate the wind park *LCOE*, and then nest profit if 4 cents per kWh is paid by the local utility and the royalties are $0.01 per generated kWh.

34. A 3.50 GWe power plant cost $6.85 billion to build in 4 years. The capacity factor is 93% and its lifetime is 36 years. Assuming a discount rate of 6.5%, the combine fuel and O&M costs is $12.5/MWh, calculate the leveled cost of the generated electricity.

35. Repeat the *LCOE* calculation for the hybrid power system in Example 5.18, assuming that there is no governmental grant, the money for the initial investment is borrowed with 7% interest, the inflation rate is 1.2% and there is fixed annual tax credit of 3.5% of the levelized annual capital cost.

References and Further Readings

1. E. DeGarmo, W. Sullivan, and J. Bondatelli, *Engineering Economy* (9th ed.), Macmillan, New York, 1993.
2. W. Short, D.J. Packey, and T. Holt, *A Manual for the Economic Evaluation of Energy Efficiency and Renewable Energy Technologies*, National Renewable Energy Laboratory (NREL), NREL/TP-462–5173 Report, 1995.
3. T. Greadel and B. Allenby, *An introduction to LCA, In Industrial Ecology* (2nd ed.), Prentice Hall, Upper Saddle River, NJ, 2003.
4. IEA Report-2005, *Statistics Information on Renewables*, Paris, France, 2005.
5. E. Hau, *Wind Turbines—Fundamentals, Technologies, Application, Economics*, Springer, New York, 2006.
6. V. Quaschning, *Understanding Renewable Energy Systems*, Earthscan, New York, 2006.
7. J. Twidell and T. Weir, *Renewable Energy Sources*, Taylor & Francis Group, London, UK, 2006.
8. R.A. Ristinen and J.J. Kraushaar, *Energy and Environment*, Wiley, Hoboken, NJ, 2006.
9. J. Andrews and N. Jelley, *Energy Science, Principles, Technology and Impacts*, Oxford University Press, Oxford, UK, 2007.
10. E.L. McFarland, J.L. Hunt, and J.L. Campbell, *Energy, Physics and the Environment* (3rd ed.), Cengage Learning, Mason, OH, 2007.
11. M. Kaltshmitt, W. Streicher, and A. Weise (eds.), *Renewable Energy—Technology, Economics and Environment*, Springer, Berlin, Germany, 2007.
12. F. Kreith and D.Y. Goswami (eds.), *Handbook of Energy Efficiency and Renewable Energy*, CRC Press, Boca Raton, FL, 2007.
13. B.K. Hodge, *Alternative Energy Systems and Applications*, John Wiley & Sons, Hoboken, NJ, 2010.
14. P. Jackson, *Getting Design Right: A Systems Approach*, CRC Press, Boca Raton, FL, 2010.
15. IFC International Administration 2010–35 Report, *Cost and Performance of Distributed Wind Turbines*, 2010.
16. D. Newman, T. Eschenbach, and J. Lavelle, *Engineering Economic Analysis*, Oxford University Press, New York, 2011.
17. K. Branker, M.J.M. Pathak, and J.M. Pearce, A Review of solar photovoltaic levelized cost of electricity, *Renewable & Sustainable Energy Reviews*, Vol. 15, pp. 4470–4482, 2011.
18. B. Everett and G. Boyle, *Energy Systems and Sustainability: Power for a Sustainable Future* (2nd ed.), Oxford University Press, Oxford, UK, 2012.
19. L. Blank and A. Tarquin, *Engineering Economics* (9th ed.), McGraw-Hill, New York, 2012.
20. G. Boyle, *Renewable Energy—Power for a Sustainable Future*, Oxford University Press, Oxford, UK, 2012.
21. E.E. Michaelides (ed.), *Alternative Energy Sources*, Springer, Berlin, Germany, 2012.
22. F.M. Vanek, L.D. Albright, and L.T. Angenent, *Energy Systems Engineering—Evaluation and Implementation* (2nd ed.), McGraw-Hill, New York, 2012.
23. IEA Report- 2013, *Key World Energy Statistics*, International Energy Agency, Paris, France, 2013.
24. M. Kaltschmitt, N.J. Themelis, L.Y. Bronicki, L. Söder, and L.A. Vega (eds.), *Renewable Energy Systems*, Springer, New York, 2013.
25. G. Petrecca, *Energy Conversion and Management: Principles and Applications*, Springer, New York, 2014.
26. R.A. Dunlap, *Sustainable Energy*, Cengage Learning, Stamford, CT, 2015.
27. V. Nelson and K. Starcher, *Introduction to Renewable Energy (Energy and the Environment)*, CRC Press, Boca Raton, FL, 2015.
28. R. Belu, *Industrial Power Systems with Distributed and Embedded Generation*, The IET Press, Stevenage, UK, 2018.

6

Distributed Generation and Microgrids

6.1 Introduction, Distributed and Dispersed Generation

Economic, technological and environmental incentives and issues are changing the face of the electricity generation and transmission. Centralized energy generation structures, needing very complex and bulk power networks to produce and to transfer energy to the consumers, have high investment, operation and maintenance costs, while the overall system efficiency is low due to the large losses in these networks and systems. In the same time, the existing electricity and energy infrastructures are becoming older, and in addition, there are new energy supply security and environmental issues with the construction of new network components (e.g., power plants or high-voltage transmission lines). Other critical issues of the power distribution are maintaining the required power quality and supply stability, while the customers have increased power quality and supply stability demands, due to the extended use of sensitive or critical loads, which may also significantly affect the power quality. For such reasons the centralized power generation is giving way to smaller and distributed energy supplies. Distributed generation (DG) is loosely defined as a small-scale electricity generation, often located at the consumption points. Usually the connection is to the distribution networks or on the meter customer side. For most DGs the customer uses all of the energy output, and any surplus is delivered to the main grid or stored. If the customers require additional power than the DG generation, the power is taken from the grid. Distributed generation has become more present into the power systems for reasons such as an alternative to the construction of large power plants, constraints on the construction of new transmission lines, higher power quality and supply stability demands. It is also economical attractive as the cost of small-scale generation keeps decreasing with technology advances, changing the economic and regulatory environment and the electricity market liberalization. Future electricity distribution networks with large DG and renewable energy systems (RES) penetration, energy storage units, electric vehicles and customers with smart meters and controllable loads are requiring the creation of new grid architecture, *the smart grid*. Distributed energy resources (DERs) are encompassing a wide range of technologies, such as wind turbines, gas turbines, micro-turbines, PV systems, fuel cells or energy storage units. These emerging technologies have lower emissions and the higher potential to lower the overall costs. Their applications include substation power support, deferrals of transmission and distribution upgrades, high efficient combined heat and power (CHP) generation units, through the capturing of the waste heat. They are also characterized by higher power quality and the possibility of smarter power distribution. In present environment it is quite unlikely that the traditional power grids can expand rapidly and perform well enough to meet the future economy needs and the expanded electricity uses. Distributed energy resources (DER) options are vital, as the economies struggle to meet the growing of the electricity uses, and are pushing the limits of affordable power quality and grid expansion. While grid dependency has intensified, smaller generation using a broad spectrum of the generation mix has emerged as competitive alternatives to large central generation stations. Waste energy recovery is one of the DER key advantages. Heat can be productively applied to many end-uses, but when used for cooling, being particularly valuable, by displacing the high priced electricity and simultaneously lowers the site peak power requirements, i.e., both saves expensive on-peak electricity, while downsizing other system requirements.

The pressure of improving the overall power system efficiency, power quality, energy supply security, stability and environmental impacts has forced the energy industries to answer to these issues. There are also significant increases in the electricity and energy demands, while the DG and RES seems one way to cope with these demands and grid issues. DER and RES integration into the existing energy networks can result in many benefits, in addition to the reduced grid losses and environmental impacts, such as relieved transmission and distribution (T&D) congestion, peak demand shavings, voltage support, reduced price fluctuations and the deferred investments to upgrade existing systems. DERs have and are expected to have even more impacts in the future on the energy market. The energy storage systems (e.g., batteries, fuel cell stacks, flywheels or thermal energy storage) are included to harness the excess of produced electricity during the off-peak and low demand periods, for the use during the peak periods or when needed, reducing the needs for the high-cost peak-load generators. The energy storage units are also providing the dispatchability of the renewable energy sources, having no dispatchability by their own, e.g., PV arrays or wind turbines. Modern power systems generate and supplies electricity through a complex process and system, consisting of electricity generation in large power plants, usually located close to the primary energy sources (e.g., coal mines, water reservoir), far away from the large consumer centers, delivering the electricity by a large passive but complex distribution infrastructure, involving high-voltage (HV), medium-voltage (MV) and low-voltage (LV) electric networks. Power distribution operates mostly radially, in which the power is flowing in one direction, from HV levels down to customers, along the distribution feeders. Nowadays, the technological advancements, environmental policies and the expansion of the electricity markets are promoting significant changes into the electricity industry. New technologies allow the electricity to be generated in smaller size units or in DG units, located in the MV and LV grid sections, closer to the users. Moreover, the increasing RES use in order to reduce the environmental impacts and diversify the supply leads to a new electricity supply schemes. In this new paradigm, the power production is not exclusive to the generation end, but is shifted to the MV and LV networks, with part of the energy supplied by the centralized generation and part is produced by the DG units, closer to the customers. Large-scale DG integration is a main trend into modern power systems. These generators are of considerable smaller size compared to the traditional generation units (Figure 6.1).

Microgrids and DER systems are not new in many respects, however are receiving increased attention because their abilities to be used in CHP applications, to provide peak power, demand reduction, back-up power, improved power quality and ancillary services, and being expected to have even more significant impacts on the future energy systems (Belu, 2014). A wide range of DG and energy storage technologies are currently in use or in development, under consideration and research. The DER benefits in to the grid T&D include the reduction in system losses, enhanced service reliability, stability and power quality, improved voltage regulation, or relieved T&D congestion. However, the DER grid integration has several issues, such as protection, control and management, besides

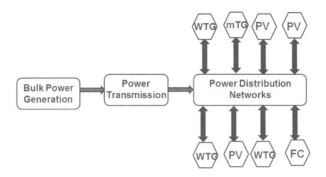

FIGURE 6.1
Distributed energy resources on the power distribution network.

other technical, economical and regulatory issues. DGs may cause the system to lose its radial power flow or the increased system fault levels. Short circuit of a power distribution network changes when its state changes or when generators in the distribution networks are disconnected, resulting in longer fault clearing time, unwanted equipment disconnection or unnecessary protection device operation. Therefore, new protection schemes for DG and power distribution networks are needed, but the issue is not yet properly addressed. A well-designed power distribution can handle the generation addition through suitable transformers, proper grounding and protection. However, there are limits to the DG and RES additions, beyond which it is critical to change the existing equipment and protection. These involve protection relays, switchgears, changes in the voltage regulation systems, revised grounding and transfer trips. The power systems supply electricity through a complex system, consisting of three major sections before the electricity reaches the users: *generation, transmission* and *distribution*. Today power systems are evolving into the *smart grid*, having several advantages and benefits, such as supply stability and security, increased efficiency and power quality, or less environmental impacts. For example, by adding CHP capabilities to DER facilities can increase the overall system efficiencies to as much as 80% or even higher, a dramatic improvement over producing electricity and heat separately, which usually has the efficiencies about 40% or lower. Among the new developments is the microgrid concept, an ideal vehicle for such addition at various facilities, spanning from residential buildings to the university campuses or military, commercial and industrial facilities. The improved efficiency results of meeting the thermal loads with the waste heat of the electricity generation. In microgrids, the CHP inclusion provides additional benefits, making them economically more attractive to the customers. The process of converting the energy input to the final output comprises often a number of intermediate conversions. Usually, several conversion stages are involved in any energy conversion process, and each stage with its conversion efficiency and the overall system efficiency is the multiple of the individual efficiencies. Clearly, the more stages are in a process the lower is the overall system efficiency, meaning higher costs, so any energy conversion process must be designed with fewer possible conversion phases. Another way to get round to the efficiency problem is to harness the process waste heat and use it elsewhere. This idea is the cogeneration basis, where the waste heat from electricity generation is used as process heat, rather than allowing it to run off into the atmosphere.

There are increased trends of the DG and RES installation, both in the utility networks and downstream of the meter. One DG key economic potential at customer premises lies into the primary fuel waste heat local conversion into electricity. Combinations of complementary generation systems based on renewable energy or mixed generators, the hybrid power system, can be further combined into a microgrid. When designing a hybrid power system, several aspects are considered, such as availability, reliability, cost or environmental impacts, while the former are the key issues. Presence of generation closer to the demand increases power quality of the electricity delivered to sensitive end-uses. The three perceived benefits, e.g., increased energy efficiency, reduced pollutant emissions and improved power quality the key drivers for DER deployment, although other benefits, such as lower transmission losses or grid expansion deferrals, are also important. While the DER applications can reduce the power system expansion needs, controlling a large number of DER units creates a daunting new challenge for operating the network safely and efficiently. This challenge can be addressed by microgrids, entities that are coordinating DERs in a consistently decentralized way, reducing the grid control burdens and providing their full benefits. A microgrid comprises a locally controlled cluster of DG and RES units, behaving from grid perspective, as a single entity both electrically and into the energy market. Microgrid is not a term that is used similarly everywhere, but more or less common all MG definitions are specifying that a microgrid is part of the future smart grid with island operation capability. Microgrid is expected to enable new opportunities for customers and business stakeholders, by using local energy resources. DER units are consisting of distributed generation, energy storages, or demand response, all having impacts on the power balance. One MG purpose is to improve power distribution reliability by reducing the amount and duration of interruptions with island operation functionality. A microgrid operates safely and efficiently within its local power distribution network, able of islanding, when needed. MG design and operation demand new skills and technologies, while power distribution containing several DER units may require considerable operational control capabilities. While not strictly compliant with the above definition, small isolated power systems are usually considered as microgrids. They apply similar technology and provide added insights into how power systems may evolve differently where they are currently rudimentary or non-existent. Usually, microgrids are spanned on the power distribution, supplying AC loads. Such microgrids with AC electric infrastructure are the AC microgrids. However, recently, the DC microgrids, exploiting a DC electric infrastructure, have gained attention due to their advantages. In either case, the energy sources are integrated into the microgrid by using power electronic interfaces. In AC microgrids, RES and DG units are integrated into the grid through AC–AC power converters (inverters) while in DC microgrids these sources are connected to the grid through DC–DC power converters. In both cases, the converters are equipped with control circuits to meet the network control objectives, such as active and reactive power flow, voltage and frequency control in AC microgrids, and power and voltage control in DC microgrids.

In a typical MG architecture (Figure 6.2), the electric network is often of radial type, connected to the power distribution through a separation device, the point of common coupling (PCC), usually a static switch. Each feeder has a circuit breaker and a power flow controller. The MG structure comprises LV network(s), controllable and non-controllable loads, micro-sources, energy storage units and hierarchical management and control schemes supported by communication, used to monitor and control micro-sources, storage units and loads. The multilevel control system core is the MG central controller (MG-CC). At a second control level, load, storage unit and micro-source controllers exchange information with the MG-CC, managing the MG operation. The exchanged data amount is usually small, e.g., messages containing controller set points, information requests sent by the MG-CC to source and storage unit controllers on the active and reactive powers, voltage levels and messages to control MG switches. MG design, operation and power distribution

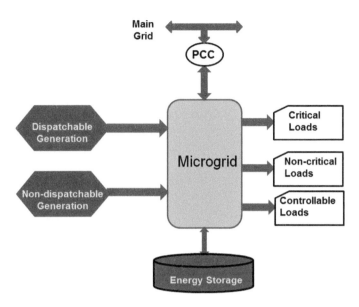

FIGURE 6.2
Microgrid architecture and structure.

with high DER penetration require considerable control capabilities. Power distribution is evolving from currently passive to an active network, in the sense that decision-making and control are distributed, while the power flow is bidirectional. This power network eases DG, RES, demand side and energy storage integration, creating opportunities for novel types of equipment and services, all of which are required to conform to new protocols and standards. The main function of active power distribution is to efficiently link generation with consumer demands, allowing both to decide how best to operate in real-time. Power flow assessment, frequency, voltage control and protection require cost-competitive technologies and advanced communication, as critical factors. The active power distribution requires the implementation of radically new system concepts. Microgrids, characterized as *the smart grid building blocks*, being perhaps the most promising approach. The MG organization is based on the control capabilities over the network operation offered by the increased DG and RES penetration at the distribution level, allowing the MG to operate when isolated from the main grid, in case of faults or other external disturbances or natural disasters, thus increasing the supply quality.

To meet the increased energy demands and the environment and natural resource preservation, the power grids have to incorporate RES and DG units to a greater degree than today. Furthermore, due to deregulations and increased dependability, today's energy systems show an increased brittleness. To cope with the issues of the flexibility and increased brittleness new supporting information infrastructures and intelligent components of the energy systems are needed. On the other hand, autonomous off-grid electrification based on the on-site RES generation has been proven to be capable of delivering reliable and acceptable quality electricity for rural locations, islands or remote areas. Microgrids are often described as a self-contained set of local generation, power distribution assets, protection and control units and smart loads that can operate in either utility grid-connected or islanded modes. In addition to providing reliable power supply, microgrids can provide a wide range of ancillary services, e.g., voltage support, frequency regulation, power factor correction, harmonic control, spinning and non-spinning reserves. Microgrids are distributive in their nature, by including DG and RES units, conventional generation, energy storage units, protection systems, critical and non-critical loads and other elements. In order to achieve an MG coordinated performance within the scope of power distribution, it is required to perform distributed or cooperative control. MG control and management components are performing sensory, communication, management, operation and control tasks, to achieve MG required reliability levels, being that the microgrids possess self-configurable networks for communication among the MG components, leading to increased needs for information, control, real-time, automated, interactive technologies, smart metering and provision of timely information and control options to the consumers.

Microgrid bases are the DER, DG and energy storage units, demand response, all of which have impact on the power balance. One MG purpose is to improve power distribution reliability by reducing the periods of interruptions, via its island operation capability. It is useful to understand the MG definition variations for the detailed MG studies. Microgrid can be: (1) a power distribution part, having islanding capability, reducing the power outages; (2) an integration platform for the supply-side (energy sources) and demand side resources (energy storage and controllable loads) located into the local distribution; and (3) an autonomous power system able to provide power and all critical services, the islanded operation being a special emergency case. This list is showing that the MG definition is incomplete, consisting of different views. However, in summary microgrids are entities that can provide improved and more reliable energy distribution, enabling easier RES and DG integration. Microgrids may also be considered as self-managed and self-serviced power systems where islanded operation is only a special case. Differences in the MG definitions are explained by the operation drivers, e.g., the operating environment and network structure variations. For example, the fact that electricity transmission and distribution are regulated as monopoly businesses has impact. In addition to variations into the MG functionalities, the way how the wanted functionalities are achieved vary significantly. Grid configuration varies from simple radial to complicated meshed grids. Control and computing capabilities (centralized or decentralized), and the generation have also impacts on the MG definitions. Basically the MG term describes a network with the following features: (1) system has islanding capability, being normally grid-connected; (2) DER units are producing large part of the consumed electricity; and (3) power flow is bidirectional. There are still uncertainties related to microgrids, such as what are the stakeholder roles, how fast the new technology is implemented, and whether microgrids are going to be a cost effective way to limit outages. The MG future is highly dependent on the smart grid features, e.g., fast, reliable and inexpensive communication, cost effective energy storage, or DER implementation. Microgrids have been extensively investigated in a number of research and demonstration projects with considerable work undertaken on their design and operation. However, to date, the microgrids have not been implemented widely in the electricity systems, and a robust commercial justification or business case for their use has yet to be developed.

Over the last two decades there is an increased interest in the DC or hybrid (AC and DC) power networks for reasons, such as cost, efficiency, energy conservation, reduced losses, increased DG and RES uses, supply reliability, security or higher power quality demands. DC networks are used in commercial applications, for practical reasons, such as the easier DC machine speed control, compatibility with digital electronics or more reliable and simpler power systems, using DC sources and batteries. If the DC electric systems will gain interest, there will be a transition period when both AC and DC are used in parallel, while the existing loads, without significant modifications are operating independent on the voltage type. It is also of interest to study within which voltage ranges existing loads can operate with the DC power systems. For example, the resistive loads operate with DC in the full tested range without any changes. Problems may arise with the load switches, which are not designed to interrupt DC currents. On the other hand, an electronic interface is needed when to connect DC supplies with an AC system. The interface design has great significance on the DC or AC power system operation. A well-designed interface needs to provide a controllable DC-link voltage, higher power quality and transient performance during faults and disturbances, and have also lower losses and cost. Moreover, bidirectional power flow capability is desired if generation is present into the DC system, to transfer power to the AC system during low-load or high-generation conditions in the DC system. Finally, galvanic isolation is required to prevent a current path between the AC and the DC systems in the case of faults. Commercial LV power systems often have sensitive nonlinear loads, which must be protected from disturbances and outages to operate correctly. A DC microgrid or nanogrid can be preferred over an AC microgrid in cases where there is sensitive electronic equipment and the power sources are connected through power electronic interfaces, having the advantages of simpler and more efficient power electronic interfaces. In DC microgrid, energy storage, power sources and the loads are normally interconnected through DC busses.

For decades, chemical plants, refineries, military installations, college campuses and other large facilities, having the ability to generate and manage their own electricity needs, while remaining connected to the grid for supplemental or emergency needs were structured as microgrids. Today, the affordable and locally produced electricity is encouraging the MG expansion, having the potential to revolutionize the power grids, by increasing efficiency and environmental sustainability. Microgrids offer an efficient energy system, based on collocating sources and loads that can operate independently in case of outages or energy crises. MG concept, a major part of the smart grid assumes an aggregation of loads and DG units operating as a single entity to provide power and heat. The MG critical elements are: (1) *embedded distributed energy resources*; (2) *advanced energy storage*; and (3) *flexible demand*. Most of the MGs combine loads, sources and storage units, allowing the intentional islanding operation, while trying to reuse the available waste heat and energy, improved efficiency, power quality and reliability, minimizing the overall energy consumption, reduced pollutant emissions, or cost efficient electricity infrastructure replacement. MGs link several DERs into a small power network serving some or all of the customer energy needs, having the potential to accommodate new electricity demands, shifting the current electricity paradigm to one in which the local, clean energy resources supplements the power grid. However, MGs are raising several challenges, such as protection and control. MG potential advantages are: more reliable energy supply to customers in the event of a major disturbance, reduced transmission losses, preventing network congestions, enhancing the system reliability and stability. One of the main MG advantages is *economic savings*, through the MG energy portfolio. The major microgrid benefits are:

- *Sustainability*: The microgrid portfolio enables a hedge against fuel cost increases.
- *Stewardship*: The MG enables deep penetration of renewables, missions reduction, green marketing.
- *Reliability*: The microgrid actively controls the network for better reliability.

As a result of the major operation differences between large power grids and microgrids, new approaches of the unit commitment, control, protection, management and economic dispatch must be developed. The major differences between these two models are outlined in this chapter. The RES and DG penetration in microgrids is much higher than in the conventional power grids. For such reason it is difficult to forecast and predict the MG power outputs. Moreover, if the microgrid is working in connected mode, it could sell to or buy electricity from the grid depending on the prices. Although in large power grids it is possible to store energy, the large part of generated power is instantly consumed. However, due to the reduced MG size a large part of the produced energy is often stored, which is a reason for the MG extended use of the distributed storage systems. Unlike the microgrids, committing a generation unit in large power systems means bringing online a large power generator, a task entailing difficulties and time than in the case of small generators. Furthermore, small MG generators have more flexibility constraints in relation to the minimum time that a unit must be down or up. These limits are set to avoid equipment fatigue and to prevent excessive maintenance and repair costs due to the frequent unit cycling. Another major difference between large power systems and microgrids is that in the latter case the number of generators involved may easily vary throughout the time. Since the search space grows exponentially with the number of units a unit commitment algorithm, working well with a limited number of generators may often not be able to handle larger number of units. Although the unit commitment and economic objectives remain the same of that in the large power systems, the MG characteristics are making necessary to modify the existing algorithms. The MG control system is composed of several levels, and specific characteristics. Firstly, the control assigns the active and reactive powers to each microgrid element according to the load demands, by fixing the voltages and frequencies references whereas the controllers are responding to the voltage and frequency deviations. Finally, the control scheme employs a *droop control* to emulate the physical behavior of each grid element. The chapter sections are focusing on the microgrid concepts, architecture, structure, benefits, advantages and applications. Other chapter sections are dedicated to the microgrid control and protection issues, DC microgrid concepts, structure, control and protection methods. A chapter summary, critical references, questions and problems are also included.

6.2 Microgrid Concepts, Configurations and Architectures

Significant changes into the regulatory and operational climate of the electric utilities and the emergence of DG and RES systems have opened new opportunities for on-site power generation, located at user sites where the electricity and heat are meeting the customer demands with an emphasis on supply security, reliability, power quality and less harm to the environment. The DER portfolio includes generators, energy storage, load control and for certain systems, advanced power electronic interfaces, while the MG concept assumes their aggregation, as a single entity to provide power and heat, an integrated platform of supply-side generation, energy storage, demand resources and controllable loads, located typically in a local power distribution. A microgrid is typically located at the LV grid sections with installed generation capacity in the MW ranges, although there are exceptions, with a microgrid located in MV section for interconnection or other purposes. Most of the MG energy sources are power electronic interfaced to provide the required flexibility and to insure operation as a single aggregated entity. The control flexibility allows microgrid to present itself to the main power system as a single controllable entity, meeting local needs for supply reliability and security, being proposed as one of the solutions to the grid reliability and stability issues. MGs offer three major advantages over traditional electricity supply involving large generation stations and long transmission through a complex electric networks, such as:

- Application of combined heat and power, as well as any heat and energy waste recovery technology;
- Opportunities to tailor the power quality delivered to suit the end-user requirements; and
- Create a favorable environment for energy efficiency, DG and RES investments and operation.

Traditional power grid is dominated by the unidirectional power flows and production based on the large synchronous generators. The developments to the RES and DG generation lead to the new aspects such as energy storage units located into the microgrids, bidirectional power flows, customers who are using and producing energy, the prosumers. The challenges to the RES and DG integration can be faced through the MG approach. MG definition includes clearly defined electrical boundaries, local control or flexible loads, the DG and RES integration issues, secured power supply, the challenges of the bidirectional power flows, demand side participation, and intermittent generation. The MG can use the waste heat through micro-CHP, implying an integrated energy system, delivering both electricity and thermal energy, having the capacity of converting over 80% of its primary fuel into useable energy. Small local energy sources can be sited optimally for heat uses, so the distributed energy systems, integrated with DGs are very pro-CHP. Power quality and reliability are used in quantifying the levels of electrical service. Both scheduled and unscheduled outages affect the availability of services to the end-users, increasing the dependence on on-site backup power supplies, quite often very expensive. The power quality degradation has more subtle, but important effects as well. Voltage sags, harmonics and imbalances are triggered by switching events and by faults. While power quality events do not lead to electrical losses, they may degrade the end-use processes, affecting the equipment operation, performances and durability. DERs have the potential to increase system reliability and power quality due to the supply decentralization. Increase in reliability levels is obtained if DG units are operating autonomously in transient conditions, when there are outages or disturbances upstream in the electrical supply. DG units located close to the load delivers electricity with minimal losses, bringing higher values than power coming from large, central generators through the grid transmission and distribution infrastructures. With the RES and DGs, the fossil fuel dependency and on the fluctuating prices, pollutant emissions and operation costs are minimized. If, in addition, DG and consumption are integrated into a single system, the power supply reliability the increases significantly. The importance and quantification of the MG benefits must be recognized and fully incorporated within the technical, commercial and regulatory framework. However, under the present grid codes, all DG types (renewable or not) must be shut-down during the power outages. This is precisely when the on-site energy sources offer the greatest values to both the generation owners and the society. A microgrid can optimize the following aspects: power quality and reliability, sustainability and economic benefits, capacity of the off-, on-grid or in-operation dual modes. MG structure can vary a lot and variables, such as the generator type and number, grid topology impact the final solution. An MG definition is related to its functionalities, capabilities and operation, and not all microgrid functionalities need to be implemented in all microgrid operation. All of the MG variables include uncertainties, making hard to predict the final, integrated microgrid model. The MG capabilities can also vary according to the needs. However, the MG main idea is islanding capability, which can reduce outages and offer independency for customers. The microgrids are similar to a small power system which encompasses various components such as DG units, loads and storage devices that are interconnected. In terms of power type, the microgrid can be classified as an AC power system, a DC power system, or a hybrid system, as shown in Figure 6.3. When used, each microgrid type presents advantages and disadvantages, and the application can set the MG type.

The DC microgrid presents several operational advantages, such as most of the DG systems employed in the microgrids, are such as PV units and fuel cells DC supply. Storage devices have a DC output voltage, so connecting them to the DC microgrid only require a voltage regulator, as compared to an AC microgrid which additionally needs to synchronize the system by matching the voltage magnitude, the phase and the frequency to the grid. Most of the DC microgrid connected loads are the conventional ones, e.g., electronic devices, TVs, computers, lights, variable speed drives and appliances, reducing or even eliminating the needs for multiple power conversions, such as AC-to-DC or DC-to-AC, as is required for AC microgrids. The DC microgrids do not use transformers, making them more efficient, smaller in size, and reliable. There is no reactive power flow in a DC microgrid, and the voltage control is concerned with only active power flow. In an AC microgrid the voltage control is related to the reactive power flow at the same time injecting the active power. On the other hand an AC microgrid has a facility to use existing infrastructure from the utility grid, the nature of its power system and its compatibility with the utility grid. When using an AC microgrid, there is no requirement to reconfigure loads or the building power system of the supply. This implies that the AC loads are connected directly to the AC microgrid without any power conversion through an AC-DC converter interface. Also, it contributes to the utility grid stability by offering reactive power support for balancing and ancillary services, being fully compatibility with the utility grid.

6.2.1 Microgrid Features and Benefits

In theory, a microgrid can be formed in any grid section, based on the control methods and/ or on different voltage levels as long as there is enough local power production and control capability. The MG operation principle remains the same despite the variations. An MG is an aggregate controllable entity able to operate in islanded mode. The MG configurations are

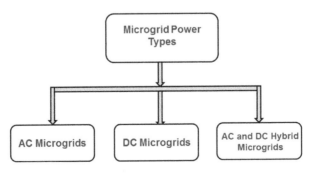

FIGURE 6.3
Microgrid classification based on the power type (AC, DC or hybrid).

separated into four cases: (1) island microgrid, (2) customer microgrid, (3) LV microgrid and (4) MV microgrid. The first type is the stand-alone microgrid where DERs and DGs provide electricity to a customer or a small community outside the grid, such as islands, isolated farms or villages far away from the grid. Second type refers to the LV customer microgrid, such as a farm or a detached building having own DER units to provide the needed electricity, operating usually in parallel with utility grid. However, in case of a fault into the grid, the microgrid is operating in island mode. In such cases, the microgrid needs energy storage to maintain constant electricity supply. Third MG type, the LV microgrid, consists of a group of customers, including anything from few consumption points to the whole LV network fed by a transformer. Power production of this microgrid type is based on several small-scale power units of different type. In the MV microgrid larger generation units are connected to the MV network section. Such microgrid can consist of a part or the whole of the HV or MV substation output. MV microgrids offer an opportunity for wind parks or other large RES facilities to produce electricity during utility grid interruption, another MG advantage. The basic principles are the same in all MG types, as well as the concepts and protocols are the same, but when it comes to more details, there are separating factors. In addition to the technical issues, separating factors are related to the market structure and the grid ownership. The solution must be feasible for all business parties, and the biggest challenge, related to all of microgrids, is how to do things economically. However, there also are several variables related to the MG operation and management. In addition to the differences in the voltage levels, there are differences in the number and type of the generators, or the type and nominal power of the DER units. Higher number of the generators increases the system complexity, requiring communication to enable the needed control capabilities. On the other hand higher number of generators increases the reliability because the system is not collapsing if one unit is not working.

A microgrid is usually connected to the grid at a PCC, appearing to the grid as a single controllable entity. The connection switch connects the microgrid to the distribution system. The microgrid enables high DG and RES penetration, without requiring power distribution re-design or re-configuration. A major MG feature is to ensure stable operation during faults and network disturbances. Autonomous operation is realized by opening the static switch, disconnecting the microgrid from the main grid. DG generations and the loads are separated from the power distribution to isolate them during faults and/or malfunctions, or by intentionally disconnecting when the grid power quality falls below certain standards. Once the microgrid is grid isolated the DG, RES and energy storage units are maintaining the prescribed voltage and frequency levels while sharing the power. Microgrids major desired features are:

- Accommodates a wide variety of generation options: distributed, intermittent or dispatchable
- Empowers the consumers to actively manage their energy use and to reduce the energy costs
- Allowing plug-and-play functionality, for switching to a suitable operation mode either into the grid-connected or islanded operation, providing the voltage and frequency protection during islanded operation and capability to resynchronize safely the MG connection to the grid.
- MGs can operate without grid connection, during islanding mode, while all loads are supplied and shared by DG units, providing solutions to supply power in the emergency and power shortages.

- Often microgrids are equipped with thermal energy units capable of waste heat recovering, an inherent by-product of fossil-based electricity generation, or energy conversion, through CHP systems.
- MGs can service a variety of loads (residential, office, industrial parks and commercial facilities).
- Provides power quality at the twenty-first-century standards, as required by the end-users.
- MGs can provide self-healing, anticipate and instantly respond to the system problems in order to avoid or mitigate power outages and power quality problems.
- Tolerant of attacks, by mitigating and standing resilient to physical and/or cyber-attacks.
- Enabling competitive, open energy market, through real-time information and lower transaction costs.
- Asset optimization, through the IT capabilities, minimizing operation and maintenance costs.

In summary, the three MG main benefits are: (1) increased power system flexibility; (2) the opportunities to tailor the power quality to the end-user requirements; and (3) an MG favorable the environment to establish higher energy efficiency and small-scale energy generation investments. Microgrids, consisting of flexible loads, energy storage and advanced control are able to integrate more renewable energy systems into the grid at the local level, being able to coordinate the DER units and to balance the power demand and supply locally and efficiently. Microgrids can power buildings, neighborhoods or entire cities, even if the grid suffers severe outages. The microgrid can provide uninterrupted power supply to customers during unexpected power outages, such as natural disasters and severe human induced faults in the utility grid, an important feature in the case of critical loads, such as hospitals, first responder units or military facilities. Less essential loads can be switched off to increase the withstand time depending on the availability of the primary energy sources. Reduction of grid interaction results in improving the self-consumption. Microgrids can reduce and control the electricity demand and mitigate grid congestion, helping to lower the electricity prices, reducing the peak power requirements. In remote areas where electricity is not available, a microgrid can reduce the investment costs for substations, transmission lines or other infrastructure components. Microgrids, with advanced control technology, can generate electricity from renewable energy sources only adding a small cost compared to the conventional grid. Additionally, as RES units are not requiring the fuel cost, the electricity tariff is not influenced by the cost fluctuations. The price is also lowered, due to the limited transmission losses and sophisticated transmission equipment requirements since the grid interaction is reduced. In remote areas microgrids can reduce the electricity bills for customers. Self-consumption and self-sufficiency coefficients are one of the parameters characterizing the generation unit performances. The self-consumption coefficient is defined as the ratio of the internal consumed energy to the total generated RES or DG energy:

$$SC = \frac{\text{DG or RES Internally Consumed Power}}{\text{Total Generated Power}} \tag{6.1}$$

Self-sufficiency coefficient is determined as the ratio of the internal consumed energy to the total load (electricity demand):

$$SS = \frac{\text{DG or RES Internally Consumed Power}}{\text{Total Electricity Demand}} \tag{6.2}$$

Flexibility is a feature of power system that is easiest to be identified in the portfolio of any power station. The inclusion of the CHP into MGs is making the power network more efficient and flexible. Since the electricity transport is more convenient than the heat, placing generation where economically attractive heat sinks exist is a desirable configuration. Optimal dispersion allows the generators to be smaller and deeply embedded with the energy demands. Unscheduled outages and malfunctions are usually more disruptive and threatening to people and property, while power quality deteriorations have mixed and less dramatic effects. While the ideal is rarely achieved in practice, the macro-grid paradigm is to provide a universal level of the power quality requirements to every load. The MG technologies and the technology selection on the demand side are holding a unique advantage, absent into the macro-grid. The RES units on both demand and supply sides have a chance at being even handed and considered, without the macro-grid issues. Since the microgrids penetration into the grids is expected to be relatively low, the initial unit commitment schedule of the centralized power generation is not expected to change dramatically in near future. However, there are changes into the economic dispatch of the most expensive (critical network units), the network losses and pollutant emissions. In order to successfully integrate RES and DG units and to fully operate microgrids several technical issues and challenges must be addressed to ensure that the present reliability levels are not significantly affected, and the DG and RES benefits are fully harnessed. In this regard, the main issues are: the schedule and dispatch of RES units under supply and demand uncertainty and the determination of appropriate levels of reserves, reliable and economical MG operation with high penetration RES levels in stand-alone operation modes; the design of appropriate demand side management schemes to allow customers to react to the power grid changes; the new market models, allowing competitive participation of intermittent energy units, and provide appropriate incentives for the investments, development of appropriate protection schemes at the distribution level to account for bidirectional power flows, voltage and frequency control methods to account for the power-electronics-interfaced DG and RES units; and last the development of market and control mechanisms that exhibit a plug-and-play feature to allow for seamless integration of DERs and DG units.

Example 6.1: A microgrid consists of four 15 kW wind turbines, 36 PV panels of 345 W each, and four battery banks with enough capacity to store the all excess energy, and the power management, control and protection units. The microgrid is supplying power to a remote atmospheric research facility having a daily power demand of 13.5 kW. Each wind turbine and PV panel has a capacity factor of 0.23 and 0.26, respectively. Each wind turbine is using internally 0.25 kW of the generated power, while each of the PV panel 7.5 W, the battery banks, power management, control and protection, is using 1.5% of the output generated power to operate. Estimate the MG self-consumption and self-sufficiency coefficients.

Solution: The average daily wind turbine output is 13.8 kW, and PV panels 1.8 kW, an overall average daily generated power of 15.6 kW. The total internally used power by wind turbines and PV panels is then 0.52 kW. The overall power output is then: 15.08 kW. The power used by the battery banks, power management, control and protection units is 0.237 kW. The MG self-consumption coefficient is then:

$$SC = \frac{0.52+0.237}{15.6} = 0.0485 \text{ or } 4.85\%$$

And the MG self-sufficiency coefficient is:

$$SS = \frac{0.52+0.237}{13.5} = 0.0561 \text{ or } 5.61\%$$

6.2.2 Summary of the Microgrid Concepts, Issues, Objectives and Benefits

A microgrid is simply defined as an aggregation of generators, energy storage units and loads that can take the form of a shopping center, industrial park, campus or a large building. To the utility, a microgrid is a controllable electric load, that can be constant, can increase at night when the electricity is cheaper, or can be held at zero during the grid stresses. An MG supersedes all the combined advantages of individual DG sources, while including at a smaller scale all the grid advantages. Microgrids are self-sufficient energy networks with autonomous control, communication and protection, being capable of providing support to the transmission network in grid-connected mode and with capacity in excess of coincident peak demand. Microgrids behave as a coordinated entity networked by advanced power electronic interfaces and control capabilities. It has been proposed that one solution to the reliability and stability issues is to take advantage of the MG technologies. A *microgrid*, a popular term within the power community, either being still vaguely defined, is offering major advantages over a traditional electricity supply involving central generation, long distance transmission and complex power distribution. One of the major MG aims is to combine the benefits of renewable energy, low-carbon generation technologies and high efficient CHP systems. The choice of a distributed generator mainly depends on the climate and the area topology. Sustainability of a microgrid system depends on the energy scenario, strategy and policy, varying from region to region. Microgrids can potentially improve the technical performance of local power distribution grid through: the reduction of the energy losses due to the decreased line power flows; voltage variation mitigation through reactive power control and constrained active power dispatch; peak loading relief of constrained network devices through selective scheduling of MS outputs; and enhancement of supply reliability via partial or complete islanding during loss of the main grid.

The waste heat recovery through CHP implies an integrated energy system, such as a microgrid delivering both electricity and heat. Small energy sources can be sited optimally for heat utilization, so a distributed system, integrated with distributed generation is very pro-CHP. The RES integration into the power system provides unique challenges to the power system engineers, due to their intermittent nature, so the central generation is still required to provide the base power supply and backup power when RES are not generating enough power. Systems with intermittent sources can experience similar problems as the systems with large and intermittent loads, but the DG can ease the burden by filling in when intermittent generation is low and by smoothing the transmission loading. This microgrid is connected to a larger system and a disconnect switch that *islands* the DG units to protect sensitive loads. A major MG component is the disconnect switch enabling the microgrid to maintain compliance with the standards, requiring the high reliability and power quality. Small electrical generators have a lower inertia and are better at automatic load following and help avoiding large standby charges, occurring when there is only a large generator.

The presence of multiple DG units makes the chance of an all-out failure very unlikely, especially when there are energy storage and backup generation units available. Such configurations create peer-to-peer networks with even no needs for a master controller to the MG operation. A peer-to-peer system implies that the microgrid continues to operate, regardless the loss of a component or a generator. When one source is lost, the microgrid regains its original functionality by new energy source additions, if available. The ability to interchange generators and components with plug-and-play capability is one of the MG main requirements. The concept can be extended to allow the generators to sit idly on when there is more capacity than needed. As the load increases, additional generators come online at a predetermined set point to maintain the power balance. Intelligent control can sense the extra generation capacity and turn-off the generators, increasing the overall efficiencies. The microgrid, is operating in a grid-connected mode. However, it is also expected to provide sufficient generation capacity, controls and operational strategies to supply at least a load portion when disconnected from the distribution system at the PCC and remain operational as an autonomous entity. The existing power utility practice does not permit accidental islanding and automatic resynchronization, due to safety reasons. However, the high DER penetration necessitates provisions for both islanded and grid-connected operation modes, smooth transition between them, to enable the best use of the MG resources. DER units, in terms of their MG interfaces, are divided into two groups. The first one includes conventional (rotary) units, interfacing the microgrid. The second group consists of electronically coupled units, using power electronic converters to provide the coupling media with the host system. The control concepts, strategies and characteristics of power electronic converters, as the DG and RES interface media are significantly different than those of the rotating machines, as listed in Table 6.1. Therefore, the MG control strategies and dynamic behavior, particularly in an autonomous operation mode, can be very different than that of a conventional power system. Figure 6.4 shows the schematic diagram of the MG building blocks that includes load, generation, storage and thermal units, implying two various component and system control levels.

TABLE 6.1

Interface and Control Types of the DER and DG Units

DER & DG Type	Primary Energy Source	DER Interface	Power Flow Control
Conventional DG	Reciprocating engines small-hydro, fixed-speed wind turbine	Synchronous or induction generators	Power flow and/or wind turbine control
Non-conventional DG	Microturbines, variable speed wind turbines, solar PVs, fuel cells	Power electronics converters	DG control, speed, output, frequency-voltage control
Energy Storage	Batteries, ultra-capacitors, flywheels, TESs	Power electronics converters	DG control, state-of-charge, speed, output, frequency-voltage control

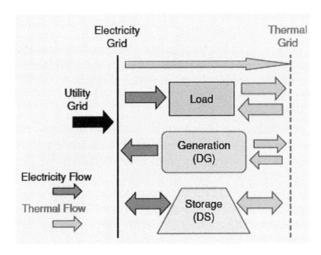

FIGURE 6.4
A general representation of the microgrid building blocks.

DG and RES units are usually connected at LV level to the host microgrid. A DG unit usually comprises energy source(s), network interface(s), and switchgear at the PCC. In a conventional DG unit (e.g., a synchronous generator driven by a diesel engine) the prime-mover is also acting also as the MG and energy source interface. For electronically coupled DG units, the coupling power converter is providing a conversion layer and control (e.g., voltage and frequency control), and is also acting as the MG interface. The input power to the interface converter from the source side is either AC at fixed or variable frequency or DC type. The MG convertor side is at the frequency of either 50 or 60 Hz for AC type. Table 6.1 outlines typical interface configurations for DG and RES power flow control for the most common primary energy sources and energy storage media. A hybrid DER unit is often interfaced to the host microgrid through an interface that includes bidirectional AC-DC and DC-DC power converters. In terms of power flow control, a DG unit is either a dispatchable or a non-dispatchable. The dispatchable DG unit output power is externally controlled, through set points provided by a supervisory control system. A dispatchable DG unit is either a fast-acting or a slow-response unit. For example, a dispatchable DG unit can use a reciprocating engine as its prime-mover, being equipped with speed control and fuel in-flow adjustment. The automatic voltage regulator (AVR) controls the internal voltage of the synchronous generator. The governor and the AVR are controlling its real and reactive power outputs, based on the dispatch strategy. In contrast, for a non-dispatchable DG unit power output is controlled by the operating conditions of its primary energy source. For example, a wind turbine is operated based on the maximum power point tracking (MPPT) concept to extract the wind maximum possible power, and its output power varies according to the wind conditions. IEEE Standards Coordinating Committee 21 supports the development of *IEEE P1547.4 Draft Guide for Design, Operation, and Integration of Distributed Resource Island Systems with Electric Power Systems*, providing approaches and practices for the MG design, operation, and integration, covering the ability to disconnect and reconnect to the grid. The guide covers the DER interconnection and microgrids, being intended for designers, operators, system integrators and equipment manufacturers. Its implementation expands the DER and MG benefits by enabling improved reliability, based on the IEEE 1547-2003 standard specifications.

6.2.3 Microgrids versus Virtual Power Plants

DG systems have been used to displace energy from conventional generating plants but not to change their capacity. Small distributed generators are not visible to the system operators and are controlled to maximize energy from renewable energy sources or in response to the heat demands of the host site and not to provide additional capacity for the power system. A *virtual power plant* (VPP) is defined as a cluster of DERs, collectively operated by a central control entity. A VPP can replace a power plant, while providing higher efficiency and flexibility, through this way of operation leads to very large generation plant margins, underutilization of assets and low operating efficiencies. The concept of VPPs has been developed to increase the DG visibility and control and to allow very large numbers of these small generation units to be aggregated, in order to take part into the markets for energy and ancillary services. In a VPP, the DG units together with controllable, responsive loads are aggregated into a controllable entity, that it is visible to the power system operator, and can be controlled to support the system operation and can effectively trade in energy markets. Such entity behaves in a power system in a similar manner as a large transmission-connected generation station. Although the microgrid and the VPP appear to be similar concepts, there are quite a few distinct differences, such as the *locality*, in a microgrid, DERs are located within the same distribution network, aiming to satisfy primarily local energy demand. In VPPs, DERs are not necessarily located on the same network, being coordinated over a wider area. The VPP aggregated production participates in trading in the energy markets. The *size* differences, installed capacities of the microgrids are typically small (from few kW to several MW), while the VPP's power ratings can be much larger. Costumer interests in microgrids are focusing on the local consumption, while VPP systems are dealing with consumption only as a flexible energy resource, participating into the aggregate power trading. Following on from the definition of a VPP as a commercial entity that aggregates different generation, storage or flexible loads, regardless of their locations, the technical VPP has been proposed, which also takes into account local network constraints. In any case, VPPs, as virtual generators, tend to ignore local consumption, except for DSI, while MGs acknowledge local power consumption and give end consumers the choice of purchasing local generation or generation from the upstream energy market. This leads to a better controllability of microgrids, where both the supply and the demand resources of a microgrid can be simultaneously optimized, leading to better DG profitability.

Through VPP aggregation of the individual DG units become visible, gain access to the energy markets, maximizing the revenue opportunities and system operation benefits from the effective DG use and increased operation efficiency. When operating alone the DG units, even in a large number, there is no sufficient capacity, flexibility or controllability to allow them to effectively take part in the system management and energy market activities. A VPP represents a portfolio of DG units, through which smaller electric units can take part in power system operation. A VPP aggregates the capacity of diverse DG units, creating a single operating entity from the parameters characterizing the generators. A VPP is characterized by a set of parameters, such as scheduled output, ramp rates, voltage regulation capability or reserve. Furthermore, as the VPP also includes controllable loads, parameters such as demand-price flexibility and load recovery patterns are characterizing a VPP. A VPP performs similarly to a large generating unit (Figure 6.5). As a VPP is composed of several DG units of various technology operating pattern and availability, its characteristics are varying significantly during operation. Furthermore, as the VPP is connected to various power distribution points, the network characteristics (topology,

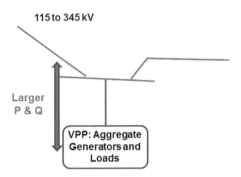

FIGURE 6.5
Aggregate VPP, larger active and reactive power transfer and control.

impedances and constraints) are affecting the VPP characteristics. A VPP facilitates the trading in the wholesale energy markets, providing services to support the transmission system management through various reserve types, frequency and voltage regulation. In the VPP concept, the activities of market participation, system management and support are described as the *VPP commercial* and *technical* activities. For the DG units in a VPP portfolio, the approach reduces the imbalance risks associated with individual unit operation and provides the benefits of resource diversity and increased capacity through aggregation. DG units can benefit from scale economies in the market participation to maximize revenue. A VPP can still represent distributed generation from various places, but aggregation of resources occurs by location, resulting in a set of generation and load portfolios defined by the geographic location. The technical VPP provides DG visibility to the system operators, allowing the DG units to contribute to the system management and to facilitate the use of controllable loads to provide system balancing at the lowest cost. The technical VPP aggregates controllable loads, generators and networks within a single electric geographical area and models its responses. A hierarchy of technical VPP aggregation can be created to characterize the DG operation connected to the grid low-, medium- and high-voltage sections, but at the grid interfaces the VPP presents a single profile representing the whole local network.

6.2.4 Microgrid Architecture, Types and Structure

Microgrids consist of several basic operation technologies, including: distributed generation, distributed energy storage, loads, interconnection switches, control subsystems and protection devices. Basically an MG structure consists of five major components: the energy micro-sources, loads, energy storage unit(s), control and protection systems and the point of common coupling. These five components are usually connected to an LV power distribution network; the MG entity incorporates a variety of micro-sources and loads that are supported by the power electronic interfaces. In order to provide synchronization and control, the mode of operation can be determined by the PCC. Distributed energy storage (DES) units are used in MGs where the generation and loads are not matching, providing a bridge in meeting power and energy requirements, and enhancing its performance in three ways. It stabilizes and permits DG units to run at a constant and stable output, despite load fluctuations. It provides the ride-through capability when there are dynamic variations of primary energy supply. It permits DG units to seamlessly operate as a dispatchable unit. The MG control is designed to safely operate the system in grid-connected and stand-alone modes. The control may be based on a central controller or imbedded as autonomous parts of each distributed generator. When the utility is disconnected the system must control the local voltage and frequency, providing the instantaneous real power difference between generation and loads, the difference between generated and the actual reactive power consumed by the load, and protecting the microgrid. In the grid-connected mode, the microgrid can be considered as a controllable load, or it can supply power and act like a generator from the grid point of view. In the islanded mode, the energy generation, storage, load control and power quality control are implemented in a stand-alone system approach. The overall energy storage denotes the total DES units within a smart microgrid that are under control of the microgrid energy management system. The microgrid concept was initially favored for its ability to provide fault isolation and ease of DG handling. Within the context of the smart grid, the MG definition is evolving into the smart microgrid, and information and communication technologies are becoming integrated to load control tasks and also being used for energy trading among communities. Currently, there is growing interest in building microgrids for campuses, military bases, or remote communities.

A traditional MG architecture consists of several energy sources and loads, with a microgrid connected to a larger system and a disconnect switch that *islands* the DG units to protect sensitive loads. A major factor is the disconnect switch, enabling the microgrid to maintain compliance with current commercial standards such as IEEE 1547. Such a switch is necessary to realize the high reliability and power quality that microgrids offer. It has been found that in terms of energy security, multiple small generators are more efficient than relying on a single large generator. Small generators have a lower inertia and are better at automatic load following and help avoiding large standby charges that occur when there is only a single large generator. Having multiple DG units available makes the chance of an all-out failure less likely, especially if there is backup generation capable of being quickly and easily connected to the system. The configuration of multiple independent generators creates a peer-to-peer network that insures that there is no master controller, and it is critical to the MG operation. Having a master controller creates a single point of failure which is not an ideal situation when the end user demands high reliability in the electrical system. A peer-to-peer system implies that the microgrid can continue to operate with the loss of any component or generator. This ability to interchange generators and create components with plug-and-play functionality is one of the main MG requirements. Plug-and-play elements imply not only that any unit is replaceable but that a unit can be placed at any system point, without re-engineering the controls. This functionality gives the benefit of placing generation close to the load, further increasing efficiency by reducing transmission losses. The concept can be extended to allow the generators to sit idly on the system when there is more electrical capacity than necessary, while when the load increases, additional generators would come at a predetermined set point necessary to maintain the correct power balance. Intelligent devices can sense when there is extra generation capacity on the system, can disconnect and turn-off the generators to save fuel and increase the efficiencies, allowing for generators to automatically drop-off the system when there is no longer a high demand of power. This is assuming a hierarchy of generators with different set points so that not all the generators drop-off at the same time.

A typical MG structure and components, with large distributed energy units, as in Figure 6.6, includes a power supplying network, several power sources, loads and energy storage units. The distributed energy sources, energy storage and the loads can be of any type, all being connected with the distributed network. The MG generated power may be

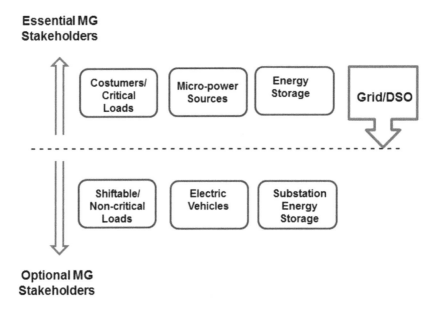

FIGURE 6.6
Microgrid stakeholders, configuration and components.

higher, equal or lower than the loads, and can have large variations. In case of larger generated power than the total loads, the excess is stored or supplied to the grid, while when
the generation is less than the loads, the energy is supplied by the energy storage and/or
grid. A microgrid is most likely to be set on a small, dense group of contiguous geographic
sites that exchange electrical energy and may be heat. The MG generators and loads are
located and coordinated to minimize the costs of the electricity and heat, while operating
safely and maintaining power balance and quality. Microgrids mange to move the power
quality control closer to the end-uses, matching the end-user needs effectively, therefore,
improving the overall efficiency of electricity supply. As the microgrids become more
prevalent, the grid power quality standards can ultimately be matched to the purpose
of bulk power delivery. Most of the MG micro-sources must be power electronic based
to provide the required flexibility, to allow the microgrid to present itself to power system as a single controlled unit, with plug-and-play simplicity to meet the customer needs.
Within microgrids, loads and energy sources can be disconnected from and reconnected
to the utility grid with minimal disruption to the loads. The required flexible load can be
met under controlled operation at higher energy efficiency, by providing both power and
heat. The key technical issues raised by microgrid include: *protection, energy management,
load control* and *power electronic interfaces*. In a basic MG architecture, the electrical system
is assumed to be radial with several feeders and several loads. The radial system is connected to the distribution system through a static switch at the PCC. Each feeder has circuit breaker and power flow controller. Besides the LV network, loads, generators, energy
storage, a microgrid has a hierarchical-type management and control schemes supported
by a communication infrastructure to monitor and control micro-sources and loads. When
the MG number reaches a large share into the LV network, similar technical benefits are
expected in the upstream grids as a consequence of multi-microgrid operation. The level
of technical benefits from a microgrid, however, depends strongly on two factors: the opti-
mality of DERs allocation and the degree of the coordination among components. Just as

effective planning of DER dimensioning and interconnection decisions can maximize unit contribution to system performance, unguided penetration of oversized units at weak grid points can create more technical problems than benefits.

6.2.5 Microgrid Key Components and Challenges

In summary all microgrids consist of DER units, power conversion equipment, communication system, controllers and energy management to obtain a flexible energy management. DER units, are involving DG and RES, distributed energy storage to meet energy demand. Controllers are needed for microgrid to apply the demands to the DER and energy storage units and to control their parameters, e.g., frequency, voltage, discharge rates and power quality. Power conversion is used to detect the microgrid running state. If the DER units are producing DC or AC voltage with other magnitude and frequency than the grid, the power electric interfaces are needed. Communication system is monitoring and controlling the MG information, interconnecting the MG elements, ensuring the management and control. Energy management system is used for data gathering and device control, state estimate and reliability evaluation of the system. It is also predicting the output power from the renewable energy systems, load forecasting and planning. The basic MG architecture presented in Figure 6.7 is showing microgrid structure, control and power flows. The power distribution networks can be classified in three types: DC line, 60 or 50 Hz AC line frequency, or in the case of special microgrids, such as ships, aircrafts, data and telecommunication centers high-frequency AC (HFAC) lines. However, most of the DERs generate DC power and the DC distribution system has no power quality issues, research on the DC microgrid system is getting importance. The DC current from DERs is transformed to the AC signal by inverters and then transmitted to the load side. There are several ways to connect DER units in an MG system. The use HFAC signals and lines in a microgrid is a new concept, still at the developmental stage. In HFAC microgrid, the DERs are connected to a common bus, and the generated electricity is transformed and HF by power converters and is transmitted to the load side, and converted back to 50 Hz or 60 AC by an AC-AC power converter. The load is connected to the distribution network,

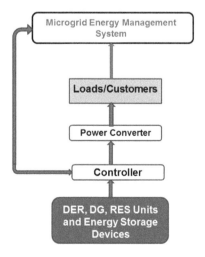

FIGURE 6.7
Microgrid operation flow and key components.

which can guarantee an effective interaction between microgrid and distribution network. At higher frequency the harmonics are filtered thus limiting power quality problems. But main disadvantage is that HFAC increases the power loss.

For power control and protection, communication systems are very important. The basic communication methods so far used in the existing communication test-beds are: power-line carrier, broadband over power line, leased telephone line, global system for mobile communication, LAN, WAN or Internet (TCP/IP), wireless radio communication, optic fiber, WiFi 802.11b, WiMAX 802.16, and ZigBee/IEEE 802.15.4 (for advanced metering infrastructure). A microgrid has to be able to import or export energy from the grid, control the power flows, and balance the voltage bus level. To achieve these objectives, small generators, storage devices and loads have to be controlled. Usually, the distributed energy sources or energy storage devices are using power electronic interfaces for MG connections. The principal control functions for non-interactive control method are the active power, reactive power, voltage and frequency control. The droop control method is often used for interactive or distributed control. In the interactive control, the power electronic converters have two separate operation modes, if they are connected to the grid follow the grid inverter, acting as a current source, while they are acting as a voltage source if the microgrid is in the islanded mode. A compromise between fully centralized and decentralized control is the hierarchical control scheme. In the context of power systems, the hierarchical control includes three control levels: primary, secondary and tertiary. These control levels differ in their response speed, the operating time frame and infrastructure requirements, e.g., need for communication.

Microgrids are not considered as a really new concept since historically first power systems where small-scale grids, while such networks already existed in remote areas, where the grid inter-connection is not possible due to technical or economic reasons. Nevertheless, combustion-based electric generators, which are fully deterministic and dispatchable, have been so far the most common generation choice. The main challenge is to make microgrids ensure the system operation without relying on fossil fuels but through an efficient coordination of different zero carbon emission technologies. Althrough microgrids may have arbitrary configurations, some elements are usually present, such as renewable energy sources, energy storage and controllable generation units. However, it should be underlined that the high RES integration, in spite of many environmental advantages, raises some technical concerns which must be solved in order to ensure system reliability. The most relevant challenges in microgrid management and control include:

1. *Intermittent power*: The power output of the renewable energy sources is variable and is determined by external factors, such as weather and different hours of the day. Therefore, there are situations where the power balance is not feasible. To overcome this issue, the microgrid is equipped with energy storage units that are charged when there is power availability and discharged when a load peak occurs. However, also the energy storage units are not fully controllable sources since they depend on their state parameters.

2. *Bidirectional power flows*: Power distribution feeders were designed for unidirectional power flows. However, the introduction of DERs to low-voltage levels can cause reverse power flows, given for instance the energy storage presence that can either absorb or deliver power, leading to complications in protection coordination, undesirable power flow patterns, fault current distribution or voltage control.

3. *Low-inertia*: Unlike the power systems where the high number of synchronous generators ensures large system inertia, microgrids are characterized by a low-inertia characteristic as most DG sources are controlled through power electronics converters. This interface is necessary since many micro-generation units generate DC power or not synchronous AC power, like wind turbines, therefore power converters, such as inverters are needed. Although such interfaces enhance the dynamic performance, the lack of synchronous and high-inertia rotating generators make the system control more critical as relevant voltage or frequency deviations can occur, especially when the microgrid is not supported by the host grid.

4. *Uncertainty*: This is another issue for the correct system coordination since neither MG generation sources nor loads are deterministic systems. Indeed, even though load profiles and weather forecasts are often available, their reliability is controversial. This factor is more critical in microgrids than in large power systems due to the reduced number of loads and the high correlation variations of the available energy resources, limiting so the averaging effect that a large electrical system has.

All these issues may be overcome through the presence of a supervising, advanced and properly designed control system that is in charge of the coordination of all microgrid systems. It has to ensure that the reliability is never compromised, especially in islanded operation, and it could also take into account the economic factors for an efficient management of the energy resources.

6.3 Microgrid Control, Protection and Management

Microgrids are defined as LV power distribution networks, consisting of DER units, energy storage, critical and non-critical loads, offering considerable control capabilities over the network operations. However, in the future the microgrid could be defined in a more general way as a part of the smart power distribution grids with islanding operation capabilities, meaning that a certain part of distribution network with DER and RES units is managed as an aggregate entity with an intelligent management system. Communication, information technologies and data management will play important roles in such smart microgrid structures. These technologies help to place a smart layer over the MG infrastructure, allowing command and control subsystem to operate and engage with the microgrid assets. Typical MG structures include: building LV microgrid (customer microgrid), separate power network, LV microgrid consisting of the feeders connected to an MV or LV distribution transformer, microgrid MV network feeder and the MV substation microgrid. The role of the MG management system can be a transition of the distribution management system into the LV levels in distribution networks. Typical control actions are network reconfiguration, by switching feeder operations, or voltage control, through capacitor switching or transformer tap changing. In addition, there are three common MG features, such as total system energy requirements are achieved efficiently, by using CHP unit(s) for heating and cooling, heterogeneous electricity security levels, quality, reliability and availability, matching various customer requirements. For the basic energy demand and supply balance, a microgrid has energy balancing capabilities and

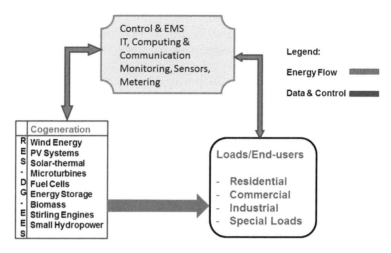

FIGURE 6.8
Microgrid structure, energy, control and data flows.

equipment, such as dispatchable loads and energy storage units (Figure 6.8), contributing to power exchange minimization or maximization of the trading profit (in case of free exchange under favorable pricing conditions). The most important MG impact consists of providing higher supply reliability and better power quality to the customers. Additional benefits are provided to the local utility by dispatchable power for use during peak power periods and postponing the distribution system upgrades. Microgrids are expected to be an essential part of the smart grids with self-healing capabilities. In addition, microgrids have a special self-healing capability, the operation in the island mode during the grid disturbances. Thereby, the MG concept can allow the reliability benefit of DER units to be realized and fulfill the future energy efficiency requirements. Technical choices made in the MG structure must be justified by the needs of normal operation, but at the same time supporting the solutions for the implementation of island operation. However, the microgrids have challenges that need to be dealt with for proper operation, such as the variable and intermittent nature of solar radiation and wind velocity, causing serious issues in voltage stability. Usually the focus is more on the MG control part to ensure stability in the system. More specifically, a control strategy based on voltage droop control that takes into account all grid dynamics is often used. There are suggestions on a power-based droop control for certain microgrids that are aiming to reduce the voltage fluctuations caused by the varying output of the solar panels or wind turbines.

Microgrids lack several attributes of the large central power systems. On a high-voltage transmission system, the reactance of the circuits is much larger than the resistance, allowing the effective decoupling of the control of the real power and reactive flows. If the resistance of a circuit is neglected (a reasonable assumption for the HV networks), the reactive power flow is controlled by the voltage magnitudes at each end of the circuit and the real power flow is determined by the relative voltage angles, P and Q flows are considered independently (Figure 6.8). This simplifying assumption cannot be made on a microgrid where the resistance of the circuits may exceed their inductive reactance. Frequency in an HV power system is determined by the balance of the generation and loads either taking kinetic energy from the spinning generators and loads or by supplying energy to the spinning machines. The speed with which the system frequency changes, when a large load is connected or a generator trips, is determined by the overall inertia of all the system spinning machines. In a microgrid, many of the micro-generators either produce DC or are connected through power converters that decouple the generator from the AC voltage. If the loads are static or connected though power converters, then no spinning masses are directly connected to the microgrid, so in islanded operation the frequency must be synthesized through the converter control systems. Controllability of intermittent RES units is limited by the physical nature of their primary energy sources. Moreover, limiting RES production is clearly undesirable due to the high investment and low operating costs of these units and their environmental benefits. Consequently, it is generally not advisable to curtail intermittent RES units, unless they cause line overloads or overvoltages. The operation strategy for intermittent RES units is therefore described as *priority dispatch*, that is, intermittent RES units are usually excluded from the unit commitment schedule, as long as they do not violate system constraints. Units with independent reactive power interfaces (decoupled from the active power output) can be included in reactive power dispatch to improve the MG technical performance.

6.3.1 Microgrid Control Methods

The two microgrid operation modes—grid-connected operation and grid-islanded operation—require specific control and protection methods and capabilities. For grid-connected operation, the microgrid is required to follow the power distribution rules and requirements, without active participation in the main power system operation, in order to maintain the power system stability. In the case of grid-connected operation, the microgrid is drawing from or supply power to the main grid, working like a controllable load or controllable energy source. In stand-alone mode, the MG components are often controlled on the basis of a decentralized approach to balance the power, ensuring each of the MG loads, the energy demands and the overall MG energy balance. From grid view, a microgrid is an aggregate entity as a generation or as a load. On the other hand, a microgrid itself is a small power system that is operating autonomously, supplying reliable and high quality electricity to customers. Therefore, a good MG control and management system is needed. The purposes of the microgrid control are: new micro-sources can be added to the system in direct ways, without any or less significant modifications of the existing equipment, the microgrid can choose operation point autonomously, the microgrid can connect to or disconnect itself from the grid in a rapid and seamless fashion, reactive and active power are controlled independently, the system imbalances and disturbances can be corrected, and the microgrid can meet the utility grid load dynamics requirements. When the grid is affected by any abnormal conditions, the microgrid is disconnected and changed to an islanded operation mode. If the microgrid is in grid-islanded mode, it must be able to handle aspects, such as supply and demand balancing, power quality as required, voltage and frequency balance, and communication among the microgrid components. The control of an isolated microgrid means balancing the generation and the demand power to keep high performance with an acceptable range of frequency and voltage amplitude. Two main control strategies are used for this purpose: a power quality inverter controller to keep the active power constant at a desirable power factor, and a voltage source inverter (VSI) controller to regulate the frequency and voltage amplitude. Notice that most of the micro-source technologies that can be installed in a microgrid are not suitable for direct connection to the electrical network because of the characteristics of the energy produced. Therefore power electronic interfaces (DC-AC or AC-DC-AC) are required in order that any microgrid to operate.

Power converters used to connect DG and RES units with the grid or other power sources must have the capability to continue to function in stand-alone mode when the other sources are unavailable, in order to supply critical loads. Power converters connected to energy storage units must be bidirectional to charge and discharge these devices. In the MG operation mode, the converter connects the power source in parallel with other energy sources to supply local loads and possibly feed power into the main grid if needed. Parallel connection of embedded generators is governed by standards, requiring that the embedded generator is not regulating or opposing the voltage at the PCC point, and the current fed into the grid is of high quality, complying with its THD content upper limits. There is also a standard limit for the maximum DC current injected into the grid. The grid injected power is controlled either by direct control of the current fed into the grid or by the power angle control. In the latter case, the voltage is set to be sinusoidal. However, using power angle control without directly controlling the output current may not be effective at reducing the THD of the output current, when the grid voltage is highly distorted. This can be an issue in the case of electric generators that are using the power angle control, raising the question of whether it is reasonable to specify current THD limits, regardless of the utility voltage quality. In practice, the converter output current or voltage needs to be synchronized with the grid, which is usually achieved by using a phase-locked loop or grid voltage zero-crossing detection. The standards are requiring that the embedded generators and power electronic converters to incorporate anti-islanding features, in order to be able to disconnect from the PCC, when the grid power is lost. It is desirable for the converter to continue to supply critical local loads when the grid is disconnected, by the anti-islanding protection schemes. In such stand-alone mode the converter needs to maintain constant voltage and frequency, regardless of load imbalance or the current quality, which can be highly distorted if the load is nonlinear. An issue in a grid-disconnected microgrid can be when two or more power converters are switching to stand-alone mode to supply critical loads. In such case, the converters need to share the load equitably, requiring additional control. There are several methods for parallel connection, broadly classified into frequency-voltage droop and master-slave methods, whereby one converter acts as a master, setting the frequency and voltage, communicating to the other converters their load share. In a microgrid, energy storage devices are needed to handle disturbances and fast load changes and to provide power in the stand-alone mode if the RES and DG units are not providing enough power. In other words, energy storage is needed to accommodate the variations of the available power or of demand, thus improving the MG reliability.

The control system of a grid-connected power converter must cater for the various operating modes. In the grid-connected mode, either a maximum power tracking system or the user specifies the power and power factor to be injected into the grid. The control system needs then to translate that into a reference demand current, if the output current into the grid is to be controlled. Alternatively, the controller needs to determine the output converter voltage and power angle, if power flow into the grid is controlled by the power angle control, and the reference signals must be synchronized with the grid. It is common to use the real and reactive power transformation to translate the measured AC voltage and current, assuming that the measured signals are pure sinusoids and the grid is balanced, which in practice is often not the case. A slight imbalance and harmonic distortions are often present, which act as disturbances that cause a deterioration of the THD of the output current. The controller must incorporate anti-islanding protection features, and seamlessly switch from grid-connected mode to stand-alone mode to supply critical loads, in parallel with other power converters, meaning the switching from a current control strategy to a voltage control strategy, a challenging procedure. An alternative is to adopt a voltage and power angle (i.e., frequency) control strategy for all operation modes, making the transition between different modes seamless. However, the cost and size of the converter and filter components remain an issue. It may also be beneficial to take an over-all system approach when designing the power converters, rather than considering the converter design in isolation. The proliferation of inverter interfaced DG units is raising issues related to the coordination of the protection relays, in both MG operation modes. The coordinated control of several RES and DG units can be achieved through various methods and techniques, ranging from a centralized control approach to a fully decentral-ized approach, depending on the share of responsibilities assumed by a central controller and the local DG controllers and flexible loads. In particular, control with limited com-munication and computing capabilities is a challenging problem favoring the adoption of decentralized approaches. Complexity is increased by the large number of distributed energy resources and the possible conflicting requirements of their owners.

The MG control functionalities can be separated into three groups, as shown in Figure 6.9. The lower level is closely related to the MG components and local control, the medium level to the overall microgrid control and the upper level to the interface to the upstream network. In a broad sense, the MG controls are classified as: (a) local controls, the basic cat-egory of the MG controls, controlling the operating points of the micro-sources and power electronic interfaces, without communication, (b) centralized controls: the micro-source and load controllers, MG central controllers and distribution management system (DMS), and (c) the decentralized controls, maximizing the autonomy of the micro-sources and loads. MG control, during islanded mode is particularly critical. In such case, the control includes voltage and frequency regulation and the load-sharing optimization. The inter-nal MG control includes all the functionalities, requiring the collaboration among the MG components: load and RES forecast, load management, unit commitment or dispatch,

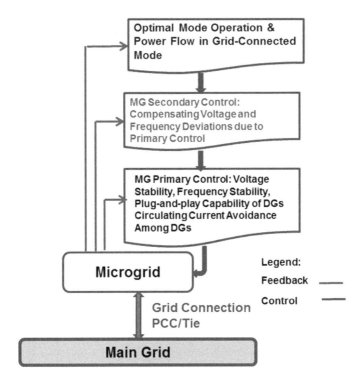

FIGURE 6.9
Microgrid hierarchical control levels.

secondary voltage and frequency control, secondary active and reactive power control, security monitoring, and black start. The local control level includes all the functionalities that are local and performed by a single DG unit, energy storage unit or controllable load, such as protection, primary voltage and frequency control, primary active and reactive power control, or energy storage management. On the other hand, the control and operation of the smart grids, including microgrids, needs to be supported by sophisticated information system and advanced communication networks. Microprocessors are used extensively within microgrids providing their ability to develop sophisticated inverter and load controllers or for other active components. An interesting characteristic of the new microprocessors is that they provide adequate processing power, communication capabilities and sophisticated software-middleware at very low costs. However, there is no general structure of MG control architecture, since the configuration depends on the MG type and the existing infrastructure. In order to analyze the MG control and management architecture a clear idea of the power distribution systems is needed, including the DMS and the automated meter reading (AMR) systems. The DMS is responsible for the monitoring of the main MV and some critical MV or LV substations. Usually the DMS is not controlling the DERs or the loads. Typical control actions are network reconfiguration, by switching operations in the main feeders, and the voltage control. The AMR system is responsible for the collection of the meter readings. By using the advanced metering infrastructure, there are capabilities of controlling loads locally, either via the meter or via the home area network, with the meter as gateway.

The proper MG control is a prerequisite and a critical condition for a stable and economically efficient operation. The main MG control functionalities are distinguished into three groups: upstream network interface, microgrid control and local component control and protection. The core interaction with the upstream network is related to market participation, specifically MG actions to import or to export energy following the MG management structure decisions. Owing to the relatively MG small size, the controller can manage several microgrids, in order to maximize the profit and provide ancillary services to the upstream network. The main MG control structure functions are: voltage and frequency regulation for grid-connected and isolated operating modes, proper load sharing, RES and DG coordination, microgrid resynchronization with the grid, power flow control and the optimizing the microgrid operating cost, and proper handling of transients and restoration of desired conditions when switching between modes. These requirements are of different significances and timescales, thus requiring a hierarchical control structure to address each requirement at a different control hierarchy level. The microgrid is usually conceived to operate within three control hierarchical levels, similar to the main power grids. The MG hierarchical control strategy consists of three levels, *primary*, *secondary* and *tertiary controls*. The primary and secondary control levels are related with the power quality operation of the microgrid while the tertiary level is related with the economic dispatch. The primary control layer constitutes the lower level of the hierarchical control architecture, having the responsibility to deal with the fastest system dynamics, operating at the fastest timescale, to maintain the MG voltage and frequency stability subsequent to the islanding process when switching from grid-connected mode. The primary control can have different configurations, but usually consist of two sequential control stages: the inverter output control and the droop control. It is essential to provide independent active and reactive power sharing controls for the DG and RES units in the presence of both linear and nonlinear loads. Moreover, the power sharing control avoids undesired circulating currents. *Inverter output control* represents the inner loop, in charge of maintaining the inverter output set points with a series of current and voltage control loops. *Droop control* is a scheme designed to quickly stabilize MG frequency and voltages, during large power variations, as well as during the islanding event. Its purpose is to set the set points for the inverter output control through a proportional action linking the variations of generated active power and reactive power to the variations of network frequency and voltages. It is essential to provide DG independent active and reactive power sharing controls, in the presence of both linear and nonlinear loads. Moreover, the power sharing control avoids undesired currents. The primary control level includes fundamental control hardware, the zero level, comprising the RES and DG internal voltage and current control loops.

The secondary control layer operates in conjunction to the primary control, allowing both to consider primary dynamics at steady state and also to have enough time to perform complex computations. The secondary control compensates for the voltage and frequency deviations caused by the operation of the primary controls and restores frequency and voltage synchronization. The purpose of the secondary control layer is not only to restore frequency and voltage deviations but it may also be responsible for the economical operation of the microgrid either in MG grid-connected and stand-alone mode. In this secondary control case, two main approaches are adopted: centralized and decentralized. The first one enables the implementation of online algorithms that can achieve relevant results in terms of efficient and secure operation. However, a centralized control is not a flexible framework since even a small change on the MG structure implies that the controller setting to be modified. On the other hand, the decentralized approach exhibits the desirable plug-and-play feature through the easy incorporation of new DER units without changing the control scheme. Nevertheless at the same time this approach cannot ensure an optimal high level coordination. Usually, in islanded mode it is preferred to implement a centralized structure since power balances must be properly managed to avoid serious frequency or voltage deviations. At the highest level and slowest timescale, the tertiary control manages the power flow between the microgrid and the main grid, facilitating optimal economic operation. This is the highest control level, typically operating in the order of minutes to hours. It has not fixed purpose, being designed to optimize power flows between different microgrids or between a microgrid and the main grid. Hence, this control layer could be needed only in grid-connected mode, while during stand-alone operation the highest coordination is usually performed by secondary control. A hierarchical control structure is shown in Figure 6.10. Notice that a droop control scheme may lead to steady-state deviations, a slower secondary control loop can be used to regulate frequency and voltages toward their nominal values.

The secondary control of AC microgrids usually is exploiting a centralized control structure. Central controllers command globally on the gathered system-wide information and require a complex and in some cases two-way communication network that adversely affects system flexibility and configurability and increases the reliability concerns by posing single point of failure. The single point of failure means that by the failure of the central controller, the whole control system fails. The distributed control is sometimes introduced to implement the MG secondary control. The distributed structure of the communication network improves the system reliability. In this control structure, the control protocols are distributed on all DG units. Therefore, the requirement for a central controller is obviated and the control system does not fail down subsequent to outage of a single unit. The secondary control of AC microgrids is similar to the tracking synchronization problem of a multi-agent system where the DG voltages and frequencies are required to track their nominal values. The dynamics of DG units in a microgrid is nonlinear and non-identical. Therefore, input–output feedback

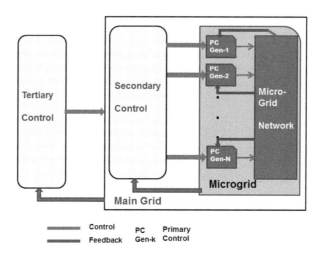

FIGURE 6.10
Hierarchical microgrid control structure.

linearization is used to transform the nonlinear heterogeneous dynamics of DGs to linear dynamics. Input-output feedback linearization transforms the secondary voltage control to a second-order tracking synchronization problem. Tertiary control is the highest level of control and sets the long-term set points depending on the requirements of an optimal power flow, e.g., based on the information received about the status of the DER units, market signals and other system requirements (Figure 6.10). The tertiary control is handled by the microgrid energy management system, which controls the power flows within the microgrid and any power exchange with the main grid. In addition to an increased network capabilities and functionality, if the DER units are allowed to operate autonomously in transient conditions, the system reliability increases. If the load suddenly increases or the system experiences a power loss, the energy sources have the capability to use local voltage, power and frequency to set a new operating set point, insuring the system stability. Grid-tied microgrids are operating in three configurations. First is the unit power control configuration where each DG unit regulates the voltage and the injected power at the connection point. Here, any load change within the microgrid is provided power through the grid since every unit regulates to a constant output power. This configuration is ideal for the CHP applications because power production depends on the heat demand. The second one is the feeder flow control, in which each DG unit regulates the voltage magnitude at the connection point and the MG power flow, while the demands are picked up by the DER units. This microgrid looks like a constant load to the grid, becoming on the utility side a dispatchable load, allowing for demand side management. The third configuration is a hybrid of the previous two with some sources regulating their output power while others regulate the MG power flow. This configuration has units operating at higher efficiency utilizing waste heat, while other units are ensuring that the power flowing from the grid stays constant under changing load conditions. Each MG control unit autonomously and effectively responds to system changes without requiring data from the loads, the static switch, or other energy sources. The basic controller uses a power and voltage feedback control, by using the real-time values for P, Q, frequency, and voltage to generate the desired voltage magnitude and angle at the inverter terminals using droop concepts. Notice, the primary control is normally operated locally, in a decentralized manner, and does not require communication. For the higher control levels (i.e., secondary and tertiary control), however, the communication plays an essential role. These communication-based control levels can be implemented with either centralized or distributed architectures.

Usually the droop control was used in within the context of controlling distributed generation focus on the idea that the inverters tie the energy source into the electrical system. In the control of parallel operation multi-converters in a distributed generator system, usually a distributed system with two distributed generation subsystems is employed. In each subsystem there are two voltage source inverters and two loads. The controller regulates the current and voltage. A droop control calculates a reference voltage and compares that operating point with the current operating conditions creating an error signal, used by the controller to adjust the control signal. The common master-slave configuration, used in droop control creates a greater chance for failure, making the configuration difficult to use as a universal controller in a microgrid with multiple energy sources. The widespread use of inverters providing power to a microgrid system can create higher-order harmonics due to internal switching. Alternative control schemes to cope with these issues are the hybrid approach that is using inverters in close proximity operate in a master-slave relationship and load sharing between distant groups using frequency droop. The master inverter uses repetitive voltage control at the common node to suppress harmonic distortion and slave inverters use repetitive control in current mode. A voltage and frequency droop control scheme for parallel inverters can be used to connect DERs to a microgrid, allowing both islanded mode operation and grid-connected operation. The droop equations are implemented by controlling an inverter in order to behave like a voltage source with virtual resistive-inductive output impedance. The droop control is used to create references for the voltage and frequency which are compared to actual values to create an error signal, used in the control methods to bring the inverter to the desired operating point. This controller introduces more complexity than the simple droop equation and any controller that includes an observer can create steady-state errors without exact parameter measurements. The energy source for the inverter is listed as a DC power supply and the experiments verified three phase operation of the control in both islanded mode and grid-connected mode. For example a voltage-power droop or frequency-reactive power boost control scheme, allowing the current controlled voltage source converters (VSC) to operate in parallel on the same microgrid was used. Each VSC in the microgrid has its own controller that sets its current references to regulate the voltage and frequency of a common microgrid bus to track the drooped references.

The control scheme also provides MG voltage and frequency regulation in islanded mode, providing over-current protection. The way that the current control, frequency estimation and droop control for the converters of a distributed system consisting of an inverter and rotating machines connected to a stiff grid are often analyzed to compare the dynamic properties of the droop methods. Inverter control in a stand-alone system may use a space vector algorithm. The advantage of using a voltage vector is that the magnitude of the fundamental voltage can control the reactive power and the angle of the vector can be used to monitor the frequency of the inverter output. In a droop-controlled DC microgrid, the power-sharing method is realized by linearly reducing the voltage reference as the output current increases. Droop control, either being widely employed as a decentralized method for load power sharing, has some limitations. The output current sharing accuracy is lowered down because of the voltage drop effect across the line impedance, an effect that is similar to the reactive power sharing of AC microgrids with inductive line impedances. To enhance the reactive power sharing accuracy in the AC microgrid with inductive line impedances, the common methods used are: the virtual impedance concept to match the unequal line impedance, the compensation method, using the remote voltage signal, in the conventional reactive power-voltage (*Q-V*) droop control, the voltage amplitude is replaced by the voltage amplitude time rate, the voltage drop across the impedance is estimated in the grid-connected operation to reach the modified slope in the *Q-V* droop control. Common loads fed by microgrids are residential, commercial, or industrial type, while the micro-sources are either conventional sources or renewable energy sources. The typical MG transmission lines and feeders are short and characterized by a non-negligible resistance component. For comparison purposes, Table 6.2 shows the typical characteristics of LV, MV and HV transmission lines. Notice that the R/X ratio in a microgrid is higher compared to their MV or HV counterparts. Faults and load imbalances are a reality within any power distribution system and the distributed generators are no exceptions. There is a need for controllers to maintain stability, and to maintain a regulated output voltage in the presence of voltage imbalances resulting from system faults or load imbalances. The controller maintains voltage within the equipment sensitivity levels, while the generators are in islanded mode, providing real and reactive power-flow control when operating in grid-tied mode. Analysis of droop control schemes for distributed generation inverters utilizing the conventional real power-frequency and reactive power-voltage control is specified by the IEEE 1547 for MG design. All microgrids are designed to operate in either grid-connected mode or islanded mode.

The diagram in Figure 6.11 is showing the main grid PCC location, the circuit-breaker which makes the microgrid transition from grid-connected mode to islanded mode, that is, the microgrid operates independently and its loads are solely supplied by its micro-sources. Energy generation sources in a microgrid are heterogeneous and can either be standard diesel generators, wind turbines, or solar panels. Given that the latter energy sources are producing either dc or variable frequency electric power, electronic power converters (i.e., inverters) are necessary for coupling the microgrid to the AC grid. Seen from the main grid, an inverter-interfaced DG unit is an actively controlled voltage source connected through impedance. During on-grid operation, the main grid ensures that the microgrid operates within the acceptable frequency and voltage ranges. The main grid manages the excess or the deficit of the MG active and reactive powers, ensuring a balance between power demand and supply. When in on-grid operation, the microgrid tracks the frequency and phase of the prevailing grid voltage at the PCC. Further, in normal grid-paralleled conditions the microgrid can be operated in a variety of control modes: based-load, load power dispatch, etc. For example, when operating in the based-load mode, a microgrid composed of inverter-interfaced DG units can be set to

TABLE 6.2

Typical Microgrid Line Parameters, Compared to the HV Systems

Line Type	R (Ω/km)	X (Ω/km)	R/X Ratio
LV Line	0.642	0.083	7.735
MV Line	0.161	0.190	0.847
HV Line	0.060	0.191	0.314

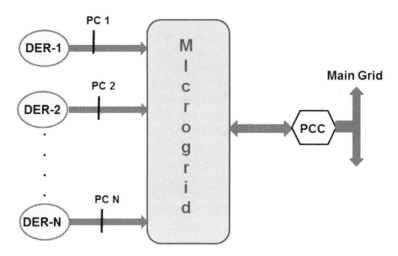

FIGURE 6.11
Conceptual design of a microgrid.

deliver a pre-specified power level to the main grid at a constant power factor. However, when operated in off-grid, the microgrid loses the reference bus at the PCC. A generalized structure of microgrid, as one shown in Figure 6.11 is often connected to the utility grid through single PCC. The isolating device is used to isolate the microgrid from the utility grid. Therefore, in off-grid mode the DG units of the microgrid must assume the role of controlling their active and reactive power supply to meet load demand since no slack bus is available anymore. Two approaches of the secondary control are: centralized control or decentralized control, a slack bus usually is connecting to the generator with the highest output power capacity, allowing the bus to compensate for any shortage or surplus of active and/or reactive power. In this context, a microgrid connected to the grid sees the latter as an infinite capacity generator. Centralized control controls all the microgrid DG units. These units are usually controlled in a topology whereby the master DG unit (typically the largest MG unit) sets the reference voltage for the slave DG units. The information shared between the DG units and the central control center requires that the communication channels have large bandwidths. The control algorithms used in this control mode ensures optimal MG operation, by optimizing renewable power input while maintaining system stability. However, the central control has several drawbacks, the most important are:

1. It has a single point failure, meaning that an eventual malfunction of the central control system leads to a system collapse, provided that the central control manages all the system micro energy sources.
2. Achieving redundancy of the central control center is expensive.
3. It necessitates expensive large bandwidth communication channels due to the high quantity of information exchanged between the DG units and the central control center.
4. The maintenance of the central control system requires complete shutdown of the microgrid.

The MG high dependency on a central control system increases its vulnerability and counteracts its sought-after distributed and independent nature. Therefore, centrally controlled microgrids do not necessarily improve the overall grid resilience. Because of the inherent limitations of central control, microgrids are usually controlled in a decentralized fashion through individual controllers associated with each DG unit. The modularity of the decentralized control provides the following advantages: (a) the individual DG controllers provide MG control redundancy, ensuring that if one controller fails, the remaining generators can compensate the loss; (2) the microgrid becomes more scalable and extendable; and (3) it is a more cost-effective solution. Microgrid decentralized control is the norm in practical applications, employing usually a droop control. Nevertheless, in an islanded microgrid, for a DG unit controlled by a droop method, only the voltage at the DG point of coupling (PC), the location where the DG unit is connected to the microgrid is controlled. However, ensuring that the voltage magnitude is controlled at the PCC does not guarantee that the voltage is within the permissible limits at downstream load buses. On the other hand, the voltage drops along the feeders, connecting micro-sources to loads, can be high enough to cause load under-voltages. For such reason, the decentralized control cannot be fully relied upon to ensure proper MG operation. By using the line phasor analysis can improve the stability of a grid with distributed generators beyond the conventional droop control. Assuming the DG and RES generators on the electrical network are three-phase voltage source inverters, they can be treated as ideal voltage sources with controllable frequency and voltage. Often the MG model is a stand-alone microgrid with low-inertia rotating prime-movers, described as the one in which the entire power is delivered to the system through energy sources that are operating without connection to a main reference grid. Usually the high-inertia generators in the main grid provide a constant reference frequency and a phase angle for a microgrid to determine if it should disconnect and reconnect. In a stand-alone system all the machines need to operate to provide a stable frequency and voltage in the presence of arbitrarily varying loads. The low-inertia of an electrical system implies that large load steps impose the changes in the generators, creating network frequency fluctuations. In the droop control approach, the frequency and the active power are closely interdependent. If a load increases, its torque increases too without a corresponding increase in the prime mover torque, which means that the rotational speed, and the generator frequency, decreases. The frequency slowing with increased load is what a droop control is trying to achieve in a controlled and stable manner. To get a understanding of the droop control, let us consider the problem of the complex power transferred by a transmission line. The transmission line is modeled as in Figure 6.12 as an *RL* circuit with the voltages at the line terminals being held constant.

The power flowing into a power line is described by the apparent power expressed by:

$$\bar{S} = P + jQ = \bar{V} \cdot \bar{I}^* = \frac{V_1}{Z}\exp(j\theta) - \frac{V_1 V_2}{Z}\exp(j\theta) \qquad (6.3)$$

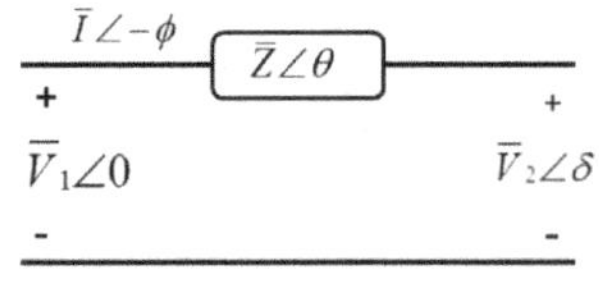

FIGURE 6.12
Power flowing through a transmission line.

Typical transmission lines are modeled with the inductance being much larger than the resistance, usually the line resistance is neglected in calculation. The active and reactive power relationships are then the well-known active and reactive power relationships, given by:

$$P = \frac{V_1 V_2}{X} \sin \delta \qquad (6.4)$$

and

$$Q = \frac{V_1^2}{X} - \frac{V_1 V_2}{X} \cos \delta \qquad (6.5)$$

Here, X is the line reactance, θ is the phase angle of the impedance, and δ is the power angle. If the power angle is small, the sine can be approximate by the angle, expressed in radians and the cosine by 1, and Equations (6.4) and (6.5) are simplified versions, showing that the power angle depends mainly on the active power and the input and the output voltage difference depends on the reactive power. In other words, if the real power is controlling the power angle, and if the reactive power is regulated, then the voltage V_1 is controlled (Figure 6.12). In the droop control method, each unit uses the frequency, instead of the power angle or phase angle, to control the active power since the initial phase values of the other units in the stand-alone power system are unknown. By regulating the real and reactive power flows, the voltage and frequency are determined. This observation leads to the common droop control equations:

$$f = f_0 - k_P \cdot P - P_N \qquad (6.6)$$

and

$$V_1 = V_0 - k_V \cdot Q - Q_N \qquad (6.7)$$

where f_0 and V_0 are the base frequency and voltage, respectively, and P_N and Q_N are the temporary set points for the real and reactive power of the machine. Coefficients k_P and k_V are the active and reactive power static droop gains. These gains are specifying the slope of the droop characteristic and hence control the reactivity speed of a droop-controlled DG unit to changes in either frequency or voltage, that is, how fast the control should react to a change in either system frequency or bus voltage magnitude. The typical droop control characteristic diagrams are shown in Figure 6.13, which shows that beyond the maximum active and reactive power limit of the DG unit, the output of the DG unit is set to that its maximum bounding limit. From the droop equations and highlighted by Figure 6.13, as the real power load on the system increases, the droop control scheme allows the system frequency to decrease. In the droop control, it should be noted that the droop method has the inherent trade-off between the active power sharing and the frequency accuracy, resulting in the frequency deviating from the nominal frequency. In summary, a droop control scheme uses only local power to detect changes in the system and adjust the operating points of the generators accordingly. The droop control uses the generator real output power to calculate the ideal operating frequency. This relaxing of a stiff frequency allows the microgrid to dampen the fast effects of changing loads, increasing the system

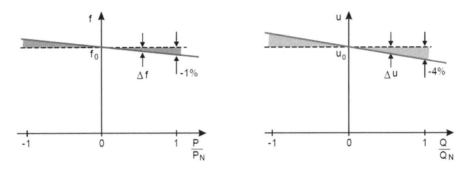

FIGURE 6.13
Drop control characteristic plots.

stability. Droop control is used to determine the effectiveness of the droop controller and its alternative forms. The frequency droop characteristic is interpreted as follows: when the frequency fall, the generating unit power output is allowed to increase the power dispatching value. A falling frequency indicates an increase in loading and a requirement for more active power. In same way when the coupling point voltage of synchronous generator decreases, the generators are allowed to increase its reactive power level. The frequency droop coefficient, k_p is defined by Equation (6.4). Typical frequency droop coefficient values for grid scale synchronous generators are 5%. A definition of k_p, frequency drop coefficient, is:

$$k_p = \frac{\text{Change in Frequency from Normal Frequency}}{\text{Change in Power from Set Point}}$$

Example 6.2: Estimate the apparent, real and reactive powers flowing into a transmission line, having the line impedance of 0.20 + j1.0 Ω, the line input voltage of 12.80 kV, the line output voltage 12.55 kV and the power angle 7.5°.

Solution: The real power and reactive power, for a configuration as one in Figure 6.12 are calculated by using Equations (6.2) and (6.3), as:

$$P = \frac{12.8 \times 12.55 \times 10^6}{1.2} \sin\left(7.5°\right) = 17.4643 \text{ MW}$$

and

$$Q = \frac{\left(12.8 \times 10^3\right)^2}{1.2} - \frac{12.8 \times 12.55 \times 10^6}{1.2} \cos\left(7.5°\right) = 2.667 \text{ MVAR}$$

The apparent power and its module (magnitude) are then:

$$S = 17.464 + j2.677 \text{ kVA}$$

$$|S| = \sqrt{17.464^2 + 2.667^2} = 17.666 \text{ kVA}$$

In the drop control methods, the line impedance angle θ largely determines the droop control law, and the usually, the inverter line impedance is considered to be inductive, due to the use of an output inductor designed to balance the output current and restrain the instantaneous circulating current. When θ is 90° or close to this value, Equations (6.4) and (6.5) are further simplified to:

$$P \approx \frac{V_1 V_2}{X} \delta \tag{6.8}$$

and

$$Q = \frac{V_1^2}{X} - \frac{V_1 V_2}{X} = \frac{V_1 (V_1 - V_2)}{X} \tag{6.9}$$

If the power angle is small, the real power is controlled by the power angle δ, while the reactive power is controlled by the source voltage, V_1. By using the droop method, each unit uses frequency to control the active power flows. The droop control method is expressed in a general form by Equations (6.6) and (6.7). Notice that the larger are the droop coefficients, the better is the power sharing. However, the frequency and amplitude regulation deteriorate. Thus, a trade-off must be considered in the droop method. From the droop equations, as the load system real power increases, the droop control allows the power system frequency to decrease. It is also worth to note the inherent trade-off between the active power sharing and the frequency accuracy, resulting in frequency deviations from the nominal system frequency. It is desirable that the controller restores the frequency to its nominal value after a disturbance. However, the frequency restoration is not practical in systems with inverters due to the inaccuracies in the inverter output frequency. The differences in the inverter frequency result in an increase of the circulating currents leading to an unstable system. However, a system with rotating machines may lend itself to a droop control with frequency restoration. If active power controllers include frequency restoration loop, the controllers are analogous to an engine governor. Engines are equipped with governors to limit the maximum safe speed when unloaded and to maintain a relatively constant speed despite changes in the loading. As the load varies, the speed may droop but over a short-time period returns to its nominal speed.

6.3.2 Protection of Microgrids

Safe operation is always major issue and concern in electricity distribution and transmission. Protection devices and careful planning are the keys to enable safe network operation under any conditions. Fuse protection is used in traditional low-voltage networks, but several studies suggested that the use of relay protection is needed to enable safe low-voltage microgrid operation. In medium-voltage network, the protection devices have been developed significantly, from old, mechanical relays to microprocessor-based relays, the so-called as numerical relays. The second-generation numerical relays have several new capabilities, such as several different measurements and setting groups, two-way data transfer and self-control. This kind of measurement, control, communication and protection units are called as feeder terminals, widely used for the MV network protection at HV-MV substation. Feeder terminals can communicate with supervisory control and data acquisition system. Large DG and RES applications are treated like all generators connected to the grid system requiring detailed impact assessments and protection studies. Smaller systems (5 MW or less) are usually connected to the sub-transmission and power distribution, being integrated into the utility protection system. The addition of DG and RES brings two perspectives for protection requirements, those of the generation owners and those of the utility. The DG or RES units require protection from short-circuits and abnormal conditions that may result into the generator damage. Abnormal conditions can be imposed on the DG units by the utility system such as over-excitation, overvoltage, unbalanced currents, abnormal frequency and shaft torque stress due to utility breaker reclosing. Utilities are concerned with the damages that the DER units may cause to the system or to customers due to unwanted fault currents or changes into the existing protection schemes. The IEEE 1547 standard only provides limited guidance and highlights only the important requirements. Newer standards are being developed toward more detailed requirements for the DG and RES integration with power distribution. The current standards are requiring not causing over-voltages or loss of utility relay coordination, the disconnection when no longer operating in parallel with utility, not to energize the utility when the utility is de-energized, not to create unintentional islands, the use of the *utility grade relays*, not to cause objectionable harmonics, loss of synchronization, and overvoltages. The micro-sources of special interest for microgrids are small (100 kW or less) units with power electronic interfaces. Power electronics provide the control and flexibility required by the MG concept, and if correctly designed power electronics and controls insure that the microgrid can meet its customers as well as the utility needs. The above characteristics are achieved using system architecture with three critical components: local micro-source controllers, system optimizer and distributed protection schemes.

The complexity of electrical power systems requires carefully protection design and planning, as well as the extended use of the smart and case sensitive protection schemes. Well-designed protection schemes are the backbone of any electrical system. DER, RES, DG and microgrids enable technologies to get more attention for their efficiency, durability, and improved reliability. There are issues relating to DER and MG protection which need to be addressed and models were developed to minimize and control the power system disturbances. The DER main purpose is to utilize local energy resources that can serve at a cost of no transmission line, as the generation is closer to the load centers. The power transmission system, with variable losses and higher cost, has made DER a more reliable and promising option that shares load jamming on conventional utility and assists localized power generation. Many of the DER systems have the capabilities to be easily controlled and handled by the distribution utilities. The DER sustainable approach is to be able to be integrated with the power systems by applying novel protective measures and approaches. The conventional power distribution system is designed as a submissive network in which large alternative energy penetration may cause bidirectional power flows that can result in voltage, current, or frequency variations. In addition, power system oscillations due to sudden fault currents could affect the protection, so requiring different protection issues in DER, needs of protection, and reliable protection schemes. The IEEE P1547-2003 is a benchmark model for interconnecting DERs with conservative electrical power systems and it also provides guidelines for general interconnection of electric utility. This involves the response to abnormal conditions when functioning, power quality issues and safety condition together with operation in utility grid-connected and islanded mode. The coordinated control of a large number of DERs can be achieved by various techniques, ranging from a basically centralized control approach to a fully decentralized approach, depending on the share of responsibilities assumed by a central controller and the local controllers of the distributed generators and flexible loads. In particular, control with limited communication and computing facilities is a challenging problem favoring the adopt ion of decentralized techniques. Complexity is increased by the large number of distributed resources and the possible conflicting requirements of their owners. The scope here is to present an overview of the technical solutions regarding the implementation of control functionalities.

The main issue when integrating the DER units into the electric networks is that the power distribution systems are intended as passive networks, carrying power, unidirectional from the HV levels (central generation) downstream to the load centers at low-voltage levels. There are several ways the protection system can fail to operate correctly, through mal-trips, without the elimination of non-faulted element or through fail-to-trip, being unable to eliminate the faulted element. A mal-trip can occur when a DER unit is feeding an upstream fault, initiated from any other DER units. Such conditions can disobey the over-current protection at the definite feeder, by disconnecting it from the system unnecessarily. Fail-to-trip occurs for downstream faults, in such case the fault current is mainly composed of the current originated from a DER unit, the fault current through the over-current protection remains passive and the faulty feeder is not disconnected from the system. The host capacity recognizes the DER percentage in the power grid that can be accepted without putting the reliability or quality of power systems in question. The system performance is seen from the host capacity definition. The estimation of the host capacity is repeated for each phenomenon in the power system operation and design, depending on the system parameters, e.g., network structure and DER type. A significant performance condition is that the DER unit must have a selective protection system with the other network protections, meaning that the DER is linked in with the system as long as possible and if the selectivity is not a prerequisite, so the system and its adjustment are kept simpler, allowing the DER separation, without taking into account other network protection devices. In normal operating conditions, the protective devices correspond to the way that the main protection works before the backup units can take any action. The DER integration into grid also increases the chance of short-circuit fault. Depending on the original protection coordination settings along with the location size and DER type, an uncoordinated condition may be acceptable, when the backup protection operates before the main, by the tripping of some loads. To make the operation stable, modifications in coordination between the protective devices and DER units must be made, which is depending on the protected energy sources (grid interfacing technology differs for every source, like power converters, induction or synchronous generators). A method, setting the DER penetration limit in terms of range, position and equipment from coordination loss viewpoint, is often used. In a given system, various coordination paths are defined, as a set of protective devices positioned along a circuit starting from the main feeder breaker to the most downstream protective device. Most of the fuses selected (lateral or sub-lateral) are similar and lead to a limited number of coordination paths. A single coordination path represents several laterals utilizing the same type of fuses. Construction of coordination charts and organizational study of protection paths leads toward a better operation. The least short-circuit current at which the loss of coordination occurs among all protective coordination paths. In a case where there is no junction between the coordination curves, the minimum current may not continue living, and the installed DER are not limited to breach the system coordination and work efficiently.

Connection to the power distribution a large DG unit or of several small DG units alters the fault current level, regardless the generator type. The changes into the fault currents can affect the fuse-breaker coordination. In the case of the inverter-connected DG units, due to the controller of these interfaces (i.e., inverters), the fault current is limited electronically to typically twice the load current or even less. Hence, an independent relay is not able to distinguish between normal operation and a fault condition without communicating with the inverter, especially in the case of large PV installations in a power distribution net-work, where there is limited increase in the phase current in the case of a fault or failure. If the fault current is not clearly distinguished from the operational current, some over-current relays are not tripping, while others can respond in seconds instead of fractions of second. The undetected fault situation can lead to high voltages despite the low fault currents. Moreover, if the fault remains undetected for longer times, it can spread out, causing equipment damages. The fault impedance is also decreasing when DG units are connected in parallel with other devices. The reduced impedance results in higher fault levels if the DG units are rotating machines or converter-interfaced units with low output impedance, without means of isolation from the distribution network. In the failure case of a fail-ure, there are unexpected high fault currents that can put the system components at risk. The fault location relative to a DG unit and substation transformer also affects the protec-tion system operation. If a fault occurs downstream of the PCC, both the main source and the DG unit contribute to the fault current. However, the relay situated upstream of the DG unit only measures the fault current supplied by the upstream source, which is only a part of the actual fault current, the relays, e.g., those with inverse time characteristics, may not function properly, resulting in coordination problems. When the fault is between the main source and a DG unit, then the fault current from the main source is not changed significantly as a DG unit is usually small. In respect to the short-circuit faults, the DG incorporation affects the amplitude, direction and duration of the fault currents. The last phenomenon happens indirectly due to the relay inverse time-current characteristics.

Low-voltage power distribution systems including DER are divided into limited pro-tective regions which are enclosed by a network consisting of transformers, generators, buses, or loads. The 3S criteria provide the requirements for basic design of a distribution protection system, holding better for the directional, distance and differential protection schemes. The abbreviation 3S stands for selectivity, sensitivity and speed of protection system. Sensitivity refers to the ability to recognize an anomalous state that exceeds a specific threshold value. The selectivity makes the protection system target oriented that disconnects only the faulted part of the system at any abnormal condition to minimize the fault implications. Speed is important for any protection scheme, and the devices should respond to system disturbances in the fastest possible time in order to avoid risks and keep the stability check. Power electronics devices are capable of supplying a smaller rated current to a fault but the rotary machines are specifically designed to supply high fault currents. The fault current depends on the current source of short-circuit in the electri-cal power system. The grid has higher short-circuit currents than the distribution inte-grated DER units. Connection to the power distribution of a single large DG unit or a large number of small DG units, using induction or synchronous generators can alter the fault current levels. As a result, the fault currents sensed by the relays is less than the fault cur-rent sensed when the power distribution is connected to transmission line electric grid. But still such values are lower than a short-circuit current supplied by the main electric grid. In this case a directional over-current relay with circuit breaker, the only practical solution is if the current is used for the fault finding. Therefore, the settings have to be constantly supervised and modified when DER generation experience changes depending upon the generation type. This change in the fault current level can disturb fuse-breaker coordination. In addition to protection of feeder outputs, protection of DER units needs to be considered because they are the foundation of microgrids. The fault impedance can also decrease when DG units are introduced into the distribution network in parallel with other devices. The reduced impedance results in high fault levels if the DG unit is a rotat-ing machine or a converter with low output impedance without means of isolation from the distribution network. In the case of a failure, there can be unexpected high fault currents that would put the system components at risk.

In conventional power systems, with generation at one end of the network, the fault current decreases with the distance of the fault location from the source increases, due to the increase in the impedance, which is proportional to the distance from the source. This phenomenon is used for discrimination of devices that use fault current magnitude. However, in the case of an islanded microgrid with DG units, as the maximum fault current is limited, the fault level at locations along the feeder is almost constant. Hence, the traditional current-based discrimination strategies may not work. New device discrimination strategies are therefore required to protect the system effectively. The *reach* of an impedance relay is the maximum fault distance, causing a relay to trigger in a certain impedance zone or in a certain time, corresponding to maximum fault impedance or minimum fault current that is detected. In the case of a fault, occurring downstream of the bus where DG units are connected to the network, the impedance measured by an upstream relay is higher than the real fault impedance (as seen from the relay), being equivalent to an apparently increased fault distance and is due to increased voltage resulting from an additional infeed at the common bus, affecting the relays, causing the delayed triggering or no triggering. This is the so-called *relay under-reaching*. The *sympathetic tripping* occurs due to the unnecessary protection device operation for faults in an outside of its zone (its operation jurisdiction). An unexpected contribution from the DG units can lead to situations when a bidirectional relay operates along with another relay, sensing the fault, thus resulting in the protection scheme malfunction. The power flow changes its direction in distribution networks with embedded DG when local generation exceeds the local consumption, which may hinder the operation of the directional relays. Moreover, reverse power flow also means a reverse voltage gradient along a radial feeder, causing additional power quality issues, the voltage limit violations, causing increased equipment voltage stress. However, in the case of highly loaded or weak networks, DG units can have a positive impact improving the network power quality. DG units also impact the role of tap changing transformers for voltage regulation in power distribution. If the DG location is close to the network infeed affects the tap changing by reducing the transformer load, resulting of a shift in tap changing characteristics, and the infeed voltage regulation is incorrect. The transformer configuration and grounding arrangements selected for DG connection to the grid must be grid-compatible, to protect the system from voltage swells and overvoltages and consequent damages.

What is expected or required from the DER protection is partially controversy to that what is required from their protection in a microgrid. However, similar challenges are confronted in both utility and microgrid operation. Following challenges, encountered in the existing protection solutions: unwanted islanding, unnecessary disconnections (lack of selectivity), reconnection failure (because fault electric arc is not go out fast enough), and protection slowness. DG units can create severe problems when a part of a distribution network with DG unit(s) is islanded. This phenomenon is the *loss of mains* (LOM) or *loss of grid* (LOG). In the case of LOM, the utility supply neither controls the voltage nor the frequency. In most cases, islanding is due to a fault in the network. If the embedded generator continues supplying power despite the disconnection, a fault might persist, being fed by the DG units. The voltage magnitude can get out of control in an islanded network as most of the embedded generators and grid interfaces are not equipped with voltage control. Frequency instability may be another result of the voltage control lack, posing a risk to the equipment. Nowadays, DG units required to disconnect from the utility when a fault occurs, to avoid an unwanted islanding and to quarantine safe operation. The DG units and the utility grid can simultaneously feed a fault current to the fault place. Due to an over-current, a circuit breaker opens and the fault current feed from utility grid stops. DG units are usually protected with over- and under-voltage and over- and under-frequency protection. Often, the islanding may cause exceeding the voltages or frequency limits and the DG disconnection. However in some cases, DER unit is able to maintain the voltage in the remaining grid. Therefore, LOM protection is needed to stop the fault current feed from distributed generator. Distributed generation increases the risk of unnecessary operation of the protection devices. Unnecessary disconnection happens if the selectivity is not realized, and the network healthy part is disconnected. In addition, DG might be disconnected because of voltage dips or spikes. Unnecessary disconnections can be avoided with selective protection settings. For instance longer disconnection times for DG units are used if are not causing problems elsewhere in the grid. If voltage dips affect the DG units, the grid fast protection settings are used. One unwanted protection feature is blinding. Protection blinding might happen if the fault current from utility grid remains small. In this case, the DERs usually feed main part of the fault current. Protection needs to be planned carefully to avoid blinding. Capability to provide fault current depends of the generation unit, and its type and features needs to be included into the protection planning. Examples show that DER units set challenges for the protection network, which are even more challenging because of the smaller fault currents and low inertia. Low inertia makes higher transients and variations possible, causing protection device malfunctions. The role of an auto-recloser is very important in the system restoring after faults lasting very short intervals. However, in the case of a distribution network with DG units, two problems may result from automatic reconnection of the utility after a short interval. The first one is that the automatic recloser attempt may fail as a result of feeding of a fault from a DG unit. The second problem is that due to active power imbalance, a change in frequency may occur in the islanded part of the grid. In such case, an attempt at reclosing the switch is coupling two asynchronously operating systems. Moreover, conventional recloser is designed to reconnect the circuit only if the substation side is energized and the opposite side is un-energized. However, in the case of DG, there are active energy sources on both sides of the recloser, hampering its operation.

6.3.3 Microgrid Protection Issues and Techniques

One of the main DG benefit is the possibility of improving the energy supply reliability and continuity, by making possible that a part of the network operates in an islanded or stand-alone mode, during a power outage. However, an MG critical challenge is to ensure stable operation during faults and various network disturbances. Transitions from interconnected to islanding mode are likely to cause mismatches between generation and loads, posing severe frequency and voltage control problems. Energy storage technologies (e.g., batteries, ultra-capacitors and flywheels) are important MG components, with the purpose to provide stable network operation during disturbances. Maintaining stability and power quality in the islanding mode requires the sophisticated control strategies, needing to include both generation and demand sides. For such purpose, there are several updates of the domestic and international regulations and standards to ensure that the DG sources have adequate short-circuit and fault protection. In order to improve system reliability and stability, the fault-ride-through capability requirements for the DG units connected in HV and MV networks are introduced and required. For the LV level similar guidelines are being considered nowadays, while different solutions are proposed to address the most common issues related to the MG protection, such as the over-current protection, in either operation mode. In designing a protection scheme of a power system, the requirements that must be considered are reliability, speed, selectivity and cost. The protection system's main function is to quickly remove from service any of the system components that started to operate in an abnormal manner. Other required functions are the personnel safety, the entire system safeguard, the supply continuity, the damage minimization and the repair cost reduction. When faults occur, the protection system is required to disrupt as fewer network sections as possible. Selective operations of the protective devices are to ensure the maximum service continuity with minimum system disconnections. Conventional power distribution is designed to operate radially, where often only a single energy source is present with unidirectional current flows from the higher voltage levels at the substation, through the distribution feeders and laterals to lower voltage levels at the end-users. The system relies on simple and low-cost protection schemes, consisting of fuses, reclosers, circuit breakers and over-current relays. Breakers and reclosers are normally installed at the main feeder to allow clearance of temporary faults before lateral fuses blow. In normal condition, they are equipped with inverse time over-current relays, usually are installed at the feeder substation. The protective devices are coordinated to operate according to selectivity criteria based on current or time, ensure that the device nearest to a fault operates first. The basic criteria, employed when coordinating time or current devices in distribution systems are:

1. The main protection clears a permanent or a temporary fault before the backup protection operate, or continue to operate until the circuit is disconnected. However, if the main protection is a fuse and the back-up protection is a recloser, it is acceptable to coordinate the fast operating recloser curve or curves to operate first, followed by the fuse, if the fault is not cleared; and

2. Fault caused supply loss should be restricted to the smallest system part for the shortest possible time.

In a conventional power distribution, the protection systems are designed assuming unidirectional power flow and are usually based on over-current relays with discriminating capabilities. For any fault situation, DG sources connected to the system are tripped off. In other words, the islanded operation of DG sources is not allowed. When a microgrid is created in a distribution network, the configuration is a complex multi-source power system, and its protection must provide system safe and secure operation, in both connected mode and islanded mode. However, the two operating modes pose additional challenges in MG protection. Therefore, two sets of protection settings are the most likely solutions to the dual operation modes. During grid-connected mode, the mains supply large fault currents to the fault point, making possible the employment of the existing protection devices. However, the protection coordination may be compromised or even entirely lost in some cases due to the presence of the DG units. The large fault currents from the microgrid, similar to ones in the main grid are not always presented, especially when the electronically coupled DG units dominate. Thus, the use of conventional overcurrent protection in the microgrid is no longer valid due to the DG low short-circuit current contributions. The protection must respond to both the main network and the microgrid faults. If a fault occurs on the main grid, the desired response is to isolate the microgrid from the main network as rapidly as needed to protect the microgrid loads, causing the MG islanding. If a fault occurs within the microgrid, the protection system is required to isolate the smallest possible faulted MG section and to eliminate the fault. Various MG protection schemes and coordination methods are available (Figure 6.14). In situations when fault current levels are drastically different between the grid-connected and the islanded operation (typically in inverter-interfaced DG units), the design of an adequate

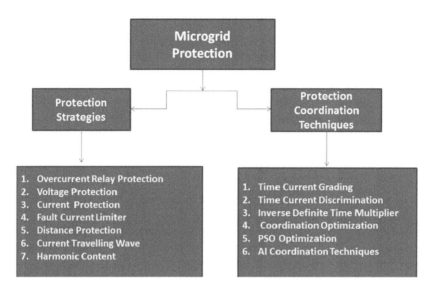

FIGURE 6.14
Different types of protection schemes for microgrid operation.

protection system, performing properly in both modes, can be a real challenge. In this regard, there is a possibility of applying a different approach which actively modifies the fault current level when the microgrid changes from grid-connected to islanded mode and vice versa, by means of certain externally installed devices, that are set to either increase or to decrease the fault level. The MG protection schemes are divided into overcurrent-based, voltage-based, current component-based, harmonic content-based, fault current limiter-based and current traveling wave-based. As for protection coordination techniques, time-current grading and optimization algorithms are often used to ensure selectivity of the protective devices. The functionality and reliability of different protection schemes need to fulfill different operation modes, different DG types, exist in a microgrid, and different issues that it has to tackle like bidirectionality of power flow and location of DG placement.

There are several possible solutions to cope with the new challenges caused by introduction of DG into the power distribution networks. For example, to solve the problems of the bidirectional power flows, the feeder main relay, fed from the same substation, is interlocked. The main relay of the feeder with DG units is equipped with an interlocking system. Once a short-circuit current is detected by a relay, a locking signal is send to the main feeder relay. Due to this locking signal, the main relay not malfunction, even if there is back feeding from the DG units to the fault. The use of directional over-current relays instead of conventional relays can also solve this problem. However, this scheme has its own limitations. Main feeder relays readjustment in terms of time settings is another solution. The feeder without DG can have faster relay settings than the relay settings of the feeder with DG units. But care has to be taken with this readjustment so that it does not hinder the coordination of these relays with the feeder downstream protection devices. There is a trade-off between the speed of reclosing and the power quality, that is, the faster the reclosing, the better the power quality. However, to ensure that the reclosing attempt is successful, instantaneous reclosing is not recommended for feeders with the DG unit. Conventional protection schemes face serious challenges when it comes to protecting an islanded microgrid with inverter-interfaced DG units, needing major revision to find methods based on the limited fault current to detect and isolate the faulty portion. The various possible solutions, as found in the literature to cope with these problems, can be broadly divided into four categories: the use of inverters having high fault current capability, that is, uprating of the inverter; communication between the inverter and protective relays; introduction of energy storage devices that are capable of supplying large current in case of a fault; and in-depth analysis of the fault behavior of an islanded microgrid with an inverter-interfaced DG unit to comprehend the behavior of system voltages and currents. This in turn helps in defining alternative fault detection and alternative protection strategies that, in case of a fault, do not rely on a large magnitude of the fault current but rely on other parameters, like change in the voltage of the system. Differential protection schemes, traditionally used for transformer protection, are also proposed for the protection of an islanded microgrid. Such schemes based on differential relays are selected as their operation, unlike over-current relays, is independent of the fault current magnitude. Such schemes solve the problem of low fault current in the case of inverter-interfaced DG units. The downside is that the protection scheme is not be able to differentiate between a fault current and an overload current, so nuisance tripping will result whenever the system is overloaded. So, traditional differential protection schemes might not be, in some instances, able to differentiate correctly between internal faults and other abnormal conditions. Also mismatch of the current transformers can be a source of malfunction. For a coordinated clearing of a fault in an islanded microgrid and to ensure selectivity, it is important that different distributed generators can communicate effectively with each other. To this end, evolving a distribution system version of the pilot wire line differential protection may be needed.

Another approach to ensure the isolated microgrid proper protection is to the use of DG units, equipped with synchronous generators, to the use of inverters having high fault current capability, or to the use a combination of both types of DG units, so that conventional protection schemes can be properly used. Such combinations are ensuring large fault currents that can be detected by the conventional protection schemes. However, for a higher rated inverter, larger size power electronic switches, inductors and capacitors are needed, making the system more expensive. Energy storage devices, like batteries, fuel cells or flywheels, can also be incorporated into the microgrids to increase the fault levels, as desired. In the case of LV networks, it is recommended that the fault current level is at least three times greater than the maximum load current for its clearance by over-current relays. Directional relays can be used to clear the fault within the microgrid provided they see a fault current exceeding the maximum load current in their tripping direction. However, this is not always the case as faults can be fed from different directions. A protection scheme for an islanded microgrid depends heavily on the inverter controller type, as the controller actively limits the available fault current from an inverter-interfaced DG unit. Thus, selection of the protection scheme can be critical for the island microgrid operation.

The inverter controller can be very important for the protection of such microgrids. A protection scheme that combines conventional OC characteristics and undervoltage initiated directional fault detection with definite time delays. A large decreasing into the network voltage cannot be used alone for detecting low levels of MG fault current, as the voltage depression have not enough gradient to discriminate the protection devices. Hence, the measurement of other parameters is needed and recommended. Adaptive protection schemes are suggested as solutions for the MG protection, both in grid-connected and in islanded modes. The basic idea is that an automatic readjustment of the relay settings is used when the microgrid changes from grid-connected to islanded mode. In an islanded microgrid, the adaptive protection strategy can be used by assigning different trip settings for different fault current levels, which in turn are linked to different magnitudes of the system voltage drops resulting from system disturbance.

6.4 DC Microgrids and Nanogrids

AC microgrids are used because the existing electrical grids are AC systems. However, a small portion of an existing AC grid, e.g., a residential community, can be converted to an AC microgrid through the installation of sufficient DG units along with an isolating switch at the grid interface for islanding purposes. Microgrids can have different configurations and can be composed of networks of different nature forming hybrid AC-DC systems interconnected through power electronic converters. A DC microgrid is more suitable for new development installations in rural areas, commercial facilities, or residential buildings. However, the DC microgrid concept can also be applied for existing installations since existing AC systems are usually three-phase systems, consisting of at least three wires, the number of wires needed for a DC network: *positive, negative* and *ground*. Nevertheless, the conversion to a DC microgrid requires a considerable amount of equipment retrofitting as well as power electronic converters. On the DC side, these power electronic converters are in charge of maintaining to voltage within certain levels and limits. Often, a droop control strategy is used to assure a fast DC voltage regulation, and how different DC microgrid elements are affecting the droop method performances, is determined through a complete DC microgrid dynamical analysis. This is often done through a detailed study on the system linearized model, while the model-based control design techniques are applied to the droop controller. DC power systems are already used in industrial and commercial applications, due to both historical and practical reasons, such as the versatile speed control of DC electrical machines and the possibility to build reliable, simple, power networks by using directly connected batteries, fuel cells and DC energy sources. Telecommunication and data centers are today supplied with 48 V DC through power converters connected to the AC grid. A low-voltage DC microgrid or nanogrid can be used to supply sensitive loads, combining the advantages of using a DC supply for electronic loads, and by using local generation units. The smaller losses due to fewer power conversion steps results in less heat which need to be dissipated, and therefore the operation costs are reduced. To ensure reliable operation of an LV DC microgrid, well-designed control and protection systems are needed. In the case of power outages, the loads are supplied from batteries or fuel cells connected to the DC bus of the DC microgrid. In some communication stations a standby diesel generator can also be used to support the network. A similar solution is also used for power supply of the control and protection equipment in power plants and substations. However, higher voltage levels (110 or 220 V DC) are used for longer distances between sources and loads, and higher power ratings. Moreover, an LV DC system is well suited for PV systems and fuel cells, which both produce DC power. Micro-turbines, small hydropower stations and wind turbines are usually generating AC with variable frequency or at different frequency than the grid, and hence such settings need an AC-DC-AC converter to meet the grid frequency requirement. These energy sources can benefit to a DC system connection, since DC-AC power converters are removed or replaced by simpler, cheaper, more robust and efficient DC-DC power converters. Also battery units can be directly connected without any converters, resulting in reduced costs and losses. Compared to an AC system, in which frequency, phase and reactive power control are required, a DC microgrid requires only the balancing of the active power in the system for proper voltage regulation. Due to the presence of the AC-DC converter, a DC microgrid does not increase the short-circuit capacity of the main grid at the interconnection point, and the interconnection converter can be used to mitigate power quality problems, e.g., harmonics, voltage sags and voltage fluctuation. All of these features result in a high-quality network. In addition, DC capacitance prevents the propagation of power system disturbances between the microgrid and the main grid.

Standards describing component modeling and calculation methods are necessary in order to analyze any power network. Available standards today for LV DC systems are IEEE Standard -399-1997 and IEC 61660, both covering the load flow and short-circuit calculations of DC auxiliary power systems. These are used, for example, in power plants and substations. Loads in these standards are modeled as constant-resistance, constant-current or constant-power loads, depending on the load characteristic. These models are adequate for load flow calculations and simplified short-circuit calculations. Historically, DC has been used in LV drive systems where speed control has been required, and AC power in applications where it has not. In the latter case simpler and more robust induction machines have been used. Speed control of DC electrical machines is obtained by changing the supplying voltage or the magnetic flux. In the early beginnings of the electric networks for DC machines a variable resistor in series with the machine, or a variable resistor in the excitation circuit were used, a simple solution, but with higher losses and a poor speed-torque characteristic. By using power electronics a better, faster and more precise control of both AC and DC machines can be obtained. Today DC machines are found in traction applications or in industrial drive systems. Although AC systems made a big breakthrough in the beginning of the twentieth-century DC solutions are adopted by a number of new applications such as distributed generation, energy storage, electric vehicles, hybrid electric vehicles, electric ships and HV DC transmission. Besides a proper control system, a well-functioning protection system is needed to ensure a reliable DC microgrid operation. It can be designed by using the techniques already used existing protection systems for higher power level DC power networks, for protection of generating stations and traction power systems. However, these DC networks utilize grid-connected rectifiers with current-limiting capability during DC faults. However, an LV DC microgrid must be connected to an AC grid through power converters with bidirectional power flow capability, and therefore a different protection-system is required.

DC microgrids can have different configurations with different RES and DG units, affecting the system control and protection in certain ways. It desirable to design a control strategy that is applicable to any microgrid configuration (with only minor changes). All components of a DC microgrid configuration are usually interconnected through power electronic converters. Focusing on the microgrid DC side, these converters are responsible for maintaining the grid voltage within reasonable limits. For this purpose, a power based droop control solution is often used to control the DC voltage fast, and to establish power sharing between the converters connected to the DC network. Apart from integrating more RES units, the DC microgrid concept has other advantages and applications. For example, DC microgrid architecture is facilitating the design of ultra-available power sources for critical loads, such as hospitals, security and first response stations, or data centers. It has been proven that, for critical loads, DC systems have a higher availability over AC systems (at least two orders magnitude higher). On top of that DC facilitates integrating majority of modern electronics since all of them internally work on DC. Coming back to renewable energy sources, a few advantages are worth mentioning when using DC microgrids:

- Reduced system power losses, by reducing the number of the AC-to-DC conversions.
- Loads are supplied through the distribution line when there is a main grid blackout, and having higher local availability is possible through local power sources and energy storage in redundant architectures.
- There is no need to synchronize distributed generators.
- Fluctuations of generated power and the loads can be compensated through energy storage modules.
- The system does not require long transmission lines, or high capacity lines.

Among so many benefits, this concept raises a couple of difficulties. As mentioned before DC systems do not experience harmonic issues because the fundamental frequency of a DC system is 0 Hz, other frequencies than the fundamental does not exist.

The functions of an LV DC microgrid protection system are to detect, isolate and clear the faults fast and accurately, in order to minimize the effects of disturbances. Its design depends on a number of factors, such as the type of faults which can occur, their consequences, the type of protection devices required, the need for backup protection, detection methods, the measures designed to prevent faults, and finally, the measures to prevent incorrect protection system operation. Possible fault types in DC microgrid are pole-to-pole and pole-to-ground faults. Pole-to-pole faults have low fault impedance, while pole-to-ground faults are characterized as either low-impedance or high-impedance faults. The location of the faults can be on the bus or one of the feeders, inside the sources or the loads. The main difference between an LV DC microgrid and other existing LV DC networks is the type of power converters, used to interconnect the DC system with the AC grid. Power converters used for example in DC auxiliary power systems for generating stations and substations and traction applications are designed to have a power flow only from the AC side to the DC side. Therefore, it is also possible to design the power converters to be able to handle faults on the DC side by limiting the current through them. However, the power flow between an LV DC microgrid and an AC grid must be bidirectional. A different type of power converter is required, and it may not be possible to limit the current through the converter during a fault in the DC microgrid or nanogrid. During a fault, all energy sources and storage units connected to the DC microgrid or nanogrid are contributing to the total fault current. The fault current from each DER unit is determined by design and the total fault impedance. The converters used in the LV DC microgrid have a limited steady-state fault-current capability due to their semiconductor switches. However, they can provide a fault current with a high amplitude and a short duration from their DC-link capacitors. Energy storage (e.g., lead-acid batteries) can provide large steady-state fault currents, but in contrast to the power converters, they have a longer rise time. The components within the DC microgrid must be protected from both overloads and short-circuits. Depending on the component sensitivity various solutions exist. Power converters are very sensitive to overcurrents, and if they are without internal current-limiting capability, they require very fast protection. Examples of such devices are fuses, hybrid circuit breakers and power electronic switches. Batteries and loads do not require fast protection, and therefore simpler and cheaper devices are used.

To achieve selectivity in the DC microgrid, it is necessary to coordinate the protection devices. Feeders and loads are preferably protected by fuses since they are simple and cheap, and it is easy to obtain selectivity.

6.4.1 DC-DC Converters

The conversion of direct current into direct current is intended to improve the power from a DC source and to match the voltage of the source and the consumers. DC–DC converters are power electronic circuits that convert a DC voltage to a different voltage level. There are different types of conversion method such as electronic, linear, switched mode, magnetic, capacitive. The purpose of DC voltage converters is not only the conversion of direct current, but also the regulation or stabilization of the voltage (or current) in the load. Converters used only for stabilization are known as stabilizers. We may distinguish between two types of DC converters: continuous converters and pulsed converters. Continuous converters stabilize the voltage in a DC circuit with variation in the source or load voltage. Continuous converters employ transistors operating in the active region of the output voltage–current characteristics. These circuits, classified as switched mode DC-DC converters are electronic devices that are used whenever change of DC electrical power from one voltage level to another is needed. In most cases the grid voltage is first rectified by a rectifier and then adjusted by a power-electronic converter, which together are called a switch mode power supply (SMPS), because the use of a switch or switches for the purpose of power conversion can be regarded as an SMPS. Even though many of the renewable energy sources and energy storage units have a DC output they need to be connected to the microgrids through DC-DC converters to regulate and improve the output DC signals. For example PV panels are connected through a DC-DC Boost converter to raise the output voltage level at the nominal microgrid voltage level and to track the maximum power point of the solar PV panel output which fluctuates depending on the solar irradiation. A power-electronic converter adjusts the grid voltage to a voltage with amplitude and frequency which is required by the load. There are different kinds of DC-DC converters. A variety of the converter names are included here: the Buck converter, the Boost converter, the Buck-Boost converter the Cuk converter, the Flyback converter, the Forward converter, the Push-pull converter the Full Bridge converter, the Half Bridge converter, the current Fed converter and the multiple output converters. For example, a Buck converter is a commonly used in circuits that steps down the voltage level from the input voltage according to the requirement. It has the advantages of simplicity and low cost. A Boost converter (step-up converter) is a power converter with an output DC voltage greater than its input DC voltage. The basic model of the Boost converter is shown in Figure 6.15. Based on the switch states, as presented in Chapter 13 the voltage equations, relating

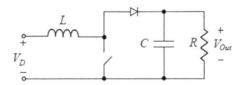

FIGURE 6.15
Diagram of a DC-DC Boost converter.

the input voltage, V_D, and the output voltage, V_{Out}, and the duty ratio, D, so the switching frequency, for Buck, Boost and Buck-Boost converters are:

$$V_{Out} = D \cdot V_D \tag{6.10a}$$

$$V_{Out} = \frac{V_D}{1-D} \tag{6.10b}$$

and

$$V_{Out} = \frac{D \cdot V_D}{1-D} \tag{6.10c}$$

Neglecting the energy consumption in the operation of the converter control subsystem, considering only the losses into the inductor and MOSFET transistor the converter efficiency can be expressed as:

$$\eta = \frac{P_{Out}}{P_{Out} + Losses} \tag{6.11}$$

Example 6.3: A Buck-Boost power converter has the input DC voltage 42 V, duty ratio 0.6, and the load resistance R_L, is 10 Ω. Calculate the output voltage, the power delivered to the load. What is its efficiency if the all converter internal losses are 31.1 W?

Solution: The output voltage is:

$$V_{Out} = \frac{D \cdot V_D}{1-D} = \frac{0.6 \times 42}{0.4} = 63 \text{ V}$$

The power delivered to the load is calculated as:

$$P_{Load} = \frac{V_{Out}^2}{R} = \frac{63^2}{10} = 396.9 \text{ W}$$

Applying Equation (6.11), the converter efficiency is:

$$\eta = \frac{396.9}{396.9 + 31.1} = 0.927 \text{ or } 92.7\%$$

Energy storage units based usually on batteries or other devices require a bidirectional DC-DC power converter, to provide power when the energy sources are not present or able to operate and to accumulate energy when the energy storage systems are not needed. The characteristics and bloc diagram of the DC-DC bidirectional converter looks like the ones in Figure 6.16.

Inverter modeling consists of two kinds of control strategies may be used to operate an inverter. The inverter model is derived according to the following control strategies: active and reactive power (*P-Q*) *inverter control*, in which the inverter is used to supply the required active and reactive power set point, and *voltage source inverter (VSI) control*, in

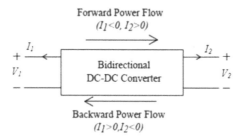

FIGURE 6.16
Bidirectional DC-DC power converter characteristics and schematics.

which the inverter is controlled *to feed the load* with predefined values for voltage and frequency. Depending on the load, the VSI real and reactive power output is defined. Notice that operating the inverter as a voltage source behind the impedance can result in real power being proportional to the phase angle across the coupling inductor, and the injected reactive power is proportional to the magnitude of the inverters output voltage.

6.4.2 DC Nanogrids, Configurations and Applications

Nanogrids are considered elements of a micro-grid, and are often referred to as small microgrids, meaning that they can be interconnected to form a larger microgrid. Nanogrids can also be separated from a microgrid and function independently with their own voltage, phase, and frequency from DC to kHz. AC, but more often are referred only to DC networks. Nanogrids play a different role to microgrids and/or main grids in the power systems hierarchy. For example, by connecting multiple nanogrids a microgrid can be formed. This introduces an alternative approach to the traditional microgrid. Interconnecting these nanogrids gives them ability to increase their range and power supply. As the nanogrid structure is often confined to a single building, the technical objectives, hardware and software often vary from that of a microgrid and/or of the main grid. A nanogrid allows a power structure to be obtained at a relatively low cost compared to microgrids and/or main grids. A nanogrid is different from a microgrid, either some microgrids can be developed for single buildings, being usually interfaced with the utility grid. However, a nanogrid, regardless whether a utility grid is present or not, it is mostly and autonomous DC-based system that is digitally connecting individual devices to one other, as well as to power generation and energy storage units within the building, being viewed as a single power domain, for voltage, capacity, reliability, administration and price. Nanogrids include energy storage internally, and the local generation is operating as a special type of nanogrid, while a building-scale microgrid can be as simple as a network of nanogrids, without any central entity. The nanogrid is conceptually similar to a ship, car or aircraft, which both house their own isolated power grid networks powered by batteries that can support electronics, lighting and internet communications. Uninterruptible power supplies also perform a similar function in buildings during grid disturbances or malfunctions. Essentially, a nanogrid allows most of the devices to plug into power sockets and connect to the nanogrid, which is balancing the energy supply with the demand from those individual loads. A nanogrid assumes and requires digital communication among its entities, DC power network, being only intended for use within or between adjacent buildings. Building-scale microgrids are built on a configuration of nanogrids and pervasive communication. The system capacity from a few kilowatts to hundreds of kilowatts, often in the lower power. There are lots of potential benefits to structuring local DC power distribution in this way, they argue. Conversion losses would be cut, investments in inverters and breakers would be reduced, and device-level controls would enable a much more nimble way to match generation or storage capabilities with demand. The building would also theoretically be immune to problems more likely to be encountered with a local microgrid or the broader centralized grid. Theoretically, such localized, autonomous power systems could be scaled up or eventually down without almost any interference with the utility grid.

The components of the nanogrid consist of a controller, gateway, load and an optional storage, as shown in Figure 6.17. The typical load sizes are in hundreds of watts and, but at times can follow in the tens of watts range. The controller is considered the core or the authority. It controls the loads, power flows, communication as well as manages the energy storage. Energy storage can be installed internally or through as second nanogrid. Solely to relieve the primary nanogrid and act as storage specifically. Gateways can be considered one way or two way, with a capacity limit. These gateways consist of two components including communication and power exchange. The communication portion should be considered generic giving it the ability to run across physical layers. The power exchange is a component that focuses on defining the various amount of voltages and capacities. Figure 6.17 shows a schematic of a typical nanogrid and its main components. The nanogrid control strategy should be extended to consider the combined consumption and production of a cluster of houses to further reduce the effects of RES generation intermittency. Nanogrid control is usually divided into two categories; supply side management and demand side management. Supply side management focuses on controlling the nanogrid power supplies and energy storage (such as PV, small-scale wind turbines, geothermal heat pumps, battery banks, fuel cells, etc.) to ensure the demand (load) is met and/or the state of charge (battery banks) is optimized. This is an important aspect of nanogrid control as often multiple energy sources exist and their integration needs to be balanced in such a way that a specific source can be selected to supply the nanogrid (e.g., with a grid tied PV system, it is favorable to supply the loads with the PV first, before supplying the unmet load with the grid). Demand side management, on the other hand, manipulates the load to meet the characteristics of the overall energy supply, balancing the nanogrid power.

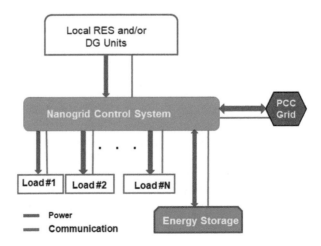

FIGURE 6.17
A typical nanogrid architecture.

There are a variety of technologies used with nanogrids, but the subject that dominates the nanogrid literature is the power converter topologies. Power converters, as presented in the previous chapter subsection, are responsible, within the nanogrid, for manipulating voltages to meet the requirements of a specific task. This is typically used, but not limited to interfacing the nanogrid energy sources with the system (nanogrid) power bus and eventually the main grid (at PCC), along with interfacing the nanogrid loads. The common categories of power converters used in nanogrids are DC-DC, DC-AC and AC-DC. There are other tasks the converters can perform within the nanogrid, such as maximum power point tracking (MPPT) of a renewable source or charge controlling a battery bank. Charge controllers make use of converters to regulate the speed of charging and ensure battery banks are not overcharged, which lengthens the life of a battery bank. The goal of MPPT is to address the nonlinearities presented to a system by a renewable source (primarily photovoltaics but can also apply to wind turbines).

6.4.3 Control of Low-Voltage DC Microgrids

The DC nature of emerging renewable energy sources (e.g., PV panels) or energy storage units (e.g., batteries, fuel cells and ultra-capacitors) efficiently lends itself toward a DC microgrid paradigm that avoids redundant conversion phases. Many of the loads are electronic DC loads and even some conventional AC loads, e.g., induction machines appear as DC loads when controlled by inverter-fed drive systems. A DC microgrid is an interconnection of DC sources and DC load through a distribution network. Given the intermittent nature of electric loads, energy sources must be dynamically controlled to supply the load power demand at any moment, while preserving a desired voltage at consumer terminals. Sources may reflect a variety of rated powers, being desirable to share the total load demand among these sources in proportion to their rated power. Such load sharing approach is known as proportional load sharing. This approach prevents overstressing of sources and helps to span lifetime of the microgrid power-generating. While the source voltages are the sole variables controlling power flow, they must be tightly managed to also ensure a desirable voltage regulation. DC microgrids are also shown to have about two orders of magnitude more availability compared to their AC counterparts, making them ideal candidates for critical applications. Moreover, DC microgrids can overcome disadvantages of AC systems, e.g., transformer inrush current, frequency synchronization, reactive power flow and power quality issues. Given the desire for developing DC microgrids, control algorithms must be tailored to account for the individual behavior of the entities, forming up a DC system, and their interactive behavior. The development of distributed control techniques that are inspired by the operation of DC systems provide global voltage regulation and accurate load sharing through a minimal communication. Proliferation of power electronics loads in DC distribution networks shifts the load consumption profiles from the traditional constant impedance loads to electronically driven loads with variable power profiles. Such fast-acting consumption patterns can destabilize the entire distribution network, given their weak nature due to the lack of the damping and generation inertia. Hardware-centric approaches focus on placement of energy storage devices or power buffers for the source decoupling, load and power distribution network dynamics. Control-centric approaches are the alternative solutions. One of the control approaches that coordinates and groups power buffers across a DC microgrid, through the information exchange in a distributed and sparse network enables power buffers to act globally, rather than locally. Commercial LV power systems often have a large amount of sensitive nonlinear loads, which must be protected from disturbances and outages in order to operate correctly. Examples of such loads are data and communication systems, control systems, safety systems and equipment for heat, ventilation and air conditioning. A way to ensure reliable power supply is to install uninterruptible power supplies (UPSs). These UPSs are used to protect the sensitive loads from transients, short interruptions with the duration up to half an hour or other power disturbances. Within this time the diesel generators are automatically started to support the UPSs. A microgrid is well suited to protect sensitive loads from power outages and power disturbances. An isolated power system with high reliability can be obtained by utilizing the local energy sources together with fast protection systems. To be able to operate the microgrid in island mode it is necessary to have an island detection system, which safely disconnects the microgrid when an AC grid outage occurs, to prevent energizing the AC grid. If a blackout in the AC power grid occurs, the microgrid is disconnected from the AC grid and used for service restoration.

Besides the advantages, the DC microgrids do not have some of the disadvantages of AC systems, e.g., transformer inrush currents, frequency synchronization, reactive power flow, phase unbalance and power quality issues. A DC microgrid is an interconnection of DC sources and DC load through a transmission and distribution network. Given the intermittent nature of electric loads, the energy sources must be dynamically controlled to provide load power demand at any moment, while preserving a desired voltage at consumer terminals. While the source voltages are the sole variables controlling power flow, they must be tightly managed to also ensure a desirable voltage regulation. A hierarchical structure is widely used to control DC sources. Such structure includes primary, secondary and tertiary levels, where the primary is the highest and the tertiary is the lowest level. Droop control is regarded as an adaptive voltage positioning method in circuit design. The principle of droop control is to linearly reduce the DC voltage reference with increasing output current. By involving the adjustable voltage deviation, which is limited within the acceptable range, the current sharing among multiple converters can be achieved. In most of the cases, the current sharing accuracy is enhanced by using larger droop coefficient. However, the voltage deviation increases accordingly, and hence, the common design criterion is to select the largest droop coefficient while limiting the DC voltage deviation at the maximum load condition, as expressed below. The controller uses droop mechanism to handle proportional load sharing, therefore a virtual resistance, R_D, is introduced to the output of the energy source. While the load sharing benefits from this virtual resistance, it is not physical impedance, thus, not causing any power losses. In this stage, the voltage controllers inside each source follow the voltage reference generated by the droop mechanism, as:

$$V_{0-i}^* = V_{Ref} - R_{D-i} \cdot i_{0-i} \tag{6.12}$$

The droop voltage deviation is then expressed as:

$$\Delta V_{DC} = |V_{Ref} - V_{DC}| \le \Delta V_{DC-\max} \tag{6.13}$$

where V_{0-i}^* is the reference voltage for the inner-loop voltage controller (converter) #i, R_{D-i} is the droop coefficient of converter #i, V_{Ref} is the MG rated voltage, and i_{0-i} is the source output current, ΔV_{DC}, and $\Delta V_{DC-\max}$, are the voltage deviation and its maximum value. In steady state, given low distribution line resistances, all terminal voltages converge to the same value. Given identical rated voltages used at all energy sources, the droop terms, R_{D-i} and i_{0-i} are sharing identical values, implying that the total load is shared among the energy sources inverse to their droop coefficients. By choosing the droop coefficients in inverse proportion to the source power ratings, the droop control successfully manages the

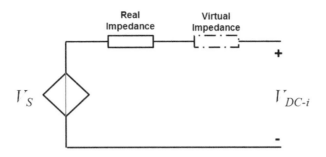

FIGURE 6.18
Thévenin equivalent circuit model of droop-controlled interface converter.

proportional load sharing. The interface converter with droop control is modeled by using the Thévenin equivalent circuit (see Figure 6.18). This virtual resistance allows additional control flexibility of the DC microgrids. Notice that it has been demonstrated the fact that the linear droop technique cannot ensure low voltage regulation and proportional current sharing. To achieve acceptable voltage regulation at full load and to ensure proportional current sharing, nonlinear and adaptive droop techniques are proposed and studied.

> **Example 6.4:** The reference voltage of a DC microgrid is 360 V (DC), the droop coefficient is 0.50 Ω and source output current is 15 A. Estimate the controller voltage and the voltage deviation.
>
> **Solution:** Applying Equations (6.12) and (6.13), the required parameter values are:
>
> $$V_{0-i}^* = 360 - 0.5 \cdot 15 = 352.5 \text{ V}$$
>
> and
>
> $$\Delta V_{DC} = |360 - 352.5| = 7.5 \text{ V}$$

Despite its benefits, the conventional droop method suffers from poor voltage regulation and load sharing, particularly when the line impedances are not negligible, as often in DC microgrid configurations. Voltage drop caused by the virtual impedance in droop mechanism, and voltage mismatch among various power converters are the main reasons. To eliminate voltage deviation induced by droop mechanism, a voltage secondary control loop is often applied to the system. The concept of secondary control, under the name of automatic generation control, has been used in large power systems to address the steady-state frequency drift caused by the droop characteristic of generation sites. It is conventionally implemented via a slow, centralized PI controller with low bandwidth communication. In AC microgrids, however, the *secondary control* term has been utilized not only for frequency regulation, but also for voltage regulation, load power sharing, grid synchronization and power quality issues. To eliminate voltage deviation induced by droop control, a voltage secondary control loop is often applied to a DC system, assigning proper voltage set point for primary control of each converter to achieve global voltage regulation. The secondary control coefficient (δV_i^κ) is changing the voltage reference of local unit(s) by shifting the droop lines up (or down), regulating the voltage to the nominal value:

$$V_{DC-i} = V_{Ref} - R_{D-i} \cdot i_{0-i} + \delta V_i^\kappa \tag{6.14}$$

where V_{Ref} is the microgrid reference voltage, V_{DC-i} is the local voltage set point for *ith* converter, i_{0-i} is the output current injection, and R_{D-i} is the droop coefficient. In the islanded operation mode, the global reference voltage, V_{Ref} is usually the rated MG voltage. However, in the grid-connected mode, a new reference voltage is set by the tertiary control in order to exchange power between grid and microgrid. On the other hand, a proper current sharing is a highly desirable feature in a microgrid operation, in order to prevent overloading of the converters. In droop-controlled DC microgrids, load power is shared among converters in proportion to their rated power. Since voltage is a local variable across the microgrid, in practical applications where line impedances are not negligible, droop control itself is not able to provide an accurate current sharing among the sources. In other words, the line impedances incapacitate the droop mechanism in proportional sharing of the load. Proper current sharing is a highly desirable feature in MG operation, e.g., to prevent circulating currents and overloading of the converters. In droop-controlled DC microgrids or nanogrids, the load power is shared among converters in agreement to their rated power. Since voltage is a local variable across the microgrid, in practical applications where line impedances are not negligible, the droop control itself is not able to provide an accurate current sharing among the sources. In other words, the line impedances incapacitate the droop mechanism in proportional sharing of the load. To improve current sharing accuracy, another secondary control loop is often employed. This current regulator generates another voltage correction term, δV_i^{cr} to be added to the droop mechanism, as expressed in the following relationships. The droop correction term forces the system to accurately share the currents among the MG components according to, for instance, the power rate of the converters. In an alternative approach, the current sharing module is updating the virtual impedance, to proper manage the current sharing and the droop correction term generated by the secondary controller, δR_{D-i} is adjusting the droop control mechanism. The relationships of these approaches are expressed by the relationships given below. Notice that although the secondary control ensures a proportional current sharing is, it can inversely affect the voltage regulation. Therefore, there is an inherent trade-off between these two control objectives, i.e., voltage regulation and current sharing.

$$V_{DC-i} = V_{Ref} - R_{D-i} \cdot i_{0-i} + \delta V_i^{cr} \tag{6.15a}$$

and

$$V_{DC-i} = V_{Ref} - \left(R_{D-i} - \delta R_{D-i} \right) \cdot i_{0-i} \tag{6.15b}$$

6.4.4 DC Microgrid Protection

The protection of DC grids, in general and the protection methods for DC microgrids and DC nanogrids in special are considered as one of the biggest challenges in the power engineering fields. Unlike in AC systems, the current has no zero crossing that extinguishes arcs, resulting in high fault currents in DC electricity systems. Moreover, series arcing can be an issue when high power loads are unplugged. Special plugs with leading pins are one of the possible solutions, but also selective load side arc detection could be implemented. Furthermore, coordination with frequency-based backup protection would need to be standardized. Traditional protection schemes in the low voltage grid rely on high short circuit currents, a radial system and unidirectional power flow for selectivity. However, since these three elements are not usually present in DC distribution systems, new short circuit protection strategies are needed. Advances in power electronics have led to a reduction of capacitor size and consequently their contribution to short-circuit currents. Oversizing the power converters is an expensive approach, sometimes employed for AC systems. A small DC nanogrid in islanded operation might not be able to produce high enough short-circuit currents, even with the oversized power converters. Since high fault currents are not inherently desirable, a new low short-circuit current protection philosophy is required. Low short-circuit currents allow for solid-state breakers to be used, enabling fast fault clearing and avoids arcing. Fast selectivity in meshed microgrids with bidirectional power flow is therefore essential and an important research challenge. Current limiting inductors, limiting the rate of change of the current are needed to be used. The significant impact of these limiting inductors on the control system needs to be taken into account. Inrush currents and ramp rates must be specified in order to allow fast fault discrimination. Grounding is another important topic to be taken into consideration since DC can cause corrosion if it flows through metallic structures in the environment for an extended period. High impedance grounding schemes have often been used in DC microgrids (e.g., data centers), since it allows to sustain operation during a single ground fault. However, selectivity is not possible and therefore this is not feasible for DC distribution grids. Solid grounding in one point allows for selective protection for residual ground currents; however, if there are multiple grounding points, ground currents would flow. Multiple grounding points would be needed in order to be able to island individual nanogrids. The development of advanced grounding schemes is therefore fundamental. Preliminary research indicates that capacitive grounding could be an interesting alternative since it provides low impedance for fault transients and blocks DC currents.

According to the fault characteristics, the DC microgrids fault types are pole-to-pole fault and pole-to-ground fault. The pole-to-ground faults are the most common in industrial systems. Usually, the fault impedances of pole-to-pole faults are low. However, the fault impedance of pole-to-ground faults can be either low or high. On the other hand, the fault types can be bus fault and feeder fault based on the fault location. The electronic equipment is vulnerable and the tolerance of the over current is finite. The fault is much severer when the fault location is closer to the energy sources, and therefore the bus fault is critical for the whole system. Faults inside VSCs and batteries may cause a pole-to-pole short-circuit fault and these are terminal faults that generally cannot be quickly cleared. In such cases, the devices have to be replaced and using fuses could be a proper choice. The fault characteristics are obtained from detailed analyses, and are helping to define protection operation time and assess the effect of any proposed protection schemes. The pole-to-pole fault is the most typical type in DC microgrids. The pole-to-pole fault response can be depicted in three stages: capacitor discharge stage, diode freewheel stage and grid-side current feeding stage. Protection devices commercially available for LV DC systems are fuses, molded-case circuit breakers (MCCB), LV power circuit breakers and isolated-case circuit breakers. Such models are specially designed for DC, but most of them can be used in both AC and DC applications. A fuse consists of a fuse link and heat-absorbing material inside a ceramic cartridge. When the current exceeds the fuse limit, the fuse link melts and an arc is formed. In order to quench the arc, the arc voltage must exceed the system voltage. This is done by stretching and cooling the arc. There is no current zero in a DC system which helps to interrupt the fault current. The fuse voltage and current ratings are given in RMS values, and are therefore valid for both AC and DC networks. An MCCB consists of a contractor, a quenching chamber and a tripping device. When an MCCB is tripped the contacts begin to separate, and an arc is formed. The arc is forced into the quenching chamber by air pressure and magnetic forces. The quenching chamber consists of multiple metal plates, designed to divide the arc into several smaller arcs, which increases the total arc voltage, decreases the arc temperature and the arc in most cases extinguishes. To improve the voltage withstands capability multiple poles can be connected in series. Molded-case circuit breakers are usually equipped with a thermal-magnetic tripping the device. Their voltage and current ratings are given in RMS values. The magnetic tripping senses the current instantaneous value, meaning that the rated current for DC is about 1.42 higher than for the AC case. There are problems associated with fuses and circuit breakers in LV DC systems such as larger time constants and longer breaker operation time. By utilizing power electronic switches such as gate-turn-on thyristors the operation speed decreases and the inductive current interruption capability are increased. However, the losses of such a solution are higher compared with a mechanical switch. Therefore a combination of one mechanical switch and one power electronic switch is often proposed to solve these issues. Grounding is a critical issue for DC microgrids protection. Different grounding options come with different fault characteristics and are influencing the configuration and setting of the protection. The purpose of grounding designs is to facilitate the ground fault detection, minimize the stray current and to ensure the personnel and equipment safety. DC microgrids can be grounded with either high resistance or low resistance, even ungrounded. The ungrounded mode has been highly recommended especially in LV applications. In that case, the common-mode voltage could not be high enough to pose a threat to personnel and equipment safety. Meanwhile the system could operate continuously when a single phase-to-ground fault occurs. However, a possible second ground fault at another pole may result in a line-to-line fault and do severe damage to the whole system. Solid grounding has rarely been adopted because of the corrosion caused by stray current. Compared with low resistance grounding, high resistance grounding limits the ground fault current so that the system could keep operating during a fault, but the detection and location of the fault becomes much more difficult. Moreover, the point of the system selected to be grounded could be the midpoint, the positive or the negative pole of the common DC link.

Similar to an AC power system, the IEC 60364 standard determines the grounding strategies.

Similar to the AC power systems, the protection schemes designed for DC systems are divided into non-unit and unit protection techniques. All the desirable characteristics like reliability, selectivity, speed, performance, economics and simplicity have to be taken into consideration when designing and setting a protection system. Non-unit protection is not able to protect a distinct zone of the power system and operates directly when the threshold is exceeded. Meanwhile, non-unit protection schemes have inherent advantages for coordinating the whole protection system. Non-unit protection realizes fault discrimination in a DC microgrid system by analyzing the current, voltage, current and voltage changing time rates (di/dt, dv/dt) and the impedance response in a range of the fault. Some protection systems are designed based on the fault current natural characteristics and its first and second order derivatives. Various faults are easily discriminated with the derivatives of the currents. The selectivity of non-unit protection methods was on the basis of complex setting values and proper time delays. Given that, current limiting methods were presented to release the time coordination tension. Because of the additional cost of the crucial communication and relay devices, the implementation of the unit protection method is closely restricted. But the development of the smart grid and microgrid suggests for an increased investment of sensors and communication infrastructures within the distribution systems to achieve advanced automatic network monitoring and management. Apparently the deployment of these devices will provide the opportunity to promote the application of the unit protection schemes. Unit protection supports a clear zone and never responds to an external circuit fault. In comparison with non-unit protection, unit protection does not have backup protection to the adjacent elements in the system, thus it is common that the non-unit protection is deployed alongside the unit protection to act as a backup protection.

6.5 Summary

Future electricity distribution networks with large amount of DERs, including DG and RES units, electricity storages, electric vehicles and customers with smart energy meters and controllable loads requiring the creation of totally new smart grid architecture. This new architecture takes advantage of the properties of DER together with new intelligent management functions and hence allows the potential of DER to be realized for different interest groups such as distribution system operators, DG producers, service providers, consumers and society. In the development of the smart grid architecture microgrids with momentary island operation possibility should be seen as basic blocks of the architecture. The term microgrid is typically used from the LV networks of the (smart) grid with an island operation capability. Microgrids allow the integration of renewable energy sources as the most optimal and reliable sources of energy into the power networks. Potential applications for microgrids are shopping or office centers, university campuses or high schools, hospitals, power for essential and critical services (police, fire, water treatment facilities), farms, located far from the grid, remote power networks (islands, rural areas, isolated villages), suburbs and blocks not connected to centralized district heating systems, military installations and facilities. The use of microgrid concept with island operation capability as part of future smart grid architecture allows the reliability benefit of DER to be realized as well as the energy efficiency aspects to be fulfilled. Also in developing countries and rapidly industrializing economies the microgrid concept can be interesting option to meet the local challenges with electricity distribution and generation, which are quite different from those for post-industrial economies. In developing economies the reach of the transmission grid is geographically limited and it is not profitable to build more transmission grid due to lack of purchasing power and low levels of average consumption among the unserved populations. Instead small remote hybrid systems based on operating diesel engine generators together with solar and/or wind power systems have been used as separate island grids. The microgrid concept highlights three essential microgrid features: local loads, local micro-power sources and intelligent control units. A microgrid can be defined as an electrical network of small modular distributed generators (micro-generators). Their generation units are usually PV systems, fuel cells, micro-turbines or small wind generators. The microgrid also includes energy storage devices and controllable loads. Most of the microgrid small-size generators are connected to the network through power electronic converters and interfaces. A microgrid can operate in grid-connected or islanded mode, and hence increase the reliability of energy supplies by disconnecting from the main distribution network in the case of network faults or reduced power quality. It can also reduce transmission and distribution losses by supplying loads from local generation and from a benign element of the distribution system. An LV DC microgrid can be preferable to an AC microgrid, where most of the sources are interconnected through a power electronic interface and most loads are sensitive electronic equipment. The advantage of an LV DC microgrid is that loads, sources and energy storage then can be connected through simpler and more efficient power-electronic interfaces. To achieve a higher quality of service, e.g., global voltage regulation, and power flow control, communication-based higher control layers must be applied to these small power systems.

In autonomous mode, each MG has its own control layers to supports its local loads. While connected, the power or current flow among microgrids may be controlled to optimize the utilization of their energy sources. It is obvious that power flow control among MGs can be achieved by adjusting their bus voltages. Thus, a trade-off needs to be taken into account between the conflicting goals of voltage regulation and power flow control. Various protection schemes for microgrids, due to their specific characteristics (bidirectional power flows, low inertia, different fault current levels, etc.) both in grid-connected and islanded mode of operation, are needed. Since microgrids operate in both grid-connected and islanded modes, the use of traditional protection schemes based on overcurrent relays would not be possible. A review of the control protection schemes and coordination techniques that have been developed or presented in the literatures is presented in this entree. All the today control and protection schemes and coordination techniques have their own merits and demerits that draw interest to their application in the microgrid operation. A good protection should be capable of fulfilling the microgrid requirements in terms of reliability, selectivity, speed and cost. For accurate selection of a control or protection scheme, a thorough study has to be done on the network characteristic and suitability of the scheme to be adopted based on the network itself. For microgrid protection, it is always better to use a scheme that utilizes communication links as it will ensure fast operation of the protective devices. As for the existing distribution system connected to a microgrid, maximizing the existing infrastructure without much investment will be the most economical decision, although not always the most desirable. This chapter has also presented a brief review of the existing energy management system (EMS) architectures for microgrids, identifying the main advantages of each approach, and has proposed a centralized EMS architecture for implementation on isolated microgrids in stand-alone mode of operation. Some relevant considerations and procedures for the model fine-tuning and performance evaluation have also been presented. Future work will concentrate on the implementation and testing of the proposed architecture.

Questions and Problems

1. List the major benefits of microgrids.

2. What is a virtual power plant? Briefly describe its attributes.

3. What are the major divers of the smart grid development?

4. List and briefly describe the smart grid major features.

5. Why a microgrid can be one of the main smart grid contributors?

6. What are the major reasons of using DC electric networks and power?

7. What are the three features, defining the microgrid concept?

8. List the attributes differentiating a VPP from a microgrid.

9. What are the major types of microgrids?

10. Define in your own words the distributed generation concept.

11. List the microgrid internal and local control functionalities.

12. Briefly discuss the microgrid protection issues.

13. List the main features of a microgrid.

14. Briefly describe a microgrid main stakeholders.

15. List and briefly discuss the microgrid protection methods.

16. Briefly describe the microgrid control methods.

17. Briefly describe the microgrid functionalities.

18. A microgrid consists of four 30 kW wind turbines, 36 PV panels of 265 W each, and four battery banks with enough capacity to store the all excess energy, and the power management, control and protection units. The microgrid is supplying power to a small remote hotel having daily energy demand of 18.5 kW. Each wind turbine and PV panel has a capacity factor of 0.26 and 0.28, respectively. Each wind turbine is using internally 0.375 kW of the generated power, each PV panel 9.5 W, while the battery banks, power management, control and protection are using 1.8% of the output generated power to operate. Estimate the MG self-consumption and self-sufficiency coefficients.

19. List and briefly discuss the major microgrid applications.

20. Compute the self-consumption coefficient of the three most common types of PV panels used in residential applications. Search the manufacturer specifications and make assumptions on your location average estimated generation.

21. Briefly describe the droop control method.

22. What are the main reasons for the use of a DC network?

23. What are the main reasons of the use of a DC nanogrid or microgrid?

24. List and briefly discuss the microgrid power types.

25. List and briefly describe the effects on the protection devices and systems, when the DG units are connected to the distribution networks.

26. What are the possible corrective measures in this case?

27. Briefly describe the microgrid protection methods.

28. List and describe the major functions of a DC microgrid protection system.

29. What is a hybrid power system?

30. Estimate the apparent, real and reactive powers flowing into a transmission line, having the line impedance of $0.12 + j1.2\ \Omega$, the line input voltage of 4.32 kV, the line output voltage 4.295 kV and the power angle 4.5°.

31. If the real and reactive powers flowing into a transmission line have the line impedance of $0.15 + j1.2\ \Omega$, the line input voltage of 13.50 kV and the power angle 5° are 36 MW and 2.8 MVAR, estimate the line output voltage.

32. A Boost power converter has the input DC voltage 28 V, duty ratio 0.6, and the load resistance R_L is 12 Ω. Calculate the output voltage and the power delivered to the load. What is its efficiency if the all converter internal losses are 24.5 W?

33. The reference voltage of a DC microgrid is 360 V (DC), the droop coefficient is changing from 0.75 to 0.25 Ω and source output current is 15 A. Estimate the minimum and maximum controller voltage and the voltage deviation.

References and Further Readings

1. IEEE Standard 399-1997, *IEEE Recommended Practice for Industrial and Commercial Power System Analysis*, 1997.
2. K. Fleischer and R. Munnings, Power systems analysis for direct current DC distribution systems, *IEEE Transactions on Industry Applications*, Vol. 32(5), pp. 982–989, 1996.
3. IEC Standard, *Short-circuit Currents in DC Auxiliary Installations in Power Plants and Substations*, IEC 61 660, Geneva, Switzerland, 1997.
4. IEEE Recommended Practice for the Design of DC Auxiliary Power Systems for Generating Stations, IEEE Std. 946-2004, 2004.
5. P. Sen, *Principles of Electric Machines and Power Electronics* (2nd ed.), Wiley, New York, 1996.
6. R.H. Lasseter, Microgrids, in *Proceedings of the IEEE Power Engineering Society Winter Meeting*, Vol. 1, January 27–31, pp. 305–308, New York, 2002.
7. R.H. Lasseter and P. Paigi, MicroGrid: A conceptual solution, in *IEEE Annual Power Electron Specialists Conference*, Vol. 1, pp. 4285–4290, 2004.
8. R.H. Lasseter, Microgrids and distributed generation, *Journal of Energy Engineering*, Vol. 1, pp. 1–7, 2007.
9. J.A.P. Lopes, C.L. Moreira, and A.G. Madureira, Defining control strategies for microgrids islanded operation, *IEEE Transactions on Power Systems*, Vol. 21, pp. 916–924, 2006.
10. G. Pepermans et al., Distributed generation: Definition, benefits and issues, *Energy Policy*, Vol. 33, pp. 787–798, 2005.
11. A.L. Dimeas and N.D. Hatziargyriou, Operation of a multi-agent system for microgrid control, *IEEE Transactions on Power Systems*, Vol. 20(3), pp. 1447–1455, 2005.
12. R.H. Lasseter, Microgrids and distributed generation, *Journal of Energy Engineering American Society of Civil Engineers*, Vol. 133(3), pp. 144–149, 2007.
13. J.P. Lopes, N. Hatziargyriou, J. Mutale, P. Djapic, and N. Jenkins, Integrating distributed generation into electric power systems: A review of drivers, challenges and opportunities, *Electric Power Systems Research*, Vol. 77(9), pp. 1189–1203, 2007.
14. D. Pudjianto, C. Ramsay, and G. Starbac, Microgrids and virtual power plants: Concepts to support the integration of distributed energy resources, *Proceedings of the Institution of Mechanical Engineers, Part A: Journal of Power and Energy (IMechE)*, Vol. 222(7), pp. 731–741, 2008.
15. A. Timbus, M. Liserre, R. Teodorescu, P. Rodriguez, and F. Blaabjerg, Evaluation of current controllers for distributed power generation systems, *IEEE Transactions on Power Electronics*, Vol. 24(3), pp. 654–664, 2009.
16. S. Chowdhury, S.P. Chowdhury, and P. Crossley, *Microgrids and Active Distribution Networks*, IET Renewable Energy Series 6, IET, London, UK, 2009.
17. D. Salomonsson, L. Soder, and L. Sannino, Protection of low-voltage DC microgrids, *IEEE Transactions on Power Delivery*, Vol. 24(3), pp. 1045–1053, 2009.
18. Z. Xiao, J. Wu, and N. Jenkins, An overview of microgrid control, *Intelligent Automation and Soft Computing*, Vol. 16(2), pp. 199–212, 2010.
19. H.J. Laaksonen, Protection principles for future microgrids, *IEEE Transactions on Power Electronics*, Vol. 25(12), pp. 2910–2918, 2010.
20. N. Jenkins, J.B. Ekanayake, and G. Strbac, *Distributed Generation*, The IET Press, London, UK, 2010.
21. J.C. Vasquez, J.M. Guerrero, J. Miret, M. Castilla, and L.G.D. Vicuna, Hierarchical control of intelligent microgrids, *IEEE Industrial Electronics Magazine*, Vol. 4, pp. 23–29, 2010.
22. J. Casazza and F. Delea, *Understanding Electric Power Systems: An Overview of Technology, the Marketplace, and Government Regulation* (2nd ed.), John Wiley & Sons, Hoboken, NJ, 2010.
23. R. Teodorescu, M. Liserre, and P. Rodriguez, *Grid Converters for Photovoltaic and Wind Power Systems*, John Wiley & Sons, Chichester, UK, 2011.

24. A. Ruiz-Alvarez, A. Colet-Subirachs, F. Alvarez-Cuevas Figuerola, O. Gomis-Bellmunt, and A. Sudria-Andreu, Operation of a utility connected microgrid using an IEC 61850-based multi-level management system, *IEEE Transactions on Smart Grids*, Vol. 3(2), pp. 858–865, 2012.
25. M. Yazdanian, G.S. Member, and A. Mehrizi-sani, Distributed control techniques in microgrids, *IEEE Transactions on Smart Grid*, Vol. 5(6), pp. 2901–2909, 2014.
26. N. Hatziargyriou (ed.), *Microgrids Architectures and Control*, Wiley, Chichester, UK, 2014.
27. R. Belu, Microgrid concepts and architectures, in *Encyclopedia of Energy Engineering & Technology* (Online) (Sohail Anwar, R. Belu et al. eds.), CRC Press/Taylor & Francis Group, New York, 2014.
28. J.M. Guerrero, J.C. Vasquez, J. Matas, L.G. de Vicuna, and M. Castilla, Hierarchical control of droop-controlled AC and DC microgrids—A general approach toward standardization, *IEEE Transactions on Industrial Electronics*, Vol. 58(1), pp. 158–172, 2011.
29. R.H. Lasseter, Smart distribution: Coupled microgrids, *Proceedings of the IEEE*, Vol. 99(6), pp. 1074–1082, 2011.
30. E. Planas, A. Gil-de Muro, J. Andreu, I. Kortabarria, and I.M. de Algeria, General aspects, hierarchical controls and droop methods in microgrids: A review, *Renewable and Sustainable Energy Reviews*, Vol. 17, pp. 147–159, 2013.
31. J.J. Justo, F. Mwasilu, J. Lee, and J.W. Jung, AC-microgrids versus DC-microgrids with distributed energy resources: A review, *Renewable Sustainable Energy Reviews*, Vol. 24, pp. 387–405, 2013.
32. R. Belu, Microgrid protection and control, in *Encyclopedia of Energy Engineering & Technology* (Online) (Sohail Anwar, R. Belu et al. eds.), CRC Press/Taylor & Francis Group, New York, 2014.
33. M. Farhadi and O. Mohammed, Adaptive energy management in redundant hybrid DC microgrid for pulse load mitigation, *IEEE Transactions on Smart Grid*, Vol. 6, pp. 54–62, 2015.
34. A.T. Elsayed, A.A. Mohamed, and O.A. Mohammed, DC microgrids and distribution systems: An overview, *Electric Power Systems Research*, Vol. 119, pp. 407–417, 2015.
35. A. Bidram and A. Davoudi, Hierarchical structure of microgrids control system, *IEEE Transactions on Smart Grid*, Vol. 3, pp. 1963–1976, 2016.
36. R.A. Huggins, *Energy Storage* (2nd ed.), Springer, Cham, Switzerland, 2016.
37. K.R. Khalilpour and A. Vassallo, *Community Energy Networks with Storage—Modeling Frameworks for Distributed Generation*, Springer, Singapore, 2016.
38. R. Bansal (ed.), *Handbook of Distributed Generation: Electric Power Technologies, Economics and Environmental Impacts*, Springer, Cham, Switzerland, 2017.
39. M.S. Mahmoud (ed.), *Microgrid: Advanced Control Methods and Renewable Energy System Integration*, Elsevier, Amsterdam, the Netherlands, 2017.
40. A. Bidram, V. Nasirian, A. Davoudi, and F.L. Lewis, *Cooperative Synchronization in Distributed Microgrid Control*, Springer, Cham, Switzerland, 2017.

7

Renewable Energy Environmental Impacts

7.1 Introduction, Renewable Energy Environmental Implications

Conventional energy conversion and its uses based on the fossil fuel gas have been proven to be highly effective drivers of the economic progress, but at the same time having quite significant adverse impacts to the environment and to human health. These conventional energy sources are facing increasing pressure from the environmental protection sector, and perhaps the most serious confronting challenge is the coal future due to the various protocols and legislations targeting significant reductions. Renewable energy sources currently supply up to 20% of the total world energy demands, while the supply is dominated by traditional biomass, used in the developing countries, mostly wood. Renewable energy, such as solar and wind energy, bioenergy, geothermal energy and small hydropower, is contributing about 2% to the energy supply. Sustainable development has become part of the national strategies and development plans in the United States and all other countries. Concerns about pollution, supply security and diversification are major drivers in the push for the RES uses. Renewable energy systems (RES) applications have significantly less pollutant emissions than fossil fuels, while most estimates of the emissions from the nuclear power plants are similar in magnitude to those from the renewable energy sources. The atmospheric CO_2 level seems to increase significantly in the last century, which can be correlated with the increased fossil fuel uses for various applications. From economic point of view the alternative energy sources are still not able to fully compete with fossil fuels. Moreover, the adding carbon capture, sequestration and storage to coal and natural gas power generation stations can significantly reduce the relative environmental advantages than that the RES systems have in terms of carbon and energy savings. It is worth to notice that the renewable energy environmental impacts are site specific, either some limited generalizations are still possible. Renewable energy is more environmentally friendly than conventional energy sources, e.g., the air emissions. The likely life-cycle emissions (taking into account harvesting, transportation and processing, power station construction, operation and decommissioning) from the main renewable energy sources used for electricity generation are shown in Table 7.1. The results are purely indicative but show the variations and relative differences between the various sources. Life-cycle emissions from renewable energy uses are small compared with those from fossil fuel plants. The nuclear power generation does have a major environmental impact; it releases no SO_2 or nitrogen oxides (NO_X) and very little CO_2, with the life-cycle emissions falling within the ranges of renewable energy sources.

Environmental and sustainability issues span a continuously growing range of pollutants, hazards and ecosystem degradation factors that affect areas from local through regional and global scale. Such concerns arise from the monitoring data, while others stem

TABLE 7.1

Life-Cycle Air Emissions of Major Renewable Energy Sources (g/kWh)

Emission	Biomass	Solar	Wind	Hydropower	Geothermal
CO_2	17.0–27.0	26.0–167.0	7.0–9.0	3.60–11.60	79.0
SO_2	0.07–0.16	0.13–0.34	0.02–0.09	0.009–0.030	0.02
NO_X	1.10–2.50	0.06–0.30	0.02–0.06	0.003–0.07	0.28

Note: IEA Report, The Environmental Implications of Renewables, OECD/IEA, 1998.

from actual or perceived environmental risks such as possible accidental releases of hazardous materials. Many environmental issues are caused by or related to the energy production, transformation and use, e.g., acid rain, stratospheric ozone depletion and climate changes. A variety of potential solutions to the environmental problems associated with the harmful pollutant emissions have been researched and developed. Renewable energy sources (RES), Distributed generation (DG) and hydrogen energy appear to be an effective solution that can play a critical role in the sustainability and environment protection. There are several indicators used to assess and evaluate the energy generation and its environmental impacts. One of them is the net energy ratio (NER), defined as the ratio of the useful grid energy output to the fossil-fuel energy consumed during the technology lifetime that represents a critical parameter in assessing whether or not a renewable energy source reduces the fossil fuel consumption. Renewable energy sources usually have an NER value greater than one. For the fossil-fuel energy technologies, the NER is referred to as the life-cycle efficiency. However, there is some inconsistency in how the NER is defined when the energy technology itself is based on fossil fuels. Often, it includes only the indirect (external) energy inputs and not the fuel (primary) energy inherent. This NER definition is not an accurate account of the total resource consumption of the specific energy technology. For example, the energy consumed by combusting coal in a coal-fired plant is not included. In such cases where the fuel primary energy is not included into the energy inputs, the NER is more accurately defined as the external energy ratio (EER), or the energy payback ratio. For renewable energy sources such as wind and solar energy, the NER and EER are the same, since such technologies are not using any fossil fuels for operation, and are not depleting any energy resource. NER values are influenced by factors such as the power plant capacity factor, life expectancy, plant construction materials (e.g., steel versus concrete for wind towers), and fuel mix during material construction. For some renewable energy sources, such as wind and solar energy, the location and the resource strength are important variables. For example, the solar installations in the areas with greater solar radiation have higher NERs. A usual approach to evaluate the environmental impacts of any generation system is to determine the unit cost of the energy production (the cost per kWh). Once the cost per kWh, or energy unit cost from production, has been determined, multiplication by the emitted quantities of the pollutants yields the cost per activity, for instance, per kWh of electricity produced by a power plant. A complete accounting of the damage costs needs to involve an life-cycle assessment (LCA), i.e., a complete inventory of emissions over the entire chain of processes involved in the activity. The total damage cost per kWh of electricity should include impacts upstream and downstream from the power plant, such as air pollution from the ships, trucks, or trains that transport the fuel to the power plant. However, in order to estimate the damage costs of a pollutant, it is needed to carry out an impact pathway analysis, tracing the pollutant from where it is emitted to the affected receptors. The principal steps of an IPA can be grouped as follows: emission, dispersion, impacts and cost. The impacts and costs are summed over all receptors of concern. For many environmental choices one needs to look not only at a particular source of pollutants but has to take into account an entire process chain by means of an LCA. For example, a comparison of power generation technologies involves an analysis of the entire fuel chain.

Sustainability is advocating that the economic development must be pursued without causing irreparable damages to the ecology and the environment. Sustainability includes all economic activities from the good and service production to the transportation and energy production. Measures of the sustainability include the calculation of pollutant emissions per unit of the desired product or service. Most notable among these measures is the carbon footprint, defined as the amount of CO_2 produced for the completion of an economic activity. The sustainability concept is a reaction to the environmental effects, caused by the anthropogenic activities and the realization that if these activities are continued unchecked and unmitigated, the planet may become uninhabitable. Sustainability encompasses ideas and concepts from several disciplines, engineering, environmental science, ecology, economics, sociology, anthropology, political science, and public policy. Central to this subject is that significant global threats, such as global warming and pollution prevention, can be tackled by a combination of technology, social awareness and public policy. Energy input and output estimates, the basic building blocks for any LCA, are used to evaluate the energy intensity and resource consumption of the technology itself. *Utilizing the energy* always means converting the energy from one form into another. For instance, in space heating, *the energy is utilized,* meaning the conversion of the wood chemical energy into heat, while in irrigation, an engine converts fuel chemical energy into mechanical energy for powering a pump, which in turn provides water for irrigation. *Generating energy* means converting energy from one form into another, e.g., a diesel engine converts the oil chemical energy into mechanical energy, or a wind turbine generates energy, converting the wind kinetic energy into mechanical energy and eventually into electrical energy through its electric generator. The *energy generation*, in fact, deals with an energy source, whereas the *energy utilization* serves an end-use of energy. In between, the energy can flow through a number of conversion phases. The words *generation* and *utilization* are a little confusing because, in fact, no energy can be created or destroyed; it is only transformed or converted from one form into another. In the generation, the energy is made available from a source, by converting it into another form. In utilizing energy, it is often converted from some intermediate form into a final useful form. In all conversion processes, part of the energy is lost, not meaning that it is destroyed, but rather that it is lost for the purpose, as is shown in Figure 7.1. The key characteristic of a conversion process is the efficiency, which is estimated as the ratio of the useful output per input. In that sense, it is a subjective value which depends on a particular purpose of a technological process and a particular input resource. Hence, the efficiency has different meanings in different disciplines. Efficiency is important in the sustainability context because it indicates how much of the

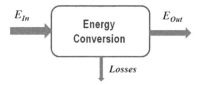

FIGURE 7.1
Energy conversion diagram.

resource is put to work, and how much of the resource is wasted. All losses are process dependent and must be analyzed specifically for each application. A part of the technological research is aimed at increasing efficiency of the conversion process via minimizing losses. The converter efficiency is defined as the quantity of energy in the desired form (the output energy) divided by the input energy into the conversion process. The efficiency, η, gives the fraction of the input energy, E_{In}, converted to a useful energy, E_{Out}, expressed as:

$$\eta(\%) = \frac{E_{Out}(\text{Useful Energy})}{E_{In}} \times 100 \qquad (7.1)$$

Notice that the efficiency of a process or technology is not necessarily measured in terms of energy units. If the useful output of the converting technology is, for example, some form of matter (e.g., water electrolyzer), the calculation can be made in terms of mass. Sometimes, efficiency analysis is used to estimate the maximum theoretical efficiency, which cannot be exceeded due to the inherent system physicochemical limitations. Finding maximum theoretical efficiency requires detailed scientific knowledge of how the process works and what unavoidable losses occur in conversion. Energy conversions can take place from any one form of energy into almost any other form of energy. Some conversions have no practical value, and which conversion is desired depends on the application. For instance, for power generation, the potential energy from hydro resources is converted into mechanical energy, whereas in the water pumping for irrigation, the reverse process. The PV cells convert the solar radiation energy into electricity, whereas the light bulbs are do the reverse. Often, in applications, the energy flows through a number of forms, as well as the conversion steps, between the source and the end-use. The energy converter can be a device, or a process, or a whole system. The overall efficiency equals the product of the efficiencies of the various components of the system. In some of the energy converters, intermediate forms of energy occur between the input energy form and the form of the output energy. For instance, with diesel engines, the intermediate form is the thermal energy.

Example 7.1: A diesel engine has an efficiency of 35% and is connected to an electric generator that has an efficiency of 93%. How much electricity is produced from 10 GJ of the fuel chemical energy?

Solution: The overall diesel engine-generator efficiency is 0.3255% or 32.55%. Applying Equation (7.1) the generated electricity is:

$$E_{Out} = \eta_{System} \times E_{In} = 0.3255 \times 10 \times 10^9 = 3.255 \times 10^9 \,\text{J or } 904.167 \,\text{kWh}$$

7.2 Life-Cycle Assessment and Analysis

One of the most important technology abilities is the conversion capability of materials and/or energy. Any technology uses inputs of energy and material(s) to create outputs of energy and/or product(s) of different qualities. A technology, as represented by the simple diagram of Figure 7.2, is typically serving as a conversion portal for specific

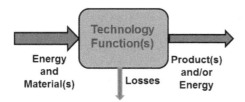

FIGURE 7.2
Energy and material(s) conversion defined as a function of technology.

purpose(s). The energy is used to produce materials, which are used to produce complex products, specific materials, or is converted into other energy forms. Notice that a conversion is also performed by natural systems, but only a technology is categorized as a human-made conversion system. Some conversion examples are: fuel chemical energy is converted by an internal combustion engine into mechanical energy, sunlight is converted by a PV panel into electricity, or electric energy is converted by a furnace into heat. The key characteristic of a conversion process is efficiency. Efficiency is estimated based on the amount of useful output per unit input, as expressed by Equation (7.11). In that sense, it is a subjective value, depending on a particular purpose of a technological process, and a particular input. Hence, efficiency has widely varying meanings in different disciplines. Efficiency is important in the sustainability context because it indicates how much of the resource is put to work, and how much of the resource is wasted in the process. The reasons for losses are process dependent and should be analyzed specifically for each application. A big part of the technological research is aimed at increasing efficiency of the conversion process via minimizing losses. Table 7.2 lists some of the efficiencies of various energy technologies, a few examples to demonstrate the variety of the converters. It should be noted that generally the efficiency of a process or technology is not necessarily

TABLE 7.2

Efficiencies of Common Energy Conversion Processes (Technologies)

Process/Technology	Process Input	Process Output	Process Efficiency (%)
Gas turbine	Gas flow	Electricity	40%
Hydropower turbine	Water flow	Electricity	90%
Wind turbine	Air flow	Electricity	20%–35%
Solar cell	Light	Electricity	15%–40%
Fuel cell	H_2(gas), O_2(gas)	Electricity	up to 85%
Electrolyzer	Electricity	H_2(gas), O_2(gas)	50%–70%
Combustion engine	Fuel (gasoline)	Mechanical energy	10%–50%
Geothermal electric plant	Heat	Electricity	10%–23%
Solar thermo-electric generator	Solar radiation	Electricity	15%
Electric motor	Electricity	Mechanical energy	60%–90%
Electric generator	Mechanical energy	Electricity	60%–96%
Power transformer	Electricity	Electricity	80%–98%
Electric heater	Electricity	Heat	up to 100%
Refrigerator	Electricity	Heat (Negative)	20%–40%
Li-ion battery	Chemical energy	Electricity	80%–90%
Fluorescent lamp	Electricity	Light	8%–15%
LEDs	Electricity	Light	Up to 53%
Photosynthesis	Light	Biomass, O_2(gas)	3%–6%

measured in terms (units) of energy. If the useful output of the converting technology is, for example, some form of matter (e.g., water), the calculation can be made in its specific terms, such as mass and volume. Sometimes, efficiency analysis is also used to estimate the maximum theoretical efficiency, which cannot be practically exceeded due to inherent physicochemical system limitations. Finding maximum theoretical efficiency requires detailed knowledge of how the process works and what unavoidable losses occur in conversion. Notice that to become part of society life, any technology needs to be adapted. Not all technologies invented are passing through the successful adaptation, and there are several critical barriers that need to be overcome in order to create a working interface between technology and society.

> **Example 7.2:** By using the values of Table 7.2, estimate the efficiency range of a power system (generation, transmission and power distribution), assuming the 93% efficiency for the overall transmission lines, and there are three power transformers in this power system.
>
> **Solution:** The power system efficiency is the product of the efficiencies of the generation, transmission and the three power transformers, while the range is set by the minimum and maximum values
>
> $$\eta_{\min} = \eta_{gen} \times \left(3 \times \eta_{\text{transformer}}\right) \times \eta_{tr-line} = .6 \times .8 \times .8 \times .8 \times .93 = 0.286 \text{ or } 28.6\%$$
>
> and
>
> $$\eta_{\max} = \eta_{gen} \times \left(3 \times \eta_{\text{transformer}}\right) \times \eta_{tr-line} = .96 \times .98 \times .98 \times .98 \times .93 = 0.840 \text{ or } 84\%$$

LCA, a *cradle-to-grave* approach, is used for assessing products, processes, or industrial systems. The *cradle-to-grave* approach begins with the gathering of raw materials from the earth to create the product, and to the end, when all materials are returned to the earth. LCA evaluates all life product stages from the perspective that they are interdependent, meaning that one operation leads to the next. There are two main types of LCA: attributional LCA (ALCA) and consequential LCA (CLCA). ALCA is used to determine or attribute the environmental burdens associated with the production and product uses at a given point of time. A CLCA is used to quantify the product environmental impacts as a result of a system change. LCA enables the estimation of the cumulative environmental impacts resulting from all stages in the product life cycle and allowing the selection of the path that is more environmentally preferable. An LCA estimates the resource requirements, the energy uses and the environmental impacts of products or services at all life stages, over its entire life time. The estimates may be derived from detailed, bottom-up analyses of mining, energy, manufacturing, transport, construction, operations and disposal processes or from top-down analyses based on national-scale economic input/output models. Overall, LCA is a useful method and tool for comparing the impacts of different technologies and for identifying points in the life cycle where improvements can be made. LCA has been developed over decades coming from product-oriented model used to evaluate environmental impact to a bigger framework that elaborates on a wider environmental, economic and social scale. At the current stage, LCA is being transformed into Life Cycle Sustainability Analysis, which links the sustainability questions with the knowledge and research needed to address them. Read the following article to learn more about the LCA background. LCA helps decision-makers select the product, process, or technology that results in the least environmental impacts. This information can be used with other factors, such as cost and performance data to find optimal solutions. LCA identifies

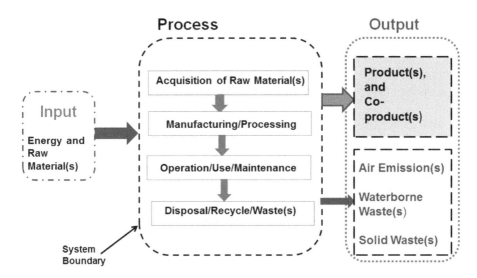

FIGURE 7.3
The main stages and typical inflows and outflows of energy, materials, products and wastes considered in an LCA process.

the transfer of environmental impacts from one media to another (for instance, a new process may lower air emissions, but creates more wastewater, etc.) and between different life-cycle stages. The diagram of Figure 7.3 illustrates the main phase or stages to be considered in any LCA analysis and application.

Any product, process, or technology requires input of some raw materials and energy at all stages: from the acquisition to manufacturing, operation and to the disposal and/or recycle. All of the life-cycle stages may produce air emissions, waterborne and solid wastes, simply because the efficiency of material uses and energy conversion is never 100%. There are losses and by-products, which sometimes are highly undesirable. LCA helps to keep track of all useful and harmful outcomes, and the diagram in Figure 7.3 provides a guideline to LCA mapping. An LCA plan includes the following main stages:

1. Goal definition and scope, consisting of the identification of a product, process, or technology, and in the establishment of the context and system boundaries

2. Inventory analysis, designed to identify and quantify energy, water and materials as inputs as well as environmental releases as outputs

3. Impact assessment, set to assess the potential human and ecological effects and to quantify the metrics

4. Data interpretation, designed to compare data from Inventory Analysis and Impact Assessment stages to select or recommend a preferred product, process, or technology

There are a few LCA limitations, such as:

- LCA thoroughness and accuracy will depend on the availability of data; gathering of data can be problematic, hence a clear understanding of the uncertainty and assumptions is important.

- Classic LCA does not determine which product, process, or technology is the most cost effective or top-performing; therefore, an LCA needs to be combined with cost analysis, technical evaluation and social metrics for comprehensive sustainability analysis.

- Unlike traditional risk assessment, LCA does not necessarily attempt to quantify any specific actual impacts, being suitable for relative comparisons and may be not sufficient for absolute predictions of risks.

Even for smaller systems, an LCA is a complex, requiring interdisciplinary technical and economic knowledge. Hence, LCA projects are usually assigned to teams of experts, being rarely performed by a single person with sufficient accuracy. The LCA is a technique for assessing aspects associated with development of a product and its potential impact throughout a product entire life. Before a new energy technology is fully implemented, the environmental superiority over competing options is asserted by evaluating its cost investments, energy and material uses, and pollutant emissions, throughout its entire life cycle. In the typical *cradle-to-grave approach* of LCA, the investigated life-cycle stages involve the exploration of materials and fuels, the product production, operation, disposal and recycling (as shown in production chain diagram of Figure 7.4). LCA approach has developed over decades from product-oriented model used to evaluate environmental impacts to a bigger framework that elaborates on wider environmental, economic and social scales. Often, an approach is to link the LCA to the sustainability, through the so-called Life Cycle Sustainability Analysis, connecting the sustainability questions with the knowledge needed to address them. The International Organization for Standardization (ISO) has standardized the basic LCA principles in four steps. The goal and scope definition describes the underlying questions, the system boundaries, and the definition of functional units for the comparison of different alternatives. The flows of pollutants, materials, and resources are investigated and recorded in the inventory analysis. The emissions and used resources are described, characterized and aggregated for various environmental issues during the impact assessment phase. Final conclusions are drawn during the interpretation. When conducting an LCA, the product, process or service design and/or development phase is often excluded, under the assumption of not contributing significantly to such impacts. However, the design or development phase decisions are highly influencing the environmental impacts in other life-cycle stages. The product, process or system design predetermines its behavior and the impacts into the later phases. Therefore, if the LCA's aim is to improve goods, processes and services, its most important goals then should be carried out in the earlier design phase and

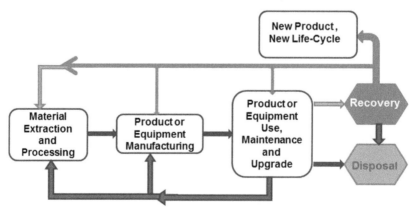

FIGURE 7.4
Product life-cycle diagram, material and energy flows.

concurrent to the other design procedures. This applies to the design or improvement within the product, process, or service life cycle, especially if interactions with other processes or stages can occur. LCA aims to analyze and compare different products, processes or services that are fulfilling the same utility, for their improvements, marketing and environmental policy. The processes within the life cycle and the associated material and energy flows or other exchanges are modeled to represent the product system and its inputs and outputs from and to the environment, resulting in system models and the inventory of environmental exchanges related to the functional unit. Besides the economic issues, of particular importance is assigned to the environmental issues associated with the choice of any energy source. The LCA is regularly applied to the generation environmental effects, helping producers to make informed decisions. The LCA purpose is not only to identify the environmental impacts, but also to create the basis for sustainable development. These impacts are analyzed during the product or service life cycle, being used for the following purposes:

1. Development of business strategy purchasing decisions
2. Product and process design
3. Product and process improvement
4. Setting eco-labeling criteria
5. Communication about environmental aspects of products
6. Comparison between different goods and services
7. Identification of places where the environmental impact is the most significant

With increasing environmental standards of modern energy systems, the up- and downstream processes, e.g., fuel supply or system production, are becoming increasingly relevant. Over the past 10 years, the use of LCA has grown rapidly. Parallel to this development, an international standardization process was started with ISO standards structuring this instrument and giving guidelines for the practitioner. The two key elements of an LCA are: the entire LCA of the investigated system and the assessment of the environmental impacts. According to the ISO, the LCA basically consists of four steps (see Figure 7.5). The first step

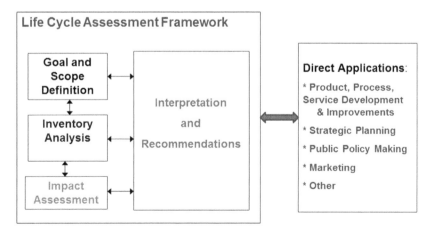

FIGURE 7.5
Phases and the LCA process structure diagram. (Adapted from ISO 14040, Environment Management–Life Cycle Assessment–Principles and Framework, 1997; ISO 14043. Environment Management–Life Cycle Assessment–Life Cycle Interpretation, 2000.)

is the goal and scope definition, in which the investigated product system, the intended application of the study, the data sources and system boundaries are described, as well as the functional unit of all related inputs and outputs are defined. The criteria for selecting input and output flows or processes are specified. In this step, the data quality requirements (e.g., time-related and geographical coverage), the data consistency, representativity and uncertainty, and the critical review procedure are described. A crucial step is the determination of the investigated impact categories. The inventory analysis (LCI) involves *data collection and calculation procedure* to quantify *the relevant inputs and outputs*. These input and output flows involve consumed or produced goods, emissions, waste streams, transportation, etc. It is essential to consider all life-cycle stages, i.e., production, operation, disposal and/or recycling. Usually, there are iterative steps leading to additional data requirements. The data collection follows the process chain, i.e., extraction, conversion, transport, production, use and disposal or recycling, respectively. Each phase can be divided into smaller phases, the *unit processes*. Every chain unit process has several incoming and outgoing material and energy flows, carefully recorded. The main product or the co-products, energy carriers, wastes and emissions into air, water or soil are the outputs leaving the system boundaries. The potential impacts of the LCI inputs and outputs are then determined by the impact assessment, which categorizes and aggregates the input and output flows to the biosphere to so-called impact categories, such as the climate changes, by multiplication with characterization factors.

The effects on the environment are assessed by the LCA analysis, encompassing all the phases of a good or service. The first LCA phase is goal and scope definition, followed by the LCI development. The LCI consists of all recognized inputs and outputs to or from the functional unit, and its results are translated into the life-cycle impact assessment, the impact categories. The analysis covers the impact categories where the environmental influence is the most significant. An LCA study involves data collection and calculation to quantify relevant inputs and outputs or the environmental load of a system. It requires backtracking for a conventional process system. LCI incorporates all foreground data, or all processes (mining, refining, transporting, plant construction, generating, distributing, decommissioning, etc.) involved in the product production, background data that support the foreground data, such as electricity, materials, transport, waste treatment, auxiliary materials and the metadata describing the foreground and the background. Using an LCA, the environmental performance indicators, including energy intensity, energy payback time and environmental impacts, can be determined for an energy technology. Performing an LCA analysis requires time and money however it can be assumed that due to new environmental regulations, the amount and quality of data increases, leading to lower costs of performing the LCA. The effective management of available resources and minimization of environmental impact due to consumption of these resources are essential. LCA due to its complex character allows one to determine how to effectively manage resources both economically and ecologically. Therefore it is a powerful tool for working out the ways of reducing natural resources consumption and maintaining a sufficient level of these at the same time. The LCA application is a tool supporting decision making where environmental issues are of the utmost importance. The LCA four steps are defined in the international standards ISO 14040 and ISO 14044. The goal and scope definition phase includes the balance object description, the system boundaries and the assumptions made. Within the impact assessment, the impact categories and their indicators are defined, and the inputs and outputs are classified by these impact categories. Using characterization factors, the impact category indicators are calculated. Finally, in the LCA interpretation step, the results are evaluated to reach conclusions. In addition, an optional sensitivity analysis can investigate responsive parameters and their influence on the results. Three different methodological approaches exist to calculate the above indicators: the energetic input output analysis, the process chain analysis and the material balance analysis. When using the input output analysis, there is a high possibility of receiving inaccurate results. However, a process chain analysis depends on detailed process data, being an extremely time-consuming balancing process.

7.2.1 LCA of Renewable Energy Technologies

LCA is a tool or method for the assessment of potential environmental impacts of products, processes and services, by systematically and adequately addressing their environmental aspects, being originally developed to assess both direct and embodied energy requirements for provision of foods and services, energy and pollutant emissions. Over the time, the LCA methods are increasingly used to analyze methods for generating, transmitting, and using energy. In particular, they are used to analyze wind turbines, PV systems, solar collectors or fuel cells. LCA aims to be an objective process when applied to a product or activity that identifies the energy and materials used and wastes released to the environment as a means of evaluating and improving environmental impact. Environmentally conscious decision making requires information on the environmental consequences of alternative products, processes or activities. An LCA is just one tool for assessing the most significant environmental impacts and benefits of such systems. The environmental impacts of conventional power plants are usually dominated by the fuel production and combustion. The construction of the power plant and the infrastructure required are usually a factor up to 10 environmentally relevant, up to 10% than the energy conversion because of the high throughput and the long lifetime of the system. Many of the RES units, such as solar PV panels, direct solar-thermal systems or wind turbines in contrast, are usually showing zero or minimal emissions during their operation. Thus, the construction of the plant, unit or station becomes the dominant factor into their environmental effects. Emissions associated with extraction, production and transportation of the fossil fuels are usually negligible over the life time of the thermal system when compared to the generation. RES for power generation do not emit significant pollutant emissions during operation. However, there may be considerable emissions associated with the material procurement, manufacture, transportation and disposal. For example, the ethanol produces lower emissions at combustion than gasoline, but the emissions to produce the crops offset these benefits or fuel cells do not emit pollutants, only water and limited heat while in operation, but hydrogen fuel production from natural gas has emission levels higher than the natural gas turbines. In the assessment of externalities related to RES technologies, it is important to include the energy-based emissions through the whole integrated system life cycle, from the manufacturing of the materials, operation and disposal. This is accomplished through a detailed LCA of all steps involved. In the energy and power sectors, including the renewable energy sources, the assessment should include the extraction, processing and transportation of fuels, building the power plant, the production of electricity and the waste disposal. The depth of the details and timeframe of an LCA study may vary to a large extent, depending on the goals and scopes. The scope, assumptions, data quality description, methodologies and the output of the LCA studies should be transparent. LCA methodology is amenable to the inclusion of scientific improvements in the state-of-the-art technology. The LCA strength is in its approach to study in a holistic manner the whole product, process, or system, enabling to avoid the suboptimization that may be the result of only few processes being focused on. The results are also related for the use of a product, which allows comparisons between alternatives.

Among RES electricity sources there is a wide variety of technologies such as hydropower, wind energy, biomass and various forms of direct solar energy utilization, e.g., PV and solar thermal power plants. Hydropower, wind energy and solar energy have special appeal, being not only inexhaustible, but with limited environmental impacts or health risks. For electricity from biomass, the main technologies that have been considered are combustion with steam turbine or gasification with gas turbine. There are significant health impacts from the air pollution emitted by the power plant and by the machinery needed for the production and transport of the fuel. The net greenhouse gas emissions from the biomass itself are zero, but there are emissions from the associated machines and vehicles, as well as chemical inputs (fertilizers, pesticides, etc.). Steam turbines are not the most efficient way of using biomass because required temperature and pressure of the steam, and thus the power plant efficiency, are relatively low. In addition, such power plants are usually too small to maximize economies and efficiencies. For hydro, PV and wind turbines, there are of course no emissions from power generation. There are upstream emissions from the material production, although the corresponding damage costs are small. However, one must not overlook the variability of wind, insolation and the amount of electricity they supply. To achieve a reliable power supply, backup capacity must be available, especially if solar and wind energy systems are providing a high fraction of the total electricity production. Of course, energy storage units are attractive solutions, but for most of the applications, the energy storage of the required magnitude and duration is still too expensive or the potential sites, for the most cost-effective options, are limited. Without sufficient energy storage the backup capacity requirement of the wind and solar energy systems implies that part of the replaced electricity comes from fossil fuels, with the costs for health and environment. Amenity impacts of renewable energy systems are dependent on local conditions, in particular the population near the site. The impacts of hydropower systems are quite variable that specific results are not recommended to be taken as general guidelines. The impacts can range from beneficial (e.g., flood control or recreational facilities) to extremely harmful if large populations are displaced without compensation or if a dam breaks. The decomposition of submerged vegetation in new reservoirs can cause significant emissions of greenhouse gases. The LCA covers the environmental and resource impacts of waste disposal processes, or those other processes which are affected by disposal strategies such as different types of collection schemes for recyclables, changed transport patterns and so on. The complexity of the task, and the number of assumptions which must be made, is showing the LCA importance in this matter. An LCA is commonly reported with an NER, the ratio of useful energy output to the grid to the fossil or nuclear energy consumed during the lifetime of the project. For renewable energy sources, NERs are expected to be greater than one, indicating a positive return over the fossil-fuel energy investment. For fossil-fuel and nuclear technologies, NERs are smaller than one and essentially represent the overall life-cycle efficiency of the project. NERs are strongly influenced by a number of underlying assumptions, such as plant capacity and life expectancy. For electricity generation from wind and solar energy, the strength of the resource (affecting the capacity factor of the installed technology) is also a critical assumption. The highest estimated NERs are generally for wind, followed by bioenergy and then solar PV. Hydropower is also expected to have high NERs, although the results shown in the graph, from a single study, are not as high as anticipated. By definition, NERs for fossil fuels are all less than one, with average estimates of 0.3, 0.4 and 0.3 for coal, natural gas and nuclear power, respectively.

7.2.2 Examples of the LCA Applications

Because operational environmental burdens are typically small for RE technologies, LCA is recognized as the most appropriate analytical tool for determining their environmental impacts of these technologies. Wind and solar energy generate electricity from the kinetic energy of the wind, and converting solar energy, respectively. Various types of wind turbines and PV systems are commercially available. Wind and solar energy are site specific energy sources. The capital cost of wind energy systems for generating electricity, for example varies from about $1.4 to $3.5/W, with larger-scale technologies being more cost effective. The generation cost of wind energy facilities varies between $0.058 to 0.24 per kWh, depending on capital cost and capacity factor. Fuel cells are electrochemical power generators with high overall and part load efficiency, operating quite with minimal polluting systems, viewed as a possible solution of the main disadvantage of wind and solar energy highly fluctuating electricity generation, due to the wind and solar radiation nature. The wind and solar energy conversion systems are used to produce hydrogen for later use in fuel cells, when the electricity is needed. However, there are significant challenges to this approach. Many hybrid power systems consist of combinations of wind turbine generator, PV arrays and energy storage devices, such as batteries and/or fuel cells. Wind power and PV array are interesting renewable energy systems with regard to development in production technologies, and thus with important updates for the inventory analysis. An integrated hybrid power system, having a 800 kW capacity is designed to deliver electricity to a local network or to a larger grid. All inflows and outflows (i.e., the materials, the energy resources, emissions and waste) were allocated to the overall amount of kWh electricity produced over the lifetime. Usually, the chosen functional unit is 1 kWh electricity output from the integrated system. The LCA main focus is to identify and quantify the emissions associated with the integrated system components (wind turbine, PV array, electrolyzer and fuel cell). The end of life of every system component, i.e., recycling, landfill and disposal are included. Notice that the fuel cell and hydrogen generation technologies are still under research and therefore it is difficult to predict the lifetime. However, most power systems are evaluated for about 20 years. The LCA steps are carried out for the entire system. LCI data are adopted for different sources, as found in the literature. LCA databanks were consulted for data regarding the environmental burden of key materials and potential recycle. When data were not available, the following assumptions can be made: (a) indirect pollutant emissions were only considered for the construction, maintenance and system disposal, and (b) the industrial, economic and energetic context is different for different components produced in different countries.

The LCA of a medium-size wind turbine begins with the calculation of its life-cycle energy requirements, pollutant emissions and energy yields. The wind turbine is a medium-size, three-balde horizontal axis, with the rated power of 500 kW. The main components of the turbine include the rotor (hub and blade), nacelle (generator, gearbox, brakes, control and power management units and transformer), tower and foundation. The assumptions are made in the analysis of wind characteristics, infrastructure, materials, maintenance, energy requirements and the savings associated with the WT decommissioning, or energy saved through the recovery of materials at the end of the turbine's life cycle. The period of the life cycle selected is 20 years, the most common WT lifetime. The main structural components of a turbine (tower and base) is capable of lasting beyond 20 years, however, more regular replacement of the moving parts, the generators, gearbox and blades is required. The LCI results for wind turbine are given in Table 7.3.

The data and the information used for the LCI of a PMFC fuel cell, found in the literature are summarized in Table 7.4. Compared to wind turbine, photovoltaics and fuel cell, very little published data exist for hydrogen production through electrolysis and additional equipment (storage, humidifier, control system, etc.). Most of the studies reported in the literature are related to hydrogen production through natural gas reforming. The hydrogen production is a critical part of this integrated power system. Using an average power mix for Western Countries, the life cycle for electrolyzer found is 11.13 g CO_2 equivalent/kWh and energy consumption is 4.32 MJ/kWh. The calculations were repeated for the production and operation of electrolyzer and accessories using different power mixes. We include in our study energy to fabricate silicon and manufacture PV

TABLE 7.3

Inventory Data for 1.5-MW Wind Turbine Adopted
and Modified from Literature

Material Used	Ton	MJ/kg	CO_2 (g/kg)
Steel	29.3	25.75	2300.6
Aluminum	0.78	39.15	3433.5
Iron	12.0	36.30	3114.0
Copper	0.19	78.20	6536.0
Plastic and Other Mat.	1.79	55.00	4522.0
Concrete	282.5	3.68	835.0

TABLE 7.4

Inventory Data of a Fuel Cell Stack

Parameters	Values
Number of required cells	2
Cell lifetime	40,000 h
Energy utilized	5100 MJ/kW
CO_2 emission	275.0 kg/kW

equipment; energy for transportation, maintenance and disposal of the system at the end of the lifetime. The result of inventory analysis and materials used for PV array of 100 kW, adapted and revised from the literature are presented in Table 7.5.

Microgrids, a type of generating systems incorporating multiple distributed generator sets linked together to provide locally generated electricity and heat are one possible alternative to the existing centralized energy system. Potential advantages of microgrids include flexibility in fuel supply options, the ability to limit emissions of greenhouse gases, and energy efficiency improvements through combined heat and power (CHP) applications. Microgrids offer the potential to utilize the best combination of power generation assets for meeting a given demand. For example, a microgrid

TABLE 7.5

Inventory Analysis of PV Array

Part	Material	Ton	Energy Rate (kWh/kg)	CO_2 (kg-CO_2/kg)
PV array	Aluminum	2.0	61.402	30.063
	Silicon	1.1	654.07	98.080
	Copper	0.3	12.745	8.983
	Insulator	1.5	28.708	21.567
	Glass	6.0	4.360	3.549
Stand	Iron/Steel	7.5	7.203	5.925
	Copper	5.0	12.745	8.983
	Insulator	3.0	28.708	21.567
Inverter	Iron/Steel	0.5	7.203	5.925
	Copper	0.3	12.745	8.983
	Insulator	0.2	28.708	21.567

may combine a solar photovoltaic (PV) array with a diesel generator to provide continuous power while utilizing renewable resources when available. Many of the benefits frequently associated with distributed generation also apply to microgrids, while microgrids also face many of the same challenges. Challenges that continue to provide barriers to the expansion of microgrids include the high capital costs of many technologies involved, the potential for more irregular power supply requiring various energy storage solutions, variable power quality and the necessity of power conditioning, limited fuel supply infrastructure and the need for specially trained technicians to provide operation and maintenance. Nevertheless, the potential advantages of distributed generation relative to centralized power have continued to drive microgrid development. These advantages include the flexibility to respond to changing conditions such as peak power requirements or changes in fuel supply, avoiding grid power outages, limiting the need for additional expansion of transmission infrastructure, the opportunity to recover some generating costs by selling power on to the grid (net-metering), the ability to limit emissions of carbon dioxide and other greenhouse gases amid growing concerns over global climate change, and the efficiency improvement opportunities offered by combined heat and power technologies. A complete understanding of the energy and environmental performance of microgrid systems requires a life-cycle perspective. LCA is an analytical technique for assessing the potential environmental burdens and impacts associated with a product system from the acquisition of the raw materials to the ultimate management of material remaining at the end of life. LCA is applied here in the evaluation of the energy and emissions performance of the building microgrid and a reference conventional system. Using the above results the life cycle and energy use are calculated for the complete integrated power system. As per inventory data, the life cycle and energy consumption are calculated at 40.6 g CO_2 equivalent/KWh and 0.372 MJ/kWh, and 37.8 CO_2 equivalent/KWh and 0.372 MJ/kWh, respectively. The results are showing that the maximum energy consumption is for electrolyzer. The results are summarized in Table 7.6. LCA is an analytical framework that is focused on the resource consumption and on the impacts to human health and the environment associated with manufacturing of specific product or service. The LCA can handle the problem-shifting by considering the full range of relevant impacts, which is one of its main advantages. For example, the LCA for the waste management systems is designed according to the same principles as LCA for products, equipment and processes.

TABLE 7.6

Results of Impact Analysis

Parameters	Numerical Values	Numerical Values	a. CO_2	b. CO_2
System component	Energy consumption (MJ/kWh)	Energy consumption (MJ/kWh)	Equivalent (g CO_2/kWh)	Equivalent (g CO_2/kWh)
Wind turbine	0.029147	0.028626	10.04509	7.0973544
PV array	NA	0.095997	NA	27.463645
Fuel cell	0.127500	0.127500	19.897455	19.897455
Electrolyzer	4.313500	4.313500	11.13987455	11.13987455
Complete integrated system	4.470100	4.565597	41.082396	65.598285
Energy payback period (yr.)	6.19×10^{-1}	6.19×10^{-1}	6.20×10^{-1}	9.80×10^{-1}

7.3 Environmental Impacts of the Renewable Energy Sources

Any energy production method impacts the environment in some ways, and renewable energy systems are no different. Like every other energy technology, wind, solar, hydropower or geothermal power plants have some environment impacts. However, the wind turbines or PV panels cause virtually no emissions during the operation and very little during the installation, maintenance and disposal. Compared to the conventional generation environmental impacts, the RES environmental impacts are relatively minor. Their human impacts include the aesthetic impacts, the impacts on cultural resources (e.g., historic, sacred, archeological and recreation sites), the human health and wellbeing impacts, such as noise or landscape changes. The environmental impact assessments and investigations are legal and technical procedures comprising a systematic testing, in which the immediate and later effects on the environment or goods affected by the decision or intent are determined, described and evaluated. The subsystem and component interactions need also to be taken into account. The environmental impact assessment is a key of the basis for verification and approvals within the approval process. Appropriate methods and procedures are used for the permitting of the exploitation of various renewable energy projects. However, there are very large variations in terms of the usage methods, technology status, environmental impacts or prospective. RES used in energy generation are helping to reduce the pollutant emissions, which may be important to mitigate the climate changes, to minimize the environmental and health complications associated with pollutants from fossil fuel energy sources. However, renewable energy such as wind, solar, geothermal, biomass and hydropower also have environmental impacts, some of which are significant. The exact type and intensity of environmental impacts varies depending on the specific technology, the geographic location and other factors. By understanding the current and potential environmental issues associated with each renewable energy source, and the steps that are taken to effectively avoid or minimize these impacts are critical for future electric supply. The developments are often restricted in areas with sensitive ecosystems, high cultural or scenic values and public involvement is invaluable in helping to identify these areas. Additional research is also needed to understand impacts of large-scale renewable energy installations on meteorology and climate. Several quantitative environmental metrics have been defined based on the energy engineering. The energy system inputs are classified as renewable resources, non-renewable resources, local energy resources and non-local (usually foreign) energy resources. The system sustainability is favored by using the renewable energy resources and local energy resources. The sum of these categories is denoted by the term R. The non-renewable (N) and non-local (F, purchased) energy resources work against sustainable development. These presumptions set the background for devising certain sustainability metrics in this study. One of such metrics, which characterizes the environmental impact of a transformational energy process, is the environmental loading ratio (*ELR*):

$$ELR = \frac{F+N}{R} \tag{7.2}$$

From this relationship, the more non-renewable and outside energy resources are involved in the energy process, the higher the ERL index. An increase in renewable energy use translates into the energy process with a lower ELR value. The lower ELR is beneficial for the environment.

7.3.1 Solar Energy

Solar energy is one of the most environmentally benign forms of alternative energy source. The most straightforward conversion of solar radiation is to thermal energy, and eventually into electricity. Besides space and water heating, high-temperature heat can be converted into electricity just as in conventional power plants, through the solar concentrating energy applications (Chapter 3). Environmental effects due to the solar thermal plant arise during the plant component manufacturing. However, the environmental effects are restricted to very limited time period of the system life cycle. Furthermore the solar thermal power plants are usually located into steppes, desert and semi-arid areas where the population is rare. Its current utilization, either as solar-thermal or as PV conversion, causes very few environmental concerns. Solar energy systems are basically generate no air or water pollution during the operation, the primary environmental, health and safety issues involve the manufacturing, installing, and their disposal and/or recycle. Materials used in the solar energy systems can create health and safety hazards for workers and anyone else coming in contact. In particular, the PV cell manufacturing often requires hazardous materials such as arsenic and cadmium. Even relatively inert silicon, a major PV cell material is hazardous to workers if it is breathed in as dust. It is also estimated that central photovoltaic-based systems require exotic inputs, some of which, such as cadmium sulfide is toxic and explosive. However, the solar energy systems can generate significant concentrations of water pollutants, including antifreeze agents, rust inhibitors and heavy metals leached from the systems. There are also indirect water pollutants via the use of herbicides to deter excessive vegetation growth around the solar collectors. Other adverse impacts of the solar energy systems are the large land area use, with no reclamation until the decommission, the generation of decommissioning non-recyclable materials (e.g., fiberglass, glass, coolant, insulations), while for PV systems, additional disposal problems are caused by the cadmium and arsenic, hazards to eyesight from reflectors and toxicants in coolant fluids. The soil erosion and compaction, wind diversion and the potential decrease in evaporation rate from soil are other solar energy environmental issues. There is an additional, likely very small danger that hazardous fumes released from PV modules attached to burning homes or buildings could injure firefighters. However, the PV systems do not have emissions into electricity generation. Building-integrated PV systems have minimal environmental and landscape impacts. Larger PV systems are usually built on free or agricultural land far from urban are restricted areas. In this case, their impact on the land is expressed in terms of flora and fauna, respectively.

Maintenance of the land covered with PV arrays involves complete or partial vegetation removal, which are drastically disrupted. There maybe also indirect wildlife impacts (e.g., habitat change, etc.). PV arrays require periodic maintenance by washing the absorptive surfaces, so water pollution. On the other hand, the PV module manufacturing requires large energy amounts. Impacts of power plants producing energy needed for PV manufacturing are also listed as negative impacts of PVs. The emissions during transport and installation are also listed under their negative impacts. However, the positive effects of the PV power installations are far more significant than the negative ones. In PV electricity generation, there are no greenhouse gas emissions, neither emission of particles that may cause respiratory problems, heavy metals or noise (no part in motion). On other hand, the solar thermal power plants occupy a large area and their impacts on the land and wildlife are similar to those of PV systems. However, these types of power plants are built for high power and they greatly reduce pollutant emissions. Unlike PV systems, large volume of water is required for cooling, as a working fluid, or for washing reflective surfaces which can affect the water quality. The reflected solar rays can cause sunburn to the birds flying near the towers. However, such power plants are often built in the desert or semiarid areas their impact on the wildlife is limited. From social point a view, these power plants can supply electricity to smaller cities or to rural areas therefore they are the basis of the regional development, providing often greater employment opportunities. Also, they may be equal to the conventional power plants of medium size, and because they have the same working principle as steam power plants, their quality of the electrical energy are the same. Time-of-day dependence can be eliminated by a hybrid version of the power plant, where natural gas is mostly used for heating water at night and cloudy days during the year and by addition of the energy storage. These power plants are built more often in desert areas, requiring an additional transmission grid to populated areas. Water is heated exclusively by solar energy, with no need for fuels and combustion, making such systems environmentally friendly. Problems occur due to higher needs for hot water where the collectors are placed on the land. The time of day is also a major problem with these systems. Insulated containers are needed for storing water in bad weather conditions.

PV panels and modules are not emitting any pollutants during their operations and uses. However, the PV cells or modules contain some toxic substances, having a potential risk of releasing these chemicals to the environment during a fire or decommissioning. The solar energy main environmental advantage is that its utilization does not involve any chemical or nuclear reactions, so no chemical emissions (e.g., CO_2, NOx, etc.) and radio nuclides are emitted, making the solar conversion a very environmentally friendly energy source. The most significant environmental impact of solar energy is associated with the production of the materials of the PV cells. Silicon, germanium and phosphorous are produced and purified with the use of significant energy amounts and involve the use of polluting chemicals such as sulfuric acid and cyanide. However, all the pollution associated with the PV cell production is localized and contained at the production facility. The pollutants at large sites are well regulated and measures are taken that all used hazardous chemicals are not released into the environment. A significant PV cell manufacturing concern is the large energy amount consumed during the manufacturing. Often, the manufacturing consumed energy of a silicon-based PV cell is equivalent to the energy that the cell is producing in about 4 years. Leaner manufacturing methods and PV cells made of new materials may reduce this significant energy requirement. Thermal pollution is also associated with all solar energy operations. Solar thermal power plants reject a significant amount of heat in their condenser. Similarly, photovoltaic arrays reject all the incident radiation that is not converted to electricity. The thermal energy (waste heat) rejected by a solar power plant, which produces the electric power dW/dt with an overall efficiency, η can be estimated by:

$$\frac{dQ_{rejected}}{dt} = \frac{dW}{dt} \times \frac{1-\eta}{\eta} \qquad (7.3)$$

Example 7.3: A 100 MW solar thermal power plant has the collector efficiency of 84%, the heat transfer efficiency of 95%, and the combined steam turbine-generator efficiency of 0.82. What the rate of the rejected waste heat?

Solution: The overall solar thermal power plant efficiency is the product of individual efficiencies, being equal with 0.65436% or 65.436%. By applying Equation (7.3) the rejected waste heat is then calculated as:

$$\frac{dQ_{rejected}}{dt} = 100 \times 10^6 \times \frac{1 - 0.65436}{0.65436} = 52.82 \text{ MW}$$

The solar radiation has a very low energy density, so the solar installations are occupying larger areas than similar fossil fuel based installations. Because the solar collectors must be accessed during the plant operation the solar collector adjacent areas are affected, the soil being compacted, levelized and pro to the erosion. On the other hand, the waste heat is dissipated over significantly larger area and does not have the same adverse impacts on the environment, as in the case of the conventional power stations. Also, the fact that solar energy installations are often located in the remote areas, their thermal pollution caused seldom affects the human population and are unnoticed. Associated with the thermal pollution is the water use, if the solar power plant is cooled by water. The used water amount by such a solar thermal power plant is the same as the water used by a similar thermal power plant. Land use is another environmental effect of solar energy utilization. Solar energy installations require large land areas. For this reason, inexpensive, unutilized and usually deserted areas are chosen for the solar energy facilities. The common land requirements, for the solar thermal power plants are summarized in Table 7.7.

Example 7.4: Estimate the minimum, maximum and average land requirements for a 50 MW solar thermal updraft power plant.

Solution: The kW equivalent of this power plant is 50,000 kW, and by using the data from Table 7.7, the land requirements are:

$$\text{Minimum land requirement: } 50,000 \times 180 = 9.0 \times 10^6 \text{ m}^2 \text{ or } 900 \text{ ha}$$

$$\text{Average land requirement: } 50,000 \times 200 = 10.0 \times 10^6 \text{ m}^2 \text{ or } 1000 \text{ ha}$$

$$\text{Maximum land requirement: } 50,000 \times 220 = 11.0 \times 10^6 \text{ m}^2 \text{ or } 1100 \text{ ha}$$

A notable beneficial environmental effect of solar energy in urban environments, when the PV panels or solar thermal collectors are sited on the building sides and rooftops, so part of the incident solar radiation is not absorbed, but is reflected back to the atmosphere or scattered. Considering the overall energy balance in the vicinity of the buildings, the solar systems are producing small but important cooling effects in the urban environment, resulting in slightly lower ambient temperatures and reducing the need for the air-conditioning. Necessary precautions should be taken for emergency situations like fire. The possibility of an accidental release of the chemicals of the PV panels to soil and groundwater poses a great threat for the environment. There also are some emissions during PV system manufacturing and transport. The emissions associated with the transport of the PV panels are quite insignificant when compared with the emissions associated with their manufacture. There are sometimes negative visual impacts depending on the type of the scheme and the surroundings of the solar PV panels and arrays. Especially for

TABLE 7.7

Land Requirements for Solar Thermal Power Plants

Solar Thermal Power Plant—Type	Land Requirement (m²/kW)
Parabolic trough and line focusing power plants	10–30
Solar updraft tower power plant	180–220
Solar pond power plant	50–60
Ground-mounted PV System	6–18

the building applications, PV modules can be used as a cladding material that could be integrated into the building during the construction phase. Solar energy system applications after the construction phase of the buildings might cause negative visual impacts. For example, due to their central tower, tower and solar updraft power plants are affecting the landscape scenery, the higher the tower the larger is the impact. However, because there is not parts in motion this type of impact can be much easier to be accepted and noticed. At the night the towers must be equipped with warning lights for aviation safety. It is recommended that solar energy system uses to be planned at the architectural phase and fitted to the building to minimize visual pollution. For the other application areas, proper sitting and design are important factors, especially for large solar energy systems. Another important factor about the control of the visual impacts is the use of color, so the usage of proper colors while assembling the PV modules or solar thermal systems. Visual burdens are defined as irregular forms that usually exist in the natural and cultural environment, which mostly appear indirectly and cause negative symbols on the human memory. Visual burdens have mostly psychological effects on the human being but could affect the physical environment and even could stop some of the environment functions. To prevent the visual burdens of the solar energy systems on the buildings, the solar energy systems should be integrated with the building during the construction phase, with proper planning, during the design phase. Solar tower, through and dish power plants are focusing the solar radiation into specific points or areas, cares and proper focusing measures must be taken to avoid any environmental or health effects of high radiation energy densities.

7.3.2 Environmental Effects of Wind Energy

Wind energy is one of the most environmentally friendly forms of electricity production, having very low emissions during the entire life cycle, but has a number of environmental effects that limit its potential. It does not involve any chemicals or produce any harmful emissions and thermal pollution, during the operation. The materials that are used to make the towers and the WT components are the commonly used structural and engineering materials. Hence, the WT construction and operation does not impose any environmental threat. There are two major ways that wind energy development may influence the ecosystem structure and functioning, through direct impacts on individual organisms and through impacts on the habitat structure and functioning. The wind energy environmental influences can propagate across a wide range of spatial scales, from the location of a single wind turbine to the landscapes, regions, and area, and a range of temporal scales from short-term noise to long-term impacts on the habitat and influences on the presence of species. The ecological effects of the wind energy facilities are complex, and can vary with spatial and temporal scale, location, season, weather, ecosystem type, species and other factors. Moreover, some effects are likely cumulative and ecological impacts can interact in complex ways at wind energy sites and at other sites associated with changed land uses and other anthropogenic disturbances. The construction and maintenance of wind energy facilities also alter ecosystem structure through vegetation clearing, soil disruption and erosion potential, and noise. Alteration of vegetation, including forest clearing, represents perhaps the most significant potential change through fragmentation and loss of habitat for some species. Changes in forest structure and the creation of openings alter microclimate and increase the amount of forest edge. Plants and animals throughout an ecosystem respond differently to these changes. There might also be important interactions between habitat alteration and the risk of fatalities, such as bat foraging behavior near turbines. There are a few rather minor environmental issues associated with wind power, the most important are:

1. Noise pollution arising due to the fact that all rotating engines always produce noise and the wind turbines are no exception; especially the ones operating at higher rotation speeds. Wind turbines produce aerodynamic noise, from air passing over the blades and mechanical noise from the turbine moving parts, e.g., the gearbox, or generator. Better designs have reduced the noise. Wind farms are developed usually far from highly populated areas, and large wind turbines are typically located in remote, rural areas with low population density. This mitigates the noise effect of to the humans, but may have significant effects on the wildlife, which can migrate due to the noise, disturbing the local ecosystem balance. Noise pollution is one of the limiting factors for the expansion and the widespread use of small wind turbines in the urban or suburban environments. The measurement, calculation and evaluation of noise generated by wind turbines are comprehensively described in the standards and guidelines.

2. Bird injuries and mortality can be one major wind energy environmental issue, because the birds may be caught by the rotating blades and be killed. Birds get killed when they collide with the rotating blades of a turbine. Migratory species are at higher risk than resident species. Siting the turbines away from migratory routes reduces this impact. The overall motion of the blade and the pressure reduction that occurs before the rotating blades, detract the birds and often kill them, which may have a significant effect on migrating bird species. Species are differing in their vulnerability to collision, in the likelihood that fatalities have larger scale cumulative impacts on biotic communities. However, the data are still inadequate to assess relative WT risks to birds. It is possible that as turbines become larger and reach higher, the risk to the more abundant bats and nocturnally migrating birds at these heights to increase.

3. Aesthetic pollution, wind turbines may affect the picturesque landscape of remote, pristine areas is often disturbed by the placement of the high towers with wind turbines on top. The aesthetic pollution has raised significant opposition to the wind energy development, especially in touristic areas. Wind turbines must be in exposed areas and are therefore highly visible. They are considered unsightly by some people, and concerns have increased with the larger size of new generation turbines.

4. Electromagnetic interference that the wind turbines, that are located near the top or the sides of hills and mountains, their operation interferes with the electromagnetic signals, TV and transmissions, radars or communication systems. Wind turbine rotors can scatter electromagnetic signals causing interference to communication systems. Appropriate siting (avoiding military zones or airports) can minimize this impact. This effect is mitigated by: (a) the location of the majority of wind turbines is in remote areas, where fewer signals need to be transmitted, and (b) the recent trends in the communication technologies.

WECS produces no air or water pollution, involves no toxic or hazardous substances (other than those commonly found in large equipment), and poses no threat to public safety. However, there are still serious obstacles facing the wind industry and public opposition reflecting concerns over the visibility and noise of wind turbines, and their impacts on wilderness areas. The WT noise pollution is sometimes similar to a small jet engine, hardly notice the noise at larger distances. Large-scale generation of electricity through wind farms can reduce wind speeds and cause stress to ecosystems. Lakes, downwind from the wind farms can become warmer because of the reduced evaporation, while the soil moisture can increase. Nevertheless, these impacts are not of great consequence except in some very sensitive areas. Another aspect of wind power is the land use, due to the required distances between wind turbines. However, the further expansion of wind power and the substitution of conventional power sources by wind energy are benevolent to the environment. Offshore wind facilities, which are currently not in operation in the United States but may become more common, require larger amounts of space because the turbines and blades are bigger than their land-based counterparts. Depending on their location, such offshore installations may compete with a variety of other ocean activities, such as fishing, recreational activities, sand, gravel, oil and gas extraction, navigation and aquaculture. Employing best practices in planning and siting can help minimize potential land use impacts of the offshore and land-based wind projects. Offshore wind turbines can have similar impacts on marine birds, but the bird deaths associated with offshore wind are minimal. Wind farms located offshore will also impact fish and other marine wildlife. Some studies suggest that turbines may actually increase fish populations by acting as artificial reefs. The impact will vary from site to site, and therefore proper research and monitoring systems are needed for each offshore wind facility.

Wind energy facilities create both positive and negative recreational impacts. On the positive side, the wind energy projects are listed as tourist sights: some offer tours or provide information about the facility and wind energy, and some of them are considering incorporating visitor centers. There are two types of potential negative impacts on recreational facilities: direct and indirect. Direct impacts can result when existing recreational activities are either precluded or require rerouting around a wind energy facility. Indirect ones include aesthetic impacts that may affect the recreational experience. These impacts can occur when scenic or natural values are critical to the recreational experience. In analyzing impacts on historic, sacred and archeological sites, the primary concern is that no permanent harm is done, affecting the site integrity. Whether or not a wind energy project would damage the resource depends on the involved site specific nature. Unlike housing developments, wind-energy projects cannot be screened from view, except behind intervening topography and vegetation. Such issues are likely to arise as wind projects are proposed in cultural landscapes, and guidance as to what constitutes an undue impact to historic or sacred sites and areas will be necessary. When it comes to aesthetics, hysteric and sacred site impacts, the wind energy projects can elicit strong reactions. To some people, WT systems are graceful sculptures, to others they are eyesores, compromising the natural landscape. Whether a community is willing to accept an altered skyline in return for cleaner power should be decided by public dialogue.

7.3.3 Geothermal Energy Environmental Issues

Geothermal energy is a clean energy source of energy, and its conversion into electricity or useful heat produces almost no pollutant emissions, CO_2 and no flue gas emissions such as soot particles, sulfur dioxide and nitrogen oxides. The geothermal power plant operation is deeply friendly to the environment. The likely adverse environmental effects of geothermal energy sources are: surface disturbances, physical effects (such as land subsidence) caused by fluid withdrawal, noise, thermal pollution and release of offensive chemicals. The various geothermal resources differ in many respects, but they raise a common set of environmental issues. Air and water pollution are two leading concerns, along with the safe disposal of hazardous waste and land subsidence. Metals, minerals, volatile species of boron, arsenic and mercury and gases leach out into the geothermal steam or hot water as it passes through the rocks. The chemicals released when geothermal fields are tapped for commercial production are hazardous or objectionable to people living and working nearby. Most geothermal power plants require large amounts of water for cooling or other purposes. In areas where the water is in short supply, there may be possible of conflicts with other water resource users. In some geothermal developments, the steam vented at the surface can contain hydrogen sulfide (H_2S), ammonia, methane and CO_2. Usually where there is volcanic activity where you find geothermal energy, there may be other tectonic activities such as earthquakes and active volcanoes. It is important to take care of a geothermal site because if the holes were drilled improperly, then potentially harmful materials could escape from underground. These hazardous materials are nearly impossible to get rid of properly. The operation of the geothermal power plants causes minimal environmental impacts. The main source of atmospheric pollution from geothermal units is the discharge of the non-condensable gases, primarily CO_2. The quantity of CO_2 released to the atmosphere is by far less than that produced by an equivalent fossil fuel plant. Geothermal power plants produce 1000–5000 times less CO_2 per kWh produced than fossil fuel plants. Hydrogen sulfide, H_2S, is another gas that is released from the condenser of a geothermal power plant. At very small concentrations, less than 1 ppm, H_2S is malodorous and leaves a distinct odor near geothermal power plants. At higher concentrations, H_2S desensitizes the olephatic nerves and may not be detected by its odor. Sulfur abatement methods may be used to eliminate the release of H_2S.The risk for harmful environmental effects are extremely low during normal operation and even during accidents. The low-risk systems result from the use of high-quality structural materials and from the mature technology with numerous safety precaution installations.

Construction of geothermal systems and power plants causes CO_2 and other pollutant emissions related to manufacturing construction materials, transport and equipment and service traffic, no different as with other types of power plants or industrial systems. Careful planning of logistics helps to minimize these emissions. Soil subsidence may become a problem in the vicinity of geothermal power plants: As steam or water is removed from the geothermal aquifers, open cracks shrink and the permeability of the aquifers decreases. The surrounding rocks and soil are displaced to fill all major voids and the soil surface subsides, usually in a uniform manner. Uneven soil subsidence may pose problems to structures and buildings. Water re-injection and rainwater seepage mitigates to a large degree the soil subsidence. Since most of the geothermal resources and power plants are located far from the major population centers, soil subsidence does not pose a significant problem to large populations. A beneficial effect of this geological volume reduction is that mechanical stresses, caused from geological movements are released, alleviating the seismic stresses and their effects on the structures. Thermal pollution, caused by the heat released to the environment as a result of the cooling system operation is another environmental effect. Finally, the noise, which is always a problem with all power plants, is also an environmental concern for geothermal units. However, since the vast majority of geothermal power plants are far from the population centers, noise pollution is not a significant environmental concern and only affects the wildlife in a similar way to the noise effect of wind turbines. Developing the underground heat exchanger of an enhanced geothermal system involves hydraulic stimulation measures that cause microseismicity. Rarely the induced microseismicity can cause irritation if physically sensed at the surface. The geothermal fluid circulates in a closed system and is not causing any damage to the environment. If a leak occurs in the installations, the fluid circulation is stopped and the leaking section replaced. The working fluid in the secondary loop of the power production circulates also in a closed system. If leaks occur, precautions on the construction and technical side help to minimize environmental pollutions. However, it may be necessary to test the thermal fluid circulation during the system development phase when the primary loop is not yet completely closed. It is only during these tests that the success of the entire effort can be visually seen from the rising white steam plumes. In binary geothermal power plants the working fluid of the secondary loop must be cooled below condensation conditions after leaving the turbines. The excess heat is released to the environment, similar to any thermal power plant. The thermal emissions of geothermal plants are, however, orders of magnitude lower than those of the conventional thermal and nuclear power plants. The emissions are considerably lower even if normalized to the power output of the units. Also geothermal projects can be plagued by very diverse failures and troubles during system development and later during operation. Careful planning, management, monitoring, educated and experienced personnel and adequate quality of materials and equipment help reducing potential troubles and thus also minimize environmental effects. In the case of EGS, where water is initially injected and then circulated through the system, not only zero CO_2 emissions are foreseen, but also none of the other problems are anticipated. However, it pays to engineer EGS within or just outside the boundaries of a hydrothermal system, in order to benefit from bonuses in temperature, depth and limited natural permeability or fluid presence. One problem that has been often reported during EGS engineering is the occurrence of microseismic activity, likely associated with the hydraulic fracturing works. These micro-earthquakes have a magnitude up to around level 2 in the Richter scale and pose no danger to existing buildings or structures. Whether they are also present during exploitation, and how they evolve overtime, remains to be identified in the field, but it is anticipated that this microseismic activity is not greater than the one encountered in hydrothermal systems. The positive environmental impact of all geothermal plants is attributed to foregone CO_2 and other emissions, had the same quantity of energy been produced by fossil fuels. However, in low enthalpy geothermal fields the deep hot water may contain dissolved CO_2. When these fluids are brought to the surface and their pressure is lowered, they tend to deposit calcite and release CO_2. Such fluids are usually located at the margins of larger hydrothermal systems. The CO_2 emissions from a geothermal field can be minimized by proper designs of the exploitation scheme. For example, maximizing the energy extracted from a given flow rate by placing the users in cascade, results in minimal CO_2 emissions. In agricultural uses, the geothermal CO_2 can be directly released within the greenhouse, in order to increase the growth rate of the plants and save fossil fuels. Adverse environmental impact may occur from low enthalpy geothermal utilization, associated with the chemistry of the geothermal fluid, which may include considerable quantities of chloride, small quantities of boron, and traces of arsenic, ammonia, mercury, or heavy metals, making it unsuitable for disposal to the surface. The problem is effectively solved by re-injecting all the produced geothermal fluid to the same deep reservoir it originated from. This practice also has benefits to the water replenishment of the deep system, improving sustainability and the economic life of the plant. Other impact from long-term low enthalpy geothermal utilization may be the dropping of water level of near surface aquifers and the flow reduction or dry-up of nearby springs and shallow water wells. The problem can be solved by reinjection, effective reservoir engineering and if the previous two measures do not solve the problem, by donating deep wells to affected water users.

The geothermal power plant operation are usually producing less noise than the equivalent produced, near leaves rustling from breeze, according to common sound level standards, and thus is not considered an issue of concern. Geothermal plants are using in average about 0.02 m³ of freshwater per MWh, while binary air-cooled plants are not using any fresh water. This compares with 1.37 m³ per MWh used by natural gas facilities. Geothermal fluids used are re-injected back into geothermal reservoirs using wells with thick casing to prevent cross-contamination of brines with groundwater systems, not being released into surface waterways. Injection reduces surface water pollution and increases geothermal reservoir resilience. Geothermal power plants can be designed to fit into their surroundings, and can be located on multiple-use lands that incorporate farming, skiing and hunting. A geothermal facility uses about 400 m² of land per GWh, while a coal facility uses over 3600 m² per GWh. Subsidence, or the slow, downward sinking of land, may be linked to the geothermal reservoir pressure decline. Injection technology is an effective mitigating technique. While earthquake activity, or seismicity, is a natural phenomenon, geothermal production and injection operations can result in low-magnitude events known as, micro-earthquakes. These events are usually not detected by humans, and are often monitored voluntarily by geothermal companies. While almost all geothermal resources currently developed for electricity production are located in the vicinity of natural geothermal surface features, much of the undeveloped geothermal resource base may be found deep under the Earth without any corresponding surface thermal manifestations. Geothermal surface features, while useful in identifying resource locations, are not used during geothermal development, and according to the laws and regulations to protect and preserve national parks and their significant thermal features. Before geothermal construction can begin, an environmental review may be required to categorize potential effects upon plants and animals. Power plants are designed to minimize the potential effect upon wildlife and vegetation, being constructed in accordance with the regulations of the host state that protect areas set for development. Geothermal power plants release very few air emissions because they avoid both environmental impacts associated with burning fuels as well as those associated with transporting and processing fuel sources. They emit only trace amounts of NOx, almost no SO_2 or particulate matter, and small amounts of CO_2. The primary pollutant is H_2S, which is naturally present in many subsurface geothermal reservoirs. With the use of advanced abatement equipment, emissions of H_2S are regularly maintained below the standards. Average life-cycle emissions at coal facilities are substantially higher than their average operational emissions which do not consider the effects of coal mining, transport, construction and decommissioning. Life-cycle emissions from geothermal facilities, in contrast, remain in the same range as operational emissions.

The stimulation related microseismicity is indispensable for the development of HDR/ EGS project. However, it is absolutely essential to avoid any surface effects that can be sensed or heard, which it is quite possible with carefully planning and conservative reservoir development strategies. It also should be kept in mind that in several documented geothermal projects massive hydraulic stimulation did not or barely result in recognizable and recordable microseismicity, but also making reservoir development difficult. Microseismic incidents are not exclusive to petro-thermal systems (EGS, HDR). Deep geothermal systems that produce hot water from fault systems or inject fluids into faults have been exposed to problematic microseismicity. There are several ways to express the intensity of microseismic events. A widely used conventional parameter is the logarithmic Richter magnitude scale, which expresses the energy released by the (micro) seismic event. By definition any seismicity below magnitude 2 is microseismicity. Minor seismicity covers the magnitude range of 2 to 4. Richter scale magnitude 3.6 characterizes an event that can be felt at the surface by most people but rarely causes structural damage. The EGS-related seismic incidents caused discussions about the cause for the unwanted seismicity above the magnitude two level and the potential risks for even stronger triggered seismicity that may cause major structural damage at the surface. Also social issues have been intensely discussed including improved communication with the local population at the project site. The general acceptance of an EGS project can vary strongly depending on complete transparency of the project and all its development phases. The perception of projects by the public depends a lot on the use of terminology. Instrumentally monitored microseismicity should definitely not be communicated to the public as *earthquakes* or *perceptible seismic activities*, the term earthquake being loaded by pictures of damage and death, prompting fears that are not backed by the microseismicity physical nature. The hydrological and geological properties and the locally existing stresses at and around the injection zone of the reservoir control the reaction on fluid injection, in some situations an earthquake may be triggered in others high-pressure fluid injection may not result in an earthquake. Anthropogenic seismicity is not exclusive to geothermal project development. It is also known from the oil and gas industry, from dammed lakes used as reservoirs for hydroelectric power systems, from underground storage of gas and compressed air, from the high-pressure injection of liquid waste in deep wells and also from underground and open pit mining.

7.3.4 Environmental Impacts of Water Energy Systems

The development of large size hydropower plants and investments has become increasingly problematic, due to several reasons. The construction of large dams has virtually ceased because most suitable undeveloped sites are under environmental protection or with very few exceptions already are used. To some extent, this slack has been taken up by the revival of the small-scale hydropower. Large hydropower projects may cause major adverse environmental impacts, especially on water quality, and large hydroelectric stations may be one of the most ecologically damaging. Large projects disturb local ecosystems, reduce biological diversity and modify water quality, causing significant socioeconomic damages. A number of projects in developing countries have been stalled or scaled down for the reason that obtaining loans from lending institutions for major projects have become very difficult. Hydropower stations are also emitting pollutant gases on a life-cycle basis (e.g., methane generated by decaying of the biomaterials in reservoirs), however far less than the burning of the fossil fuels. Large-scale hydroelectric dams are unlikely to continue to be built in many countries, instead, the future of hydroelectric power in the United States and all developed countries is likely to involve only an increased capacity at current dams and small hydropower projects. There also are a few other environmental impacts at all types of hydroelectric plants. The reservoir size created by a hydroelectric project varies widely, depending largely on the size of the hydroelectric generators and the topography. Hydroelectric plants in flat areas tend to require much more land than those in hilly areas or canyons where deeper reservoirs can hold more volume of water in a smaller space. Flooding land for a hydroelectric reservoir has extreme environmental impacts: destroys and severe damages the forest, the wildlife habitat, agricultural lands and scenic lands. On the other hand, the dammed reservoirs are used for multiple purposes, such as agricultural irrigation, flood control, recreation and even there may be beneficial wildlife impacts. All hydroelectric facilities can have major impacts on the aquatic ecosystems. However, though there are a variety of methods to minimize the impacts, fish and other organisms can be injured and killed by the hydro turbine blades. Apart from direct contact, there can also be wildlife impacts both within the dammed reservoirs and downstream from the facility. Reservoir water is more stagnant than river water, the hydroelectric plant reservoir has higher than normal amounts of sediments and nutrients, which can cultivate an excess of algae and other aquatic weeds, which can crowd out the river animals and plants. They must be controlled through harvesting or by introducing fish that are eating them. In addition, the water is lost through evaporation in dammed reservoirs at higher rates than in the flowing rivers, and if too much water is stored into the reservoir, the downstream river can be dried out. Thus, the hydroelectric operators are required to release a minimum amount of water at certain times of year. If not released appropriately, water levels downstream drops, the animal and plant life can be harmed. In addition, the reservoir water is typically low in dissolved oxygen and colder than the river water. When released, it may have negative impacts on downstream plants and animals. To mitigate these impacts, aerating turbines are installed to increase dissolved oxygen levels and multi-level water intakes can ensure that the released reservoir water comes from all reservoir levels, rather than just the bottom, the coldest with the lowest dissolved oxygen.

Pollutant emissions are produced during the development, the installation and the dismantling of the hydroelectric plants, but there may also be significant emissions during the facility operation. Such pollutant emissions are varying greatly depending on the reservoir size and the flooded land nature. Small hydropower plants emit up to 0.012 kg of CO_2 equivalent per kWh. Life-cycle emissions from large-scale hydroelectric plants built in semi-arid regions are also modest: approximately 0.02 kg of CO_2 equivalent per kWh. However, the estimates for life-cycle global hydropower plant emissions built in tropical areas or temperate peat-lands are much higher. After the area is flooded, the vegetation and soil decomposes and releases both CO_2 and methane. The exact amount of emissions depends greatly on the site specific characteristics. However, current estimates suggest that life-cycle emissions can be over 0.2 kg of CO_2 equivalent per kWh, while the estimates of life-cycle global warming emissions for natural gas generated electricity are between 0.25 and 0.9 kg of CO_2 equivalent per kWh and estimates for coal-generated electricity are 0.6 and 1.4 kg of CO_2 equivalent per kWh. Environmental hydropower plant effects can occur during the construction, building, plant operations and at the system end-life. For example, although the water used in small hydropower plants is returned to the river, the plant can affect the river flow, flora and fauna, especially in the case of small rivers. Usually, each hydropower station where water is diverted requires a license from proper authority to be developed, installed and operated. Before the license is granted, the agency also has to be convinced that the diverted water is not changing the river character. Some projects may be restricted in when operate, to ensure river flow levels are maintained, meaning that some units do not operate during summer. Even in the run-of-river plants where there is no water storage or diversion, there may be costly requirements, such as the *fish passageways*, allowing the migrating fish to pass the turbine. Once in operation, hydropower plants are welcome additions to the system because they have mostly predictable power generation and controllability. They are also extremely long-lived, system, over 50 years of reliable life, if the plant is well maintained.

Tidal power is renewable, clean and does not emit any pollutants, having significant environmental benefits, being non-polluting, and replacing fossil fuels. A barrage can provide protection over coastal areas during very high tides by acting as a storm surge barrier or breakwater. However, any proposed large-scale tidal energy project has environmental impacts that must be weighed and considered. Estuaries often have high volume of sediments moving through them, from the rivers to the sea. The barrage construction into an estuary may result in sediment accumulation within the barrage, requiring frequent dredging, while the fish migration can also be affected heavily by the tidal barrages.

Some fishes such as the salmon must migrate from saltwater to freshwater to spawn and migrate back, multiple times during their life. When the sluices are open, they can get through. Otherwise they have to seek way out through the turbines, which may end their life. Marine mammal migration is also affected in some areas and, therefore, fishing industry is affected, too. Barrages may also cause less water exchange between the sea and the basin, resulting in the decrease in the basin average salinity, which may affect the food chain of the creatures inside the basin. There are tidal energy adverse environmental effects, which are listed here:

1. For the building of the pool systems, significant construction is required, using a great deal of materials, especially cement.
2. Dams and barrages can obstruct the passages of fish and other marine life.
3. Because the tidal systems are constructed in coastal environments, they invariably interfere with fish spawning and have an adverse effect on the populations of the fish and other aquatic life.
4. Because of the large water flows involved, there are daily water turbidity and chemical composition fluctuations, which may have adverse effects on the aquatic life of bays and estuaries.
5. Local water navigation is adversely affected by the construction of barrages, dams and locks.
6. Additional pool sedimentation is necessitating dredging and disposal of the contaminated silt.

On the positive side, the construction of a tidal system may have multiple purposes to add to the quality life in the surrounding communities. For example:

1. The upper surface of a barrage that connects two sides of an estuary may be paved and used as a road that connects communities.
2. Recreational activities (sailing, boating and swimming) may be promoted within the enclosed estuary.
3. The additional construction may be used for flood control and avoidance of sea erosion.

Wave energy is clean, renewable, almost continuous, and does not pollute the atmosphere. Because the wave systems are built into the open seas, are not affecting the coastal area environment and the ecology, as tidal systems. The following are the most important environmental effects of the wave power facilities:

1. Because the wave power density is low, the wave conversion systems are massive and require large quantities of materials for their construction.
2. Most wave power systems are lengthy, which may obstruct the navigation of the ships.
3. All wave power systems include moving parts, which can kill fish and other sea animals if they are trapped into the systems. Mechanical systems for wave energy conversion operate in adverse conditions; the systems must operate and maintain structural integrity under heavy and low seas, storms and calm weather. This imposes a large constraint on the size, the strength of the materials and the design of the systems, which must withstand very high stress ranges. Because of

the water density, the forces that sea-wave systems are enduring are much higher than those of the wind energy systems. The salt in the sea water is corrosive to most materials, so any wave energy system must be designed around this constraint. The anchoring subsystem of a wave energy system is not 100% fail-safe, and if it fails, may cause the failure or damage of the wave power system. Some wave energy systems are located far from the shore, making the electric power transmission problematic and expensive, being also more susceptible to damage from the strong storms and high waves, occurring in the high seas. These are the main reasons why wave power systems are expensive and there are not any systems in operation, other than experimental or pilot ones. An advantage of future wave power systems is the high correlation between the predominant wind and high energy waves. A promising wave power site is also a good candidate for the construction of towers that produce wind power simultaneously. The construction of both types of electric production units at the same location would bring economic advantages, for example, common generator types, transmission lines and maintenance crews.

Ocean thermal energy conversion (OTEC) is another form of clean and renewable water energy system, which does not cause atmospheric pollution. Because the relative temperature differences between surface and sea-bottom water are present throughout the year, OTEC may become a continuous and rather reliable electric energy source. OTEC generation units have some negative impacts on the natural environment, but overall they are relatively clean and environmentally benign sources of electricity when are compared to conventional options such as fossil fuels or nuclear power. Moreover, the hazardous substances are limited to the working fluid (e.g., ammonia), and no noxious by-products are generated. In an open-cycle OTEC plant, cold water, released at the ocean surface releases trapped carbon dioxide, but its emission is less than 1% of those released from a thermo-electric fossil fuel power plant. The value is even lower in the case of a closed-cycle OTEC plant. Sustained discharging of the nutrient-rich and bacteria-free deep ocean cold water at the ocean surface could change the local concentrations of nutrients and dissolved gases. However, this impact could be minimized by discharging the cold water at depths greater than 50 m. In closed-cycle OTEC plants, chlorine might be used to protect the heat exchangers from microbial fouling. According to the EPA, a maximum chlorine discharge of 0.5 mg/L and an average of 0.1 mg/L are allowed. The amount of chlorine in a closed-cycle OTEC plant is within these limits. There are low-impact and rather limited environmental and ecological effects of the OTEC electricity production, such as the local aquatic life disturbance and small impediments in the local ship navigation. Because the warm OTEC resources exist primarily in smaller bays, all environmental and ecological effects are usually localized in the OTEC power plant vicinity. However, all OTEC cycles suffer from the inherent disadvantage of the low temperature differences and the implied low thermal efficiencies. For this reason, all processes must be optimized and the equipment must be maintained in good condition. This becomes problematic in the ocean and coastal environment, where salty water causes equipment corrosion and deterioration, while the growth of algae tends to obstruct the heat exchanger surfaces, pipelines and submerged water pumps. These are the reasons why the few OTEC pilot plants that were built on shore or on barges failed and were decommissioned after a short operation. OTEC power plants may cause major adverse impacts on the ocean water quality. Such plants require entraining and discharging large quantities of seawater. Marine biota can be impinged on the screens covering the OTEC plant warm and cold-water intakes. Small fishes and crustaceans may be entrained through the system, where they are experiencing rapid changes of temperature, salinity, pressure, turbidity and dissolved oxygen. A major change occurring in the cold water pipe is the depressurization of up to 10^7 Pa in water coming from a depth of 1000 m to the surface. Sea surface temperatures in the OTEC plant vicinity is lowered by the discharge of effluent from the cold water pipe, having impacts on organisms and microclimate. The pumping of large volumes of cold water from sea depths to the surface releases the dissolved gases (e.g., CO_2, oxygen and nitrogen) to the atmosphere, influencing the water pH and dissolved oxygen levels, causing stress to marine life.

7.3.5 Bioenergy Environmental Analysis

Biomass, due to its rich composition, is a versatile raw material for the production of chemicals that can be used as biofuels. There are a number of alternatives to substitute crude-based gasoline and diesel, but most importantly, the technologies are becoming available and industrially feasible. Furthermore, the production costs are becoming competitive. Biomass power plants are based on the same principle as conventional power plants, what is different are the type of fuel. It uses various remains of forestry, agriculture and livestock, which are directly burned or gasified. Combustion of biomass and biomass-derived fuels produces air pollution; beyond this, there are concerns about the impacts of using land to grow energy crops. Although ethanol use as fuel results in less CO_2 emissions when compared to gasoline, it is important to notice that avoided emissions comes to a cost in other environmental impacts. Soil erosion, water quantity and quality and loss of biodiversity are some of the environmental concerns associated with ethanol production. How serious these impacts are, depend on how carefully the resources are managed. The analysis is further complicated because there is no single biomass technology, but a wide variety of production and conversion methods, each with different environmental impacts. The stress such a massive plantation would cause on soil moisture, through uptake and evapotranspiration of precious water, has not been estimated but would obviously be quite large. Large amounts of water are also used for sugarcane or other biomass washing, processing and distillery operations. There are other impacts of a larger magnitude, on soil productivity, microclimate, or wildlife, some of which would be disastrous to the ecology of the region. Unlike other RES, these power plants have direct emissions of greenhouse gases and particulate matters, but they take into account continuous renewal of fuel, the amount of produced CO_2 is basically equivalent to consumption in the process of photosynthesis. Biomass has lower nitrogen content than coal, but it also has lower heating values (15 to 21 MJ/kg for biomass compared to 23 to 35 MJ/kg for coal). Therefore, some biomass fuels can produce more NO_X emissions than coal does for the amount of heat it produces. The production of NO_X from the nitrogen in biomass is not well understood, because the forms of nitrogen are different from those in coal. Therefore, research is needed to minimize pollutant emissions during biomass power generation processes. Research is also needed to understand the impacts that such installations have on plants and wildlife in various areas, and to develop effective ways to mitigate these impacts. Land-use impacts can be reduced by employing previously developed sites, co-occupation with other land uses, using military and government sites, and encouraging DG technologies to minimize the transmission line needs.

Large land areas are needed to grow any biomass which has a direct impact on the eco-world in those areas (herbicides, pesticides, fertilizers). For the production of biofuels (biodiesel and bioethanol), a large area of cultivable land is needed. For the production of biodiesel waste, edible oils are also used. If on one side the gas emissions are reduced through the use of biofuels, the most important consequence of biomass production for energy is the use of agricultural land, a scarce planetary resource. Agricultural crops and forests use a very large area and the development of energy crops would be very costly in the use of land. On the other hand, agricultural waste, such as bagasse, rice husk, or forestry waste is a very welcome source of energy because it is normally considered unwelcome waste and, in most cases, their producers will even pay for their removal. A significant environmental effect of the growth of biomass for energy is the high requirements of fresh water for irrigation and processing. For example, the production of a single gallon of ethanol from corn requires approximately 1000 gallons of water. If all the available land were to be used for the growth of energy crops, huge quantities of fresh water are required, which are not available in all regions. Unless the energy crops are planted close to large rivers with unused or underutilized agricultural land, the competition for the existing water resources would result in a significant strain to local populations. The situation is entirely different if sea water were to be used for the production of energy crops: there are vast areas of land close to the ocean, which are not currently used for the production of food. A technological breakthrough, allowing irrigation of energy crops with sea water will change this situation and may make feasible and more socially acceptable the widespread production of energy crops in several coastal regions. The expected globally widespread use of fertilizers and pesticides is another significant environmental effect of more biomass production. The energy crops are fast-growing plants, needing large amounts of fertilizers, pesticides and insecticides. These chemicals contain phosphorous, sulfur, nitrates, arsenic and trace metals such as zinc, lead and manganese. Many of these elements are toxic to humans or harmful to the environment. The widespread biomass combustion is another pathway that introduces these toxic elements into the atmosphere and living organisms. In addition, all the chemical compounds in the fertilizers, pesticides and insecticides invariably find their way to the hydrosphere, rivers, lakes, estuaries and oceans, following the runoff waters from rainfall. The addition of these chemicals alters the chemical composition of the freshwater and seawater. This is especially significant for the fragile ecosystems in lakes, estuaries and bays with low water circulation.

Another environmental biomass effect is the potential accelerated methane production. If left untreated for certain periods, the biomass decomposes naturally, producing CO_2 and methane, which diffuses into the atmosphere and contributes significantly to global warming. If biomass is not used promptly and is left to decompose, either in the field or in storage facilities, can cause significantly greater damage than the environmental benefits. Less significant environmental effects of the expanded biomass and energy crops include: soil erosion and depletion of soil nutrients, biodiversity losses, deforestation, which may lead to desertification, growth of monocultures, which are highly vulnerable to agricultural diseases and bacteria, higher river silt concentration and enhanced siltation rates that may lead to river or estuary eutrophication, changes in land use and irrigation patterns, which may change the region microclimate, and dust production from plowing and harvesting. It must be noted that almost all these adverse effects of extensive biomass production are associated with the energy crops and not with the agricultural, human and animal waste treatment. The expanded energy production from any wastes and the waste volume reduction on a global scale has beneficial environmental and ecological effects, in addition to producing needed energy. It is apparent that the use of food crops for energy production, electricity or biofuels, is detrimental to the environment, placing strains on water supply, being economically inefficient and socially undesirable. Providing high subsidies for the energy crops in the industrialized countries is economically inefficient. There are other impacts of a large magnitude on soil productivity, microclimate and wild life, some of which would be disastrous to the ecology of the region if bioenergy production is not properly. From aquatic weed farms, it would be difficult to prevent percolation of sewage from such large ponds to the underground aquifers and the dangers of groundwater contamination would be very real. There would be such other problems to contend with as mosquito menace and propagation of pathogens. Further, disposal of spent water hyacinth, after energy is extracted from it, would be a major problem. The removal of biomass from land and water for energy production programs may increase soil and water degradation, flooding and removal of nutrients. It might also affect wildlife and the natural biota. These and other threats to the environment from the production of biomass do not seem to have been widely understood. Conversion of natural ecosystems into energy-crop plantations changes both the habitat and food sources of wildlife and other biota. Alteration of forests and wetlands reduces many preferred habitats and mating areas of some mammals, birds and other biota. Bioenergy plants have lower emissions of SO_2 than do coal and oil plants, but they may produce more particulate matter.

> **Example 7.5:** Imagine this amount of energy demanded by a world region is to be produced by burning coal. Assuming that the energy demands for a region is about 2.464 TWh, by employing a 10% of PV, geothermal power and wind energy for electric power generation, how much CO_2 emissions are mitigated?
>
> **Solution:** The general reaction of carbon combustion is: C (coal) + O_2 = CO_2, and the average coal energy content is ~24 MJ/kg, while the average efficiency of coal-fired power plants is approximately 33%, roughly, three times as much energy must be generated by coal combustion for obtaining the specified amount of electric energy:
>
> $$\text{Combusted Energy} = \frac{\text{Target Energy}}{\eta} = \frac{2.464}{0.33} = 8.0182 \text{ TWh or } 2.887 \times 10^{19} \text{ J}$$
>
> Now, we can find how much coal we need to generate this amount of energy by dividing the amount of energy by coal energy content:
>
> $$\text{Amount of coal needed} = 2.89 \times 10^{19} \text{ J}/24 \times 10^6 \text{ J} = 1.2 \times 10^{12} \text{ kg } (=1200 \text{ million ton})$$
>
> Now, based on the assumed 95% content of carbon in the coal (let us consider anthracite coal), we calculate the amount of pure carbon burnt in the process:
>
> $$1.2 \times 10^{12} \times 0.95 = 1.14 \times 10^{12} \text{ kg C}$$
>
> Because for each mole of carbon in the coal, there is 1 mole of CO_2 produced, we can find the amount of CO_2, taking into account the molar masses of the components – C 12 g/mol and CO_2 44 g/mol, the amount of CO_2 released by cola combustion is:
>
> $$\text{Released } CO_2 \text{ Amount} = 1.14 \times 10^2 \times \frac{44}{12} = 4.18 \times 10^{12} \text{ kg } CO_2$$
>
> If the 30% is replaced by wind, PV and geothermal, quite substantial mass of CO_2, about 1.125×10^{12} kg CO_2 *is not* released into the atmosphere due to the operation of the PV systems around the world.

7.3.6 Environmental Implications of Energy Storage Systems

As energy storage technologies are emerging as one of the potential solutions for addressing the constraints caused by the high deployment of renewable energy systems, efforts are underway to identify key environmental impacts of large-scale energy storage systems. A comprehensive understanding of the environmental impacts of the energy storage systems can help the energy storage industry to develop environmentally friendly energy storage solutions and help decision makers craft sustainable energy storage policies. It is worth to mention that the term carbon-free literally means for the utilization of the high grade heat energy being derived from the natural and renewable solar radiation for achieving TES requirements in the applications ranging from solar heating to solar thermal power systems. Energy storage systems cannot store electricity itself, but can convert electricity into other forms of energy, which can be stored for later use and then be converted back to electricity when demand is high. The conversion processes can have environmental implications, which must be addressed and quantified. Regarding the solar TES applications, there are two important factors that needed to be considered: requirement of larger area for collection of heat energy, and the size of the heat storage facility. The aforementioned factors are most vital in the sense that the maximum quantity of heat energy has to be captured from the incident solar radiation during sun brilliance periods by means of efficient collectors. This can facilitate the provision of more heat energy to the TES system during the heat storage process. Thus, proper sizing of the heat storage facility also contributes to enhancing the energy efficiency of solar TES systems. When considering the environmental impacts of the energy storage technologies, they also should be compared with the impacts of other low-carbon energy system options, such as RES of DG units. The health impact of renewable electricity sources have been less comprehensively assessed, but the research and academic communities are expecting the impact of solar, wind and tidal power to be less severe than either coal or nuclear power. While carbon capture and storage technologies have the potential to reduce the emission of some air pollutants from fossil fuel combustion, they reduce the overall efficiency of electricity production and remain immature so their total effect is yet to be determined. In assessing the effect technologies have on the environment, it is usually to consider the energy required to build them (the *embedded energy*), any toxic components used, and how they can be recycled. One metric to measure the potential environmental impact of different bulk energy storage technologies is the energy stored on investment (ESOI), where a higher number indicates a better capacity to store energy, a long operating life time (cycle life), or a smaller amount of embedded energy. Mechanical energy storage technologies, such as pumped hydropower and compressed air energy storage, have a much higher ESOI than electrochemical energy storage technologies, such as batteries, by a factor of 10 up to 100. This difference is a result of the mechanical technologies, having lower embedded energy per unit capacity, and higher cycle life. Among batteries, cycle life varies greatly and this has a significant effect on their ESOI. It follows that driving innovation to improve cycle life could be a good way to reduce the environmental impacts. Further to this, the ESOI of a technology can be improved by extending its useful life, for example giving spent electric vehicle batteries a *second life* in stationary applications such as grid or domestic energy storage. The ESOI is usually defined by this ratio:

$$\text{ESOI} = \frac{\text{Overall Lifetime Stored Energy}}{\text{System Embedded Energy}} \tag{7.4}$$

Additional environmental considerations for electrochemical technologies are the scarcity and toxicity of materials used in their production. The lead used in these types of batteries is toxic and therefore must be recycled and handled properly. In addition, the sulfuric acid typically used as the electrolyte is corrosive and when overcharged the battery generates hydrogen which presents an explosion risk. Effective recycling procedures exist for lead-acid batteries in Europe and the United States, where more than 95% of lead-acid batteries are recycled at the end of their lives. This success has been attributed to the profitability of reclaimed recycled materials, the illegality of disposing of batteries, the simplicity of disassembling the standard batteries and the ease of recycling the components. However, a high incidence of lead poisoning in regions of developing world has been attributed to widespread improper recycling. The most significant drawback of Ni-Cd batteries is the highly toxic cadmium used within them. Although this metal is highly recyclable, it is exceedingly toxic. Most Nickel is recovered from end-of-life batteries since the metal is reasonably easy to retrieve from scrap and can be used in corrosion resistant alloys such as stainless steel. For example, the EU legislation is in part responsible for the replacement of Ni-Cd batteries by Ni-MH batteries and represents a significant issue to any future development of Ni-Cd battery technologies. Avoiding fumes and dust is essential for workers in this industrial sector. There are limited environmental concerns associated with NaS batteries, since the materials used in their construction are relatively environmentally inert. There is a small risk associated with the high temperature at which the battery must be operated in order to maintain the sulfur in its molten form. Li-ion batteries have a limited environmental impact since the lithium oxides and salts can be recycled. The main impact is the resource depletion, human toxicity and eco-toxicity mainly associated with copper, cobalt, nickel, thallium and silver). Lithium-ion batteries can be hazardous without proper recycling at the end-life. Recycling procedures are not fully established and are even more challenging than for lead-acid batteries, owing to a more complex design and a wider range of materials used. While there are a number of proposed solutions, an insufficient number of lithium-ion batteries have reached the end of their lives for recycling to become commercially viable. In addition, lithium-ion battery technology is still evolving, so recycling procedures developed for a specific design or chemistry could quickly become obsolete. Broad commitment from industry and government will be required to meet the challenge of developing effective recycling procedures before large numbers of electric vehicle lithium-ion batteries reach the end of their useful lives. The rare earth metal cobalt is currently used in most lithium-ion batteries, while its extraction is been associated with serious and systematic human rights violations and environmental negligence. For this reason, some lithium-ion battery manufacturers are seeking to use cobalt from other sources, and make more use of lithium-ion cell types that do not use cobalt. Similar challenges could arise in other electrochemical storage technologies, and careful attention should be paid to recyclability and resource use. Environmentally, metal-air batteries are relatively inert since no toxic materials are involved in their construction. Like other battery types, metals such as zinc or aluminum used within the battery should be recycled. The potential size and scale of these systems are likely to determine the extent to which environmental impacts are significant. Significant quantities of space may be required for holding tanks containing the electrolytes and although these substances may not be specifically toxic this requires care at the design stage. A major advantage of the technology is the ability of the technology to perform discharge cycles indefinitely so there are no significant waste products associated with operation.

Thermal energy systems can help to increase the efficiency and to reduce the environmental impacts for the energy systems, particularly in the heating and cooling of the buildings and industrial facilities and into the power generation management and control. TES allows a solar thermal power plant to extend operational hours into the evening and can help operators manage variability during extended cloudy periods. Reduced variability enables the solar thermal power plant to become closer to a dispatchable generator.

By reducing the energy use, TES systems can provide significant environmental benefits by conserving fossil fuels through increased efficiency and/or help for a significant fuel substitution, and by reducing emissions of such pollutants as CO_2, SO_2, NOx and CFCs. TES can impact air emissions in buildings by reducing quantities of ozone depleting CFC and HCFC refrigerants in chillers and emissions from fossil fuel fired heating and cooling equipment. TES helps reduce CFC use in two main ways. First, since cooling systems with TES require less chiller capacity than conventional building and industrial process thermal energy systems, they use fewer or smaller chillers and correspondingly fewer refrigerants. Second, using TES can offset the reduced cooling capacity that sometimes occurs when existing chillers are converted to more benign refrigerants, making building operators more willing to switch refrigerants. The potential aggregate air-emission reductions at power plants due to TES can be significant. TES systems can contribute to increased sustainability as they can help extend supplies of energy resources, improve costs and reduce environmental and other negative societal impacts. Sustainability objectives often lead local and national governments to incorporate environmental considerations into energy planning. TES systems are attracting increasing interest in several thermal applications, e.g., active and passive solar heating, water heating, cooling and air conditioning. Also, TES is presently identified as an economic energy storage technology for building heating, cooling and air-conditioning applications. The implementation of proper energy storage remains crucial to achieve energy security and to reduce energy generation environmental impacts. It is difficult to compare different energy storage methods using only few factors, such as storage capacity or lifespan. In fact, no single storage method can be universally used. For specific situations, geological locations and existing facilities, different energy storage methods are possible and need to be considered. For some storage methods, such as batteries, super-capacitors or solar fuels, their advancements depend on discovery of new materials and characterization of materials and properties. Less-expensive, high-capacity and environmentally friendly materials are some of the criteria in choosing the right materials. However, TES may increase operation and maintenance costs as well since the TES system itself will require maintenance and additional workers, and any increase in solar collector numbers requires additional mirror washing and potential repairs. Additional collectors also mean an increase in land use, which is especially important in the ecologically sensitive desert regions that are particularly well-suited for CSP technology.

Extremely low temperatures are required for the superconducting system to operate, can represent a safety issue. Larger scale SMES systems could require significant protection to deal with magnetic radiation issues in the immediate vicinity of the plant. Supercapacitors can be employed to enhance the energy performance of cars, for example, through regenerative braking systems, which can therefore lead to emissions benefits. In addition, supercapacitors have a long life cycle. Potential negative environmental impacts of supercapacitors arise from the materials and compounds used within their construction and operation. With no chemical and disposal issues to consider, flywheels have few environmental advantages over the batteries described earlier. Subject to stringent safety safeguards applied to the operation of heavy, rapidly rotating objects, a flywheel system should not cause problems to the local area. Compressed air energy storage, as a means of mitigating intermittency in wind power production can be favorable in certain conditions, especially when geological conditions are well suited for implementation and no energy has to be spent on the creation of a cavern. In general, ACAES have a lower impact compared to the CAES, due to fact that no fossil fuel is combusted. However, the impacts for a CAES plant are lower than those from natural gas power plant. For ACAES, an important part is the thermal energy storage and developing of the thermal mass with high heat transfer capabilities and low environmental impact is crucial to improve overall performance of these systems.

7.3.7 Waste-to-Energy Recovery Environmental Implications

Waste materials are an important source of energy, if they are properly used, so the results of assessments of the damage costs of waste treatment and the benefits of energy recovery, for the principal technologies for municipal solid waste, such as incineration and landfill are important. Overall, managing municipal solid waste (MSW) is challenging task rather than an opportunity to obtain other commodities such as recycling materials, heat or energy. Depending on economic development, climate, culture and energy sources, MSW composition varies from one country to another. Waste-to-energy (WTE) conversion has now been considering one of the optimal methods to solve the waste management problem in a sustainable way. Different advanced mechanical biological and thermochemical waste-to-energy technologies have now been applying for managing municipal waste. Energy can be recovered from biodegradable and non-biodegradable matter through thermal and biochemical conversions. The utilization of municipal solid waste as a renewable energy source could overcome waste disposal issues, while displacing the power and heat generated for fossil fuel displacement and mitigate pollutant emissions from waste treatment by converting CH_4 to CO_2. This option has therefore been considered an important and crucial factor to successful and environmental friendly municipal waste management. The main air emissions from landfill are CH_4 and CO_2. Depending on waste treatment technology, showed in the damage assessment graph. The environmental burdens are also varies for different the damage assessment result of WTE technologies in technology must carefully analyzed. Landfill is the old but still an extensively used technology for municipal waste management and advance sanitary landfill has the option to collect landfill gas and to use as biofuels or electricity production. Landfill gas is generated from the landfill site. Usually, modern landfill is divided into a large number of individual compartments, and they are filled one after another and sealed when full. In practice, it is not possible to capture all of the CH_4, the common capture rates are around 70% (although measured data are difficult to find). A capture rate of 70% for the first 40 years is assumed after closure of a compartment, after 40 years, and all the remaining CH_4 escapes to the atmosphere. These assumptions reflect uncertainty in the way that sites are managed around the time of closure is approached. One possibility is that regulators will simply sign off a site as inert after a certain period of time. Another is that regulators are requiring operators to provide evidence, such as data on emission rates, to show that a site is inert. The latter provides a higher level of environmental protection. The damage costs and the comparison between landfill and incineration turn out to be extremely sensitive to assumptions about energy recovery. Raw waste can be used as a feed stock in the advanced thermal Incineration process. Incineration processes has taken place in the presence of air and at the temperature of 850°C and waste are converted to carbon dioxide, water and noncombustible materials with solid residue. Incineration is very popular technology in Europe due to heat and electricity generation option from the waste management facilities. Primarily, two process stages are involved in the incineration processes like, combustion of the waste where flue gas is produces and cleaning of the flue gas and emitted to the atmosphere. Pyrolysis gasification (PG) is an advanced hybrid thermal waste treatment processes and is an emerging technology since the processes has been recently used for municipal waste management system. Thermal degradation (between 400°C to 1000°C) of the pyrolysis process is taking place in the absence of air and it produces syngas, oil and char. Gasification takes place at higher temperatures than pyrolysis (1000°C to 1400°C) in a controlled amount of oxygen. The gaseous product contains CO_2, CO, H_2, CH_4, H_2O trace amounts of higher hydrocarbons]. Pyrolysis is favorable to reduce heavy metal emission and also sulfur dioxide and particulates; however, oxides of nitrogen and dioxins emissions might be similar with the other thermal waste treatment technology. Thermal WTE conversion has the higher energy conversion potentials that the other technology.

7.4 Energy Management and Energy Conservation

Energy management, a paradigm and theory, having concepts, principles and methods that are recently fully accepted and employed is an important and comprehensive framework, part of the sustainable development, attempting to plan and manage the energy uses and applications on the past experience and future needs. The energy management fundamental goals are to produce goods and to provide services with the minimum energy uses and environmental impacts. The term *energy management* has different meaning to different people and in different industrial areas. The objectives of the energy management methods and techniques are to achieve and maintain optimum energy procurements and uses, throughout an organization and to minimize energy costs, energy losses and wastes without affecting production levels and quality, while minimizing in the same time the environmental energy use effects. Energy efficiency and conservation, and the renewable energy uses are the pillars of a sustainable energy policy. Above definitions are rather broad definition covers many operations from the services, product and equipment design through the product shipment and delivery. Waste minimization and disposal, which are important aspects of energy management plans, are also presenting important energy management opportunities and solutions. Energy savings and waste reductions constitute primary measures for the protection of the environment and, in addition, for the reduction of exchange effluxes, which are used to purchase the polluting fossil fuels, coal, oil and natural gas. Energy efficiency concept or energy saving concept is not to be mistakenly considered as lack of consumer's comfort. Optimal energy consumption defines a large range of solutions: reduction of energy losses at production, transmission, distribution and consumption phases, replacement of the obsolete equipment, and rather energy consuming, elimination of the unjustified energy consumption, etc. Ecological aspects should not be neglected. It is an important topic of energy saving and conservation programs envisaged by the newest energy policy. Notice that the *energy conservation* and *energy efficiency* are separate, but related concepts. *Energy conservation* is achieved when growth of energy consumption is reduced, being measured in physical terms. Energy conservation can be the result of several processes or developments, such as productivity increase or technological progress. On the other hand the *energy efficiency* is achieved when energy intensity in a specific product, production process, service or consumption is reduced without affecting the output or comfort levels. Promotion of energy efficiency contributes to energy conservation and is therefore an integral part of energy conservation policies. Energy efficiency is often viewed as a resource option like coal, oil or natural gas. It provides additional economic value by preserving the resource base and reducing pollution. Nature sets some basic limits on how efficiently energy can be used, but in most cases our products and manufacturing processes are still a long way from operating at this theoretical limit. Very simply, energy efficiency means using less energy to perform the same function. Although, energy efficiency has been in practice ever since the first oil crisis, it has today assumed even more importance because of being the most cost-effective and reliable means of mitigating the global climatic change.

From physics point of view the energy is always conserved, being only converted from one form to another. Efficiency involves using less energy to achieve the same ends or results, while the energy conservation are putting the stress on simply using less energy, even the ends may be compromised. The energy efficiency involves technical solutions, while the conservation involves mainly behavioral approaches. On the other hand, even the energy efficiency and energy conservation are often mutually reinforcing, they can be in many situations in conflict, depending mostly on the individual viewpoints. The most common definition of the energy efficiency is the fraction of the extended energy used to produce a desire result. For example, an old incandescent lamp and an LED may produce the same light intensity but using quite different input electricity, and similar an old induction motor and new more efficient one are producing the same mechanical power by using different input electrical energy. However, the energy efficiency must be estimated over the all process or system stages or phases in order to avoid wrong estimates. If for example the electricity is used to heat water in an electric heater, the efficiency is close to 100%, while the efficiency of generating electricity is included the overall process efficiency is about 30% or so. Furthermore, the energy use is not just about efficiency, an important parameter, but also about cost, time spent, feasibility, safety, convenience, the environmental implications and even the culture. On the other hand, the energy conservation problems involve many and often alternative solutions, and different decision-makers and are very complicated and complex to solve. However, the decentralized nature of the energy conservation can be an advantage into the implementation conservation methods and procedures, offering different courses of action.

Energy management system lies at the heart of all infrastructures from communications, economy and transportation. This has made the system more complex and more interdependent, increasing number of disturbances occurring that has raised the priority of energy management system infrastructure which has been improved with the aid of technology and investments. Modern industrial and commercial facilities operate complex and interrelated power systems. Energy conservation and facilities/equipment are only part of the approach to improve energy efficiency. Most energy efficiency in industry is achieved through changes in how energy is managed in a facility, rather than through installation of new technologies. Systematic management and the behavior approach have become the core efforts to improve energy efficiency today. An energy management system provides methods and procedures for integrating energy efficiency into existing industrial or commercial management systems for continuous improvements in the energy usage. Energy management is defined as a system, methods and procedures for an effective and optimum energy use in industrial processes and in operation of residential, commercial and industrial facilities, maximizing the profits, enhancing the competitive positions through organizational measures and optimization of energy efficiency in the industrial and commercial processes. Profit maximization is also achieved through the reduction in energy costs during each production and operational phase, while the three most important operational costs are those for materials, labor and energy (fuels, electrical and thermal). Moreover, the competitiveness improvement is not limited to the reduction of sensible costs, but can be achieved with an opportune energy cost management, increasing the flexibility and compliance to the changes of market and international environmental regulations. Energy management is a well-structured process, both technical and managerial in nature. In this chapter, we discuss the structure, methods and techniques used in energy management, as well as new approaches and developments in the field. The process of energy management, regardless the application size or specifics consists of the following major steps: collecting the detailed data on the energy uses, identifying the energy savings opportunities and the most accurate possible estimates of the energy amounts that can be saved, applying the most cost-effective energy saving measures, based on the opportunities found before, and tracking and monitoring the impacts of the actions and measures and eventually going back the first step and repeat the iteration. Energy management is showing the organization energy performance ratings, helping to track the energy and water consumption, to set targets and indicators and, as a result, to improve policies, identify areas that need improvements, and measure the investments required for facility upgrades. The energy management systems aim at reduced energy uses and costs, are representing a key element in any company energy management program. Energy management must be based on real time information obtained from process monitoring and control systems, and on production plans received from production planning systems. It is also useful to compare each unit and site with its own past performance to determine energy savings, but the internal view is not very useful in determining how the process compares with its competitors. Often total and comprehensive solutions include planning and scheduling methods to optimize energy uses and supply, energy balance management tools to support the real-time monitoring and control, and reporting tools to evaluate and report energy consumption, costs, efficiency and other energy-related information. Opportunities for cost reduction are greatest when both electricity consumption and prices vary over time, which is common in process industries, and open electricity market environments. Energy management concepts are built upon the plan-do-check cycle, being used as a basis for many management systems. Figure 7.6 shows how the use of the management action process leads to continuous improvement. An EMS is a collection of procedures, methods and tools designed to engage staff within an organization in managing energy uses on an ongoing basis. It allows all types of facilities and entire organizations to systematically track, analyze and plan their energy use, enabling greater control of and continual improvement in energy performance.

In the past, most of the manufacturing, industrial and large commercial organizations and facilities have lacked complex energy monitoring and control systems when compared to the ones frequently found in electrical generating plants. The reason is that the primary goal of any manufacturing organization is to cost effectively produce high quality products and thus the energy management often takes a somewhat secondary role to a production objectives. However, with increased energy prices and market competition, the companies are taking a closer look at energy saving opportunities and the potential for onsite electrical generation from alternative energy sources. Manufacturing firms are re-examining the energy management functional requirements and internal energy systems to increase efficiency and integration with the future power grid. Energy efficiency, conservation and cost are top priorities all over the world, in particular for heavy energy

FIGURE 7.6
Energy management flow chart and structure.

consumers. Energy conservation technology and facilities or equipment are only part of the approach to improve energy efficiency. Systematic management and the behavior approach have become the core efforts to improve energy efficiency today. Energy management represents a significant opportunity for organizations to reduce their energy use while maintaining or boosting productivity. Industrial and commercial sectors jointly account for approximately 60% of global energy use. Organizations in these sectors can reduce their energy use 10%–40% by effectively implementing an energy management system. The modern manufacturing plant has a large set of IT equipment including EMS and manufacturing execution systems which are used to control processes and infrastructure. The energy management key question is how to provide the best case for successful energy management within their organization, achieve the desired buy-in at top management level, and implement a successful energy management system. The energy management purpose is to provide an organizational framework to integrate energy efficiency into management practices, including fine-tuning production processes and improving the system energy efficiency. It also seeks to apply to the energy use the same culture of continual improvement that is successfully used by companies to improve quality and safety practices. Its guidelines recommend that companies need to track energy consumption, benchmark, set goals, design action plans, evaluate progress and performances, and create energy awareness throughout the organization, enabling the energy efficiency and conservation integration into existing management system for continuous improvements and to reach the requirements for high-efficient companies and ultimate to reduce cost and increase revenue. Among the high-efficient company requirements are: efficiency is a core strategy, leadership and organizational support is real and sustained, company has energy efficiency goals, there is in place a robust tracking and measurement system, resources are put for energy efficiency and energy efficiency strategy is working and results are communicated. The key feature of a successful EMS is that it is owned and fully integrated as an embedded management process within an organization or company, its implications are considered at all stages of the development process of new projects, and that these implications are part of any change control process. Standards should lead to reductions in energy cost, pollutant emissions, minimizing negative environmental impacts. A change in the organizational culture is needed in order to realize industrial energy efficiency potential.

Energy management is required because it influences a number of aspects of company operation and activities including the following aspects, energy costs, which affect the company profitability, energy costs which affect the competitiveness in the world market, national energy supply-demand balance, national trade and financial balance, local and global environments, occupational safety and health, loss prevention and waste disposal reduction, improved productivity and quality. Energy management in the form of implementing new energy efficiency technologies, new materials and manufacturing processes and the use of new technologies for business and industry is also helping companies improve their productivity and increase their product or service quality. Often, the energy savings is not the main driving factor when companies decide to purchase new equipment, use new processes and new high-tech materials. However, the combination of increased productivity and quality, reduced emissions and energy costs provides powerful incentives for companies and organizations to implement these new technologies. Energy management is all about reducing the cost of energy used by an organization, now with the added spin of minimizing carbon emissions as well. Reducing energy costs has two facets: price and quantity. Improving energy efficiency is, in part, a technical pursuit with a scientific basis. The knowledge and skills of the energy management report impart guidance for development, generate opportunities for collaboration among developing or expanding training programs, facilitate greater consistency among existing professional programs and increase awareness about the energy efficiency potentials. Steps needed during implementation of an energy management system are:

1. *Initiating an energy management program*: Understanding basic concepts and requirements, getting organization higher management commitment, establishing an energy management team and developing an energy policy and planning;

2. *Conducting an energy review*: Collecting energy data, analyzing energy consumption and costs, identifying major energy uses, conducting energy assessments and identifying potential energy saving opportunities;

3. *Energy management planning*: Setting a baseline, determining performance metrics, evaluating energy saving opportunities and selecting best projects, and developing action plans;

4. *Implementing energy management*: Obtaining resource commitments, providing training and raising energy saving awareness, communicating to all stakeholders and executing action plans;

5. *Measurement and verification*: Providing the knowledge and skills required to monitor, measure, verify, track and document energy use and savings; and

6. *Management review*: Reviewing progress, modifying goals, objectives and action plans as needed.

These steps are embedded in the *Plan-Do-Check-Act* cycle, which in addition to the common knowledge areas, includes the identified ancillary knowledge and skills that are enhancing the understanding of key energy management topics. To manage the energy a complete understanding of the company energy use, structure and demand is critical. This is based on a comprehensive energy review, consisting of the analysis of all energy use and consumption, determining the significant energy uses and then identifying and prioritizing potential opportunities for the improvements and energy savings. An energy review is an essential element of energy management planning and must be performed by personnel with a broad range of knowledge and skills. An energy review requires the collection of energy consumption data from utility bills, energy meters and other information sources. The data must be analyzed and interpreted within the context of the sites, facilities, processes, business units and equipment. For example, in many of the smaller facilities, lighting and space heating and cooling are often the dominant energy users, while larger facility large energy parts are used in process or large equipment. These are just a tiny percentage of the cost at large petrochemical sites, where the dominant energy users are associated with moving, transforming and separating feed and product materials. This analysis requires personnel that not only understand buildings, processes, energy using equipment and other process or facility factors but also possess the knowledge and skills necessary to identify viable improvement opportunities. Some of the fundamental tasks involved in conducting an energy review include: data logging and collection, metering, monitoring, measurement, and verification, facilitating and managing the process for identifying the energy efficiency and conservation opportunities.

7.5 Summary

The supply of sufficient energy has become very important to the affluence of nations and the well-being of the global society. Securing energy supplies has become a prime consideration of all the governments. The environmental impact of energy conversions inevitably disturbs the environment. For this reason it is imperative that aggressive, scientifically based monitoring, analysis and mitigation efforts be considered an integral part of any energy development. Decision making as for to deploy or not to deploy a particular technology, on what scale, at which location and when, would rely on the technology readiness analysis (not discussed in this chapter) and on technology adaptation to the social, economic and environmental spheres, which is assessed through the LCA analysis. Both types of assessments are complex, require significant amount of factual data, and must be system-specific and location-specific. Developing RES technologies that exploit the Sun, the wind, the water and geothermal energy is critical to addressing concerns about climate change and some environmental issues. The RES advantages are inexhaustibility, and an impact on the environment, while the main RES disadvantage is low energy density and the technology is yet in the development phase. However, using renewable energy sources are not eliminating all environmental concerns. Although renewable energy sources produce relatively low levels of GHG emissions and conventional air pollution, manufacturing and transporting them will produce some emissions and pollutants. The production of some PV cells, for instance, generates toxic substances that may contaminate water resources. Renewable energy installations can also disrupt land use and wildlife habitat, and some technologies consume significant quantities of water. The RES importance is that their environmental impacts can be minimal, if properly managed. Alternative energy sources are starting to substitute the fossil fuel consumption and provide the human society with sufficient energy to maintain the desired standards of living. In addition, societal concerns about global climate change may impose regulatory limitations on the consumption of fossil fuels in the near future. For this reason it is very important to know: what are the alternative energy sources that may be harnessed with the current technology, what are the engineering systems that harness this energy, how the alternative energy sources may satisfy the world energy demands, and what is the potential of these sources to supply energy on a large scale. Electrical energy storage will have a number of benefits as the electricity system becomes increasingly reliant on intermittent renewables. A range of storage technologies exist, that have different performance characteristics and costs, ranging from low-cost, large-scale mechanical technologies to higher cost electrochemical technologies. In many cases even the higher-cost technologies represent good propositions for a low-carbon electricity system, provided that their value can be realized across their multiple capabilities. No matter the magnitude of the energy storage system, certain environmental aspects should be considered. In the context of promoting smart grids as in small scale and distributed energy generation in remote communities, the major aspects of batteries' environmental impacts must be considered. Consumers must have the means and proper information on recycling of the batteries, thus a proper recycling network must operate flawlessly. In general, energy storage systems are environmentally inert during their operation and the major impacts are found during their construction. As in all chemical and mechanical devices, the installation must be made in secure rooms with all the necessary precautions, lighting, accident prevention automations and emergency exits. The concepts of energy conservation, energy efficiency or energy optimization became the major modern concerns. Energy sustainability is the cornerstone to the health and competitiveness of industries that produce and manufacture in our global economy, being more than the process of being environmentally responsible and earning the right to operate as a business. It is the ability to utilize and optimize multiple sources of secure and affordable energy for the enterprise, and then continuously improve utilization through systems analysis and through an organizational drive for continuous improvement as a core principle. Management commitment to ensure the best energy efficiency management in existing process operations, as well as a dedicated pursuit of new system technologies and processes, is the only recipe for excellence.

Life-cycle analysis is a sophisticated way of examining the total environmental impact of a product through every step of its life, from obtaining raw materials all the way through making it in a factory, selling it in a store, using it in the home and disposing of it. Disposal options include incineration, burial in a landfill, or recycling. LCA is a tool for comparing goods, processes, services and products and for identifying opportunities for reducing the impacts attributable to associated wastes, emissions and resource consumption. LCA is a valuable method of comparing environmental impacts of alternative electricity generating technologies and identifying where improvements are most likely to pay off. LCA shows that increasing system efficiencies and operating lifetimes will reduce environmental impacts for all renewable energy technology. LCA is a *cradle-to-grave* approach for assessing products, processes, industrial systems and the like. *Cradle-to-grave* begins with the gathering of raw materials from the earth to create the product, and ends at the point when all materials are returned to the earth. LCA evaluates all stages of a product's life from the perspective that they are interdependent, meaning that one operation leads to the next. LCA enables the estimation of the cumulative environmental impacts resulting from all stages in the product life cycle and allowing selecting the path or process that is more environmentally preferable. LCA helps decision-makers select the product, process, or technology that results in the least impact to the environment. RES negative impacts can be minimized with appropriate measures. There are miscellaneous precautions that can be taken to minimize the environmental impacts of the renewable energy systems, during manufacturing, transportation and operation phases. Wind turbines, for example must adhere to certain spacings from individual houses and communities. The impacts from geothermal energy are primarily from emissions, water disposal, seismicity, ground subsidence, water use and land use. LCAs are an important tool for industry and policy makers, used to determine the actual emissions of a product or technology throughout its whole life cycle. In case of energy production systems or power plants, analysis of energy required to produce the materials and processes; emissions resulting from various processes for materials production and processes resulting into their cumulated energy demand and global warming potential become important parameters when making decisions on further research, development and deployment of any technology. Some actions are required, such as:

a. Development of rigorous process and activities integration

b. Rigorous and quantitative analysis

c. Study of the rates of removal of hazardous materials from atmospheric agents.

d. Development of reliable regional models for the atmospheric dispersion and dry and wet settling of particulates generated from solar modules

e. Contingent analysis to identify the locations suitable for the construction of wind turbine towers

f. Optimize the production of electricity, fuels and chemicals in integrated bio-refineries.

g. Use wastes, refuses and lignocellulosic products or energy crops harvested in off-years.

h. Identify areas that are not included in migratory paths.

Questions and Problems

1. Briefly describe the energy conversion process.

2. The efficiency of the hydro turbine ranges from 0.70 to 0.98, while the efficiencies of the electric generators are in the range of 0.80 to 0.95. If the penstock losses are assumed to account for about 8%, what are the efficiency limits for a hydropower station?

3. How do the renewable energy sources differ from fossil fuels, i.e., what are the most common characteristics of renewable energy sources that are different from characteristics of fossil fuels used in the past?

4. By using the values of Table 7.2 or found in the literature estimate the efficiency range of a power system (generation, transmission and power distribution), assuming a reasonable transmission, distribution and transformer steps.

5. What are the four stages of an LCA?

6. Briefly describe the LCA inventory analysis.

7. What are the LCA limitations?

8. List and compare the environmental impacts and benefits of solar and wind energy.

9. What are the environmental effects that must be considered for geothermal power projects? How do they compare with those from conventional power generation systems?

10. Discuss the relative merits of the most common methods of reducing environmental impacts of energy production.

11. By using data found in the literature, develop an LCA analysis for a micro-wind turbine, power range 10 kW.

12. Repeat the above problem for a large wind turbine, power range 1 MW or higher.

13. What are the environmental benefits of wind energy?

14. What are the major negative impacts of the solar energy?

15. Briefly describe the geothermal energy environmental issues.

16. Briefly discuss the seismic-induced activities by the deep geothermal projects.

17. What are the environmental benefits of the use of thermal energy storage?

18. Briefly describe the negative environmental impacts of wind energy.

19. What are the negative impacts on the environment of the wave energy, Tindal energy and ocean thermal energy conversion?

20. Briefly describe the measures to reduce the noise generated by the wind turbines.

21. List and briefly describe the major negative effects of the biofuels.

22. Briefly describe the environmental benefits of the direct domestic solar thermal energy systems?

23. Is energy conservation likely to be more important in a renewable energy economy than it has been in the past? Explain.

24. Hydropower is the world largest source of renewable energy generated electricity, and there is potential for expansion. In some of the world regions, a renewable energy portfolio could be based only on hydropower and energy conservation. However, the hydropower is also controversial, mainly because of associated negative externalities. Describe some of these. Do you think more hydropower should be developed in the world? How would you decide whether or not to develop a particular hydro project?

25. By using data found in the literature, develop the LCA of a flat-plate solar collector.

26. Repeat the above analysis for an evacuate tube solar collector.

27. Define and briefly describe the energy stored on investment (ESOI) ratio.

28. List the environmental implications of main grid energy storage technologies.

29. Estimate the minimum, maximum and average land requirements for a 40-MW parabolic trough updraft power plant.

30. How much energy can be produced in a year if a 100-MW geothermal power plant is operated 75% of the time?

31. Repeat the previous problem of a wind park of the same power, assuming 0.26 capacity factor?

32. How much energy can be produced in one year if a 100-kW PV array is located in your area?

33. Repeat the above problem for a 200-kW solar pond power plant.

34. Determine the land required for a 75-MW photovoltaic power plant.

35. List and briefly discuss the environmental benefits of the PV systems.

36. List and briefly discuss the environmental benefits and impacts of the small hydropower systems.

37. Three coal power plants with a total power producing capacity of 1000 MW and average thermal efficiency 36% are substituted by one nuclear power plant with 33% thermal efficiency. What is the annual amount of CO_2 that is not emitted to the atmosphere? What is the increase of the waste heat produced? The heating value of the coal that was used is 28,000 kJ/kg and contains 80% carbon.

38. Compare the environmental benefits and disadvantages of geothermal power plant, solar thermal power plant and a natural gas conventional power plant.

39. If in Example 7.5 it is assumed that between 10% and 40% of the demanded energy is generated by using renewable energy sources, how much CO_2 emissions are mitigated? Assuming that initial 50% of the electrical energy is generated by coal and other 50% is generated by natural gas.

40. Repeat the previous problem for the electrical energy demanded in the United States.

41. Briefly discuss energy efficiency and conservation.

42. What are the main points of an energy management plan?

Further Readings and References

1. L.A. Vega, Ocean thermal energy conversion, in *Encyclopedia of Energy Technology and the Environment* (A. Bisio and S. Boots eds.), Vol. 3, pp. 2104–2119, Wiley, New York, 1995.
2. OECD/IEA, 1998. Benign energy? In *The Environmental Implications of Renewable*. International Energy Agency, Paris, France, 1998. Available from www.iea.org.
3. R.A. Ristinen and J.J. Kraushaar, *Energy and Environment*, Wiley, Hoboken, NJ, 2006.
4. V. Quaschning, *Understanding Renewable Energy Systems*, Earthscan, London, UK, 2006.
5. J. Andrews and N. Jelley, *Energy Science, Principles, Technology and Impacts*, Oxford University Press, Oxford, 2007.
6. J. Twidell and T. Weir, *Renewable Energy Sources* (2nd ed.), Taylor & Francis Group, Boca Raton, FL, 2006.
7. B. Sorensen et al., *Renewable Energy Focus Handbook*, Academic Press, Amsterdam, the Netherlands, 2009.
8. E.L. McFarland, J.L. Hunt, and J.L. Campbell, *Energy, Physics and the Environment* (3rd ed.), Cengage Learning, Mason, OH, 2007.
9. F.M. Vanek and L.D. Albright, *Energy Systems Engineering: Evaluation and Implementation*, McGraw Hill, New York, 2008.
10. J.F. Manwell and J.G. McGowan, *Wind Energy Explained: Theory, Design and Application*, John Wiley & Sons, Chichester, UK, 2009.
11. J.B. Guinee et al., Life cycle assessment: Past, present, and future, *Environmental Science & Technology*, Vol. 45, pp. 90–96, 2011.
12. B. Everett and G. Boyle, *Energy Systems and Sustainability: Power for a Sustainable Future* (2nd ed.), Oxford University Press, Oxford, 2012.
13. National Renewable Energy Laboratory Report, Life cycle greenhouse gas emissions from electricity generation, NREL/FS-6A20-57187, 2013.
14. R.A. Dunlap, *Sustainable Energy*, Cengage Learning, Stamford, CT, 2015.
15. V. Nelson and K. Starcher, *Introduction to Renewable Energy (Energy and the Environment)*, CRC Press, Boca Raton, FL, 2015.
16. ISO 14040. Environment Management–Life Cycle Assessment–Principles and Framework, 1997.
17. ISO 14041. Environment Management–Life Cycle Assessment–Goal and Scope Definition and Inventory Analysis, 1998.
18. ISO 14042. Environment Management–Life Cycle Assessment–Life Cycle Impact Assessment, 2000.
19. ISO 14043. Environment Management–Life Cycle Assessment–Life Cycle Interpretation, 2000.
20. L. Gagnon, C. Belanger, and Y. Uchiyama, Life-cycle assessment of electricity generation options: the status of research in year 2001, *Energy Policy*, Vol. 30, pp. 1267–1287, 2002.
21. J.B. Guinée et al. (eds.), *Handbook on Life Cycle Assessment: Operational Guide to the ISO Standards*, Kluwer Academic Publishers, Dordrecht, the Netherlands, 2002.
22. T. Greadel and B. Allenby, *An Introduction to LCA, In Industrial Ecology* (2nd ed.), Prentice Hall, Upper Saddle River, NJ, 2003.
23. M. Goralczyk, Life cycle assessment in the renewable energy sector, *Applied Energy*, Vol. 75, pp. 205–211, 2003.
24. N. Jungbluth, C. Bauer, R. Dones, and R. Frischknecht, Life cycle assessment for emerging technologies: Case studies for photovoltaic and wind power, *International Journal of Life Cycle Assessment*, Vol. 10(3), pp. 24–34, 2005.
25. M. Carpentieri, A. Corti, and L. Lombardi, Life cycle assessment (LCA) of an integrated biomass gasification combined cycle (IBGCC) with CO_2 removal, *Energy Conservation and Management*, Vol. 46, pp. 1790–1808, 2005.

26. M. Pehnt, Dynamic life-cycle assessment of renewable energy technologies, *Renewable Energy*, Vol. 31, pp. 55–71, 2006.
27. US EPA Document/Report, *Life Cycle Assessment: Principles and Practice*, EPA/600/R-06/060, 2006.
28. H. Chen, T. Cong, W. Yang, C. Tan, Y. Li, and Y. Ding, Progress in electrical energy storage system: A critical review, *Progress in Natural Science*, Vol. 19, pp. 291–312, 2009.
29. B. Fleck and M. Huot, Comparative life-cycle assessment of a small wind turbine for residential off-grid use, *Renewable Energy*, Vol. 34(12), pp. 2688–2696, 2009.
30. IEC 61400 Standards (International Electrotechnical Commission), *Wind Turbines*, Part 1, 2, 3, 4, 5, 11, 12, 13, 14, 17, 21, 22, 23, 24, and 25; IEC Central Office, Geneva. Available at http: //www. iec.ch (accessed in 2012, 2014, 2017).
31. National Academy of Engineering and National Research Council, *Power of Renewables: Opportunities and Challenges for China and the United States*. The National Academies Press, Washington, DC, 2010.
32. S. Davidsson, M. Höök, and G. Wall, A review of life cycle assessments on wind energy systems, *The International Journal of Life Cycle Assessment*, Vol. 17, pp. 1–14, 2012.
33. P. Brunner and H. Rechberger, Waste to energy–key element for sustainable waste management, *Waste Management*, Vol. 1, pp. 1–10, 2014.
34. F.R. Spellman, *Environmental Impacts of Renewable Energy*, CRC Press, Boca Raton, FL, 2014.
35. K.R. Khalilpour and A. Vassallo, *Community Energy Networks with Storage: Modeling Frameworks for Distributed Generation*, Springer, Singapore, 2016.
36. R. Bansal (ed.), *Handbook of Distributed Generation: Electric Power Technologies, Economics and Environmental Impacts*, Springer, Cham, Switzerland, 2017.

Appendix A: Common Parameters, Units and Conversion Factors

TABLE A.1

Physical Constants in SI Units

Quantity	Symbol	Value
Avogadro constant	N	$6.022169 \cdot 10^{26}\,\text{kmol}^{-1}$
Boltzmann	k	$1.380622 \cdot 10^{-23}\,\text{J/K}$
First radiation constant	$C_1 = 2 \cdot \pi \cdot h \cdot c$	$3.741844 \cdot 10^{-16}\,\text{Wm}^2$
Gas constant	R	$8.31434 \cdot 10^3\,\text{J/kmol K}$
Planck constant	h	$6.626196 \cdot 10^{-34}\,\text{Js}$
Second radiation constant	$C_2 = hc/k$	$1.438833 \cdot 10^{-2}\,\text{mK}$
Speed of light in a vacuum	c	$2.997925 \cdot 10^8\,\text{m/s}$
Stefan-Boltzmann constant	σ	$5.66961 \cdot 10^{-8}\,\text{W/m}^2\text{K}^4$
Speed of light	c	$299{,}792.458\,\text{m/s}$
Elementary charge	e	$1.602176 \cdot 10^{-19}\,\text{C}$

TABLE A.2

Multiplication Factors

Multiplication Factor	Prefix	Symbol
10^{12}	Terra	T
10^9	Giga	G
10^6	Mega	M
10^3	Kilo	K
10^2	Hecto	H
10	Deka	da
1	N/A	—
0.1	Deci	d
0.01	Centi	c
10^{-3}	Mili	m
10^{-6}	Micro	μ
10^{-9}	Nano	n
10^{-12}	Pico	p
10^{-15}	Femto	f
10^{-18}	Atto	a

TABLE A.3

System of Units and Conversion Factors

U.S. Unit	Abbreviation	SI Unit	Abbreviation	Conversion Factor
Foot	ft	Meter	m	0.3048
Mile	mi	Kilometer	km	1.6093
Inch	in	Centimeter	cm	2.54
Square feet	ft^2	Square meter	m^2	0.0903
Acre	acre	Hectare	ha	0.405
Circular mil	cmil	μm^2	—	506.7
Cubic feet	ft^3	Cubic meter	m^3	0.02831
Gallon (U.S.)	gal(U.S.)	Liter	l	3.785
Gallon (UK)	gal(UK)	Liter	l	4.445
Cubic feet	ft^3	Liter	l	28.3
Pound	lb	Kilogram	kg	0.45359
Ounces	oz	Gram	g	28.35
US ton	ton (U.S.)	Metric ton	ton (metric)	0.907
Mile/hour	mi/h	Meter/second	m/s	0.447
Flow rate	ft^3/h	Flow rate	m^3/s	0.02831
Density	lb/ft^3	Density	kg/m^3	16.020
lb-force	lbf	Force	N	4.4482
Pressure	lb/in^2	Pressure	kPa	6.8948
Pressure	bar	Pressure	Pa	10^5
Torque	lb.force ft	Torque	Nm	1.3558
Power	ft.lb/s	Power	W	1.3558
Power (horsepower)	HP	Power	W	745.7
Energy	ft.lb-force	Energy	J	1.3558
Energy (British thermal unit)	Btu	Energy	kWh	3412

TABLE A.4

Common Energy Conversion Factors

Energy Unit	SI Equivalent
1 electron volt (eV)	$1.6021 \ 10^{-19}$ J
1 erg (erg)	10^{-7} J
1 calorie (cal)	4.184 J
1 British thermal unit (Btu)	1055.6 J
1 Q (Q)	10^{18} Btu (exact)
1 quad (q)	10^{15} Btu (exact)
1 tons oil equivalent (toe)	$4.19 \cdot 10^{10}$ J
1 barrels oil equivalent (bbl)	$5.74 \cdot 10^9$ J
1 tons coal equivalent (tce)	$2.93 \cdot 10^{10}$ J
1 m^3 of natural gas	$3.4 \cdot 10^7$ J
1 liter of gasoline	$3.2 \cdot 10^7$ J
1 kWh	$3.6 \cdot 10^6$ J
1 ft^3 of natural gas (1000 Btu)	1055 kJ
1 gal. of gasoline (125,000 Btu)	131.8875 kJ

TABLE A.5

Refractive Index of Some of the Common Materials (at 20°C)

Material	Index of Refraction	Material	Index of Refraction
Diamond	2.419	Benzene	1.501
Fluorite	1.434	Carbon disulfide	1.628
Fused quartz	1.458	Carbon tetrachloride	1.461
Glass (crown)	1.520	Ethyl alcohol	1.361
Glass (flint)	1.660	Glycerin	1.473
Ice	1.309	Oil, turpentine	1.470
Polystyrene	1.590	Paraffin (liquid)	1.480
Salt ($NaCl_2$)	1.544	Water	1.333
Teflon	1.380	Air (0°C, 1 atm)	1.000293
Zircon	1.923	Carbon dioxide (0°C, 1 atm)	1.00045

TABLE A.6

Properties of Dry Air

Temperature (K)	ρ (kg/m³)	C_P (kJ/kg K)	k (W/m K)
293	1.2040	1.0056	0.02568
300	1.1774	1.0057	0.02624
350	0.9980	1.0090	0.03003
400	0.8826	1.0140	0.03365
450	0.7833	1.0207	0.03707
500	0.7048	1.0295	0.04038
600	0.5879	1.0551	0.04659
700	0.5030	1.0752	0.05230
800	0.4405	1.0978	0.05779
900	0.3925	1.1212	0.06279
1000	0.3525	1.1417	0.06752

Source: U.S. National Bureau Standards (U.S.), Tables of Thermodynamic and Transport Properties, Circular 564, 1955.

Symbols: T absolute temperature, degrees Kelvin; ρ density; C_P specific heat capacity; μ viscosity; $\nu = \mu/\rho$; and k thermal conductivity. The values of ρ, μ, ν, k and C_P are not strongly pressure-dependent and may be used over a fairly wide range of pressures.

TABLE A.7

Properties of Water

Temperature (°C)	ρ (kg/m³)	C_P (kJ/kg K)	k (W/m K)
0.00	999.8	4.225	0.566
4.44	999.8	4.208	0.575
10.0	999.2	4.195	0.585
20.0	997.8	4.179	0.604
25.0	994.7	4.174	0.625
100.0	954.3	4.219	0.684

TABLE A.8

Temperature Conversion Formulas

	Degree Celsius (°C)	Degree Fahrenheit (°F)	Kelvin Degree (K)
Degree Celsius (°C)	—	$\frac{9}{5}(^\circ C + 32)$	$K - 273.15$
Degree Fahrenheit (°F)	$\frac{5}{9}(^\circ F - 32)$	—	$1.8 \times K - 459.67$
Kelvin degree (K)	$^\circ C + 273.15$	$(459.67 + {}^\circ F) / 1.8$	—

TABLE A.9

U.S. Standard Units

Quantity	Type	Symbol	SI Conversion
Foot	Length	ft	0.305 m
Inch	Length	in	0.0254 m
Mile	Length	mi	1609.34 m
Second	Time	s	1 s
Pound	Mass/weight	lb	0.454 kg
Ounce	Mass/weight	oz	0.02835 kg
Gallon	Volume	gal	3.7854 l
Fluid ounce	Volume	fl oz	29.574 mL
Fahrenheit	Temperature	°F	$(9/5)\ ^\circ C + 32$
Ampere	Current	A	1 A
Volt	Potential/voltage	V	1 V

Appendix B: Design Parameters, Conversion Factors and Data for Renewable Energy Conversion and Energy Storage Systems

TABLE B.1

Common Units and Conversion Factors

Type	Name	Symbol	Approximate Value
Pressure	Atmosphere	atm	1.013×10^5 Pa
Pressure	Bar	bar	10^5 Pa
Pressure	Pounds per square inch	psi	6890 Pa
Mass	Ton (metric)	t	10^3 kg
Mass	Pound	lb	0.436 kg
Mass	Ounce	oz	0.02835 kg
Length	Ångström	Å	10^{-10} m
Length	Inch	in	0.0254 m
Length	Foot	ft	0.3048 m
Length	Mile (statute)	mi	1609 m
Volume	Liter	l	10^{-3} m^3
Volume	Gallon (U.S.)	gal	3.785×10^{-3} m^3

TABLE B.2

Useful Design Conversion Factors SI and American System

Quantity	Symbol	Conversion
Area	A	1 ft^2 = 0.0929 m^2 1 acre = 43,560 ft^2 = 4047 m^2 1 hectare = 10,000 m^2 1 square mile = 640 acres
Density	ρ	1 lb$_m$/ft^3 = 16.018 kg/m^3
Energy, heat or work	E, Q or W	l Btu = 55.1 J 1 kWh = 3.6 MJ 1 Therm = 105.506 MJ l Cal = 4.186 J 1 ft lb$_f$ = 1.3558 J
Force	F	1 lb$_f$ = 4.448 N
Heat flow rate (debit)	q	1 Btu/h = 0.2931 W 1 ton (refrigeration) = 3.517 kW l Btu/s = 1055.1 W
Heat flux	q/A	1 Btu/h ft^2 = 3.1525 W/m^2
Heat-transfer coefficient	h	1 Btu/h ft^2 = Z5.678 W/m^2 K
Length	L or l	1 ft = 0.3048 m 1 in = 2.54 cm 1 mi = 1.6093 km
Mass	m	1 lb$_m$ = 0.4536 kg 1 ton = 2240 lbm 1 tonne (metric) = 1000 kg

(Continued)

TABLE B.2 (*Continued*)

Useful Design Conversion Factors SI and American System

Quantity	Symbol	Conversion
Mass flow rate	dm/dt	1 lb_m/h = 0.000126 kg/s
Power	P	1 HP = 745.7 W
		1 kW = 3415 Btu/h
		1 ft. lbf/s = 1.3558 W
		1 Btu/h = 0.293 W
Pressure	P	1 lbf/in^2 (psi) = 6894.8 N/m^2 (Pa)
		1 in Hg = 3,386 N/m^2 (Pa)
		1 atm = 101,325 N/m^2 = 14.696 psi
Radiation	SR	1 langley/min = 697.4 W/m^2
Specific heat Capacity	C	1 Btu/lb_m °F = 4187 J/kg K
Enthalpy	H	1 Btu/lb_m = 2326.0 J/kg
		1 cal/g = 4184 J/kg
Thermal conductivity	K	1 Btu/h ft °F = 1.731 W/m K
Velocity	v	1 ft/s = 0.3048 m/s
		1 mi/h = 0.44703 m/s
Dynamic viscosity	μ	1 lb_m/ft. s = 1.488 N s/m^2
		1 cP = 0.00100 N s/m^2
Kinematic Viscosity	ν	1 ft^2/s = 0.09029 m^2/s
		1 ft^2/h = $2.581 \times 10^5 m^2$/s
Volume	V	1 ft^3 = 0.02832 m^3 = 28.32 l
		1 barrel = 42 gal (U.S.)
		1 gal (U.S. liq.) = 3.785 l
		1 gal (U.K.) = 4.546 l
Volumetric flow Rate	dQ/dt	1 ft^3/min (cfia) = 0.000472 m^3/s
		1 gal/min (GPM) = 0.0631 l/s

TABLE B.3

Standard Thermodynamic Properties for Selected Substances

Substance	Symbol	ΔH_f° (kJ· mol^{-1})	ΔG_f° (kJ· mol^{-1})	S_{298}° (J $K^{-1} mol^{-1}$)
Aluminum	Al(s)	0	0	28.3
	Al(g)	324.4	285.7	164.54
Antimony	Sb(s)	0	0	45.69
	Sb(g)	262.34	222.17	180.16
Arsenic	As(s)	0	0	35.1
	As(g)	302.5	261.0	174.21
Bromine	Br(l)	0	0	152.23
	Br2(g)	30.91	3.142	245.5
	Br(g)	111.88	82.429	175.0
Cadmium	Cd(s)	0	0	51.76
	Cd(g)	178.2	144.3	167.75
Carbon	C(s)/Graphite	0	0	5.74

(*Continued*)

TABLE B.3 (*Continued*)

Standard Thermodynamic Properties for Selected Substances

Substance	Symbol	ΔH_f° (kJ· mol^{-1})	ΔG_f° (kJ·mol^{-1})	S_{298}° (J K^{-1} mol^{-1})
	C(s)/Diamond	1.89	2.90	2.38
	C(g)	716.681	671.2	158.1
	CO(g)	−110.52	−137.15	197.7
	CO_2(g)	−393.51	−394.36	213.8
	CH_4(g)	−74.6	−50.5	186.3
	CH_3OH(*l*)	−239.2	−166.6	126.8
	CH_3OH(g)	−201.0	−162.3	239.9
	C_2H_2(g)	227.4	209.2	200.9
	C_2H_4(g)	52.4	68.4	219.3
	C_2H_6(g)	−84.0	−32.0	229.2
	C_2H_5OH(l)	−277.6	−174.8	160.7
	C_2H_5OH(g)	−234.8	−167.9	281.6
	C_6H_6(g)	82.927	129.66	269.2
	C_6H_6(l)	49.1	124.50	173.4
	CH_2Cl_2(l)	−124.2	−63.2	177.8
	CH_2Cl_2(g)	−95.4	−65.90	270.2
Chlorine	Cl_2(g)	0	0	223.1
	Cl(g)	121.3	105.70	165.2
Chromium	Cr(s)	0	0	23.77
	Cr(g)	396.6	351.8	174.50
Copper	Cu(s)	0	0	33.15
	Cu(g)	338.32	298.58	166.38
	$CuSO_4$(s)	−771.36	−662.2	120.9
Fluorine	F_2(g)	0	0	202.8
	F(g)	79.4	62.3	158.8
	HF(g)	−273.3	−275.4	173.8
Hydrogen	H_2(g)	0	0	130.7
	H(g)	217.97	203.26	114.7
	H^+(aq)	0	0	0
	OH^-(aq)	−230.0	−157.2	−10.75
Iron	Fe(s)	0	0	27.3
	Fe(g)	416.3	370.7	180.5
	Fe_2^+(aq)	−89.1	−78.90	−137.7
Lead	Pb(s)	0	0	64.81
	Pb(g)	195.2	162.	175.4
	Pb_2^+(aq)	−1.7	−24.43	10.5
Lithium	Li(s)	0	0	29.1
	Li(g)	159.3	126.6	138.8
	Li^+(aq)	−278.5	−293.3	13.4
Magnesium	Mg^{2+}(aq)	−466.9	−454.8	−138.1
Mercury	Hg(l)	0	0	75.9
	Hg(g)	61.4	31.8	175.0
Nickel	Ni_2^+(aq)	−64.0	−46.4	−159
Nitrogen	N_2(g)	0	0	191.6

(Continued)

TABLE B.3 (*Continued*)

Standard Thermodynamic Properties for Selected Substances

Substance	Symbol	ΔH_f° (kJ·mol^{-1})	ΔG_f° (kJ·mol^{-1})	S_{298}° (J K^{-1}mol^{-1})
	N(g)	472.704	455.5	153.3
	NO(g)	90.25	87.6	210.8
	NO$_2$(g)	33.2	51.30	240.1
	NO$_3^-$(aq)	−205.0	−108.7	355.7
	NH$_3$(g)	−45.9	−16.5	192.8
	NH$_4^+$(aq)	−132.5	−79.31	113.4
Oxygen	O$_2$(g)	0	0	205.2
	O(g)	249.17	231.7	161.1
	O$_3$(g)	142.7	163.2	238.9
Phosphorus	P$_4$(s)	0	0	164.4
	P$_4$(g)	58.91	24.4	280.0
	P(g)	314.64	278.25	163.19
Potassium	K(s)	0	0	64.7
	K(g)	89.0	60.5	160.3
	K$^+$(aq)	−252.4	−283.3	102.5
Silicon	Si(s)	0	0	18.8
	Si(g)	450.0	405.5	168
	SiO$_2$(s)	−910.7	−856.3	41.5
	SiH$_4$(g)	34.3	56.9	204.6
Silver	Ag(s)	0	0	42.55
	Ag(g)	284.9	246.0	172.89
	Ag+(aq)	105.6	77.11	72.68
Sodium	Na(s)	0	0	51.3
	Na(g)	107.5	77.0	153.7
	Na$^+$(aq)	−240.1	−261.9	59
Sulfur	S8(s) (rhombic)	0	0	256.8
	S(g)	278.81	238.25	167.82
	S$_2^-$(aq)	41.8	83.7	22
Tin	Sn(s)	0	0	51.2
Water	Sn(g)	301.2	266.2	168.5
	H$_2$O(l)	−285.83	−237.1	70.0
	H$_2$O(g)	−241.82	−228.59	188.8
	H$_2$O$_2$(l)	−187.78	−120.35	109.6
	H$_2$O$_2$(g)	−136.3	−105.6	232.7
Zinc	Zn(s)	0	0	41.6
	Zn(g)	130.73	95.14	160.98
	Zn^{2+}(aq)	−153.9	−147.1	−112.1

Note: Adapted from U.S. National Bureau Standards (U.S.), Tables of Thermodynamic and Transport Properties, Circular 564, 1955.

Legend: (g) gas; (l) liquid; (aq) aqueous solution; and (s) solid.

TABLE B.4

Thermal Capacities of Common TES Materials at 20°C

Material	Density (kg/m³)	Specific Heat (J/kg K)	Volumetric Thermal Capacity (10^6 J/m³ K)
Clay	1458	879	1.28
Brick	1800	837	1.51
Sandstone	2200	712	1.57
Wood	700	2390	1.67
Concrete	2000	880	1.76
Glass	2710	837	2.27
Aluminum	2710	896	2.43
Iron	7900	452	3.57
Steel	7840	465	3.68
Gravely	2050	1840	3.77
Earth Magnetite	5177	752	3.89
Water	988	4182	4.17

TABLE B.5

Thermodynamic Properties of Water

Temperature (K)	$\widehat{G}(T)$ (kJ·mol⁻¹)	$\widehat{H}(T)$ (kJ·mol⁻¹)	$\widehat{S}(T)$ (J·K⁻¹ mol⁻¹)	$C_P(T)$ (J·K⁻¹ mol⁻¹)
273	−305.1	−287.73	63.28	76.10
298.15	−306.69	−285.83	69.95	75.37
320	−308.27	−284.18	75.28	75.27
340	−309.82	−282.68	79.85	75.41
360	−311.46	−281.17	84.16	75.72
373	−312.58	−280.18	86.85	75.99

TABLE B.6

Water Properties at Different Temperatures

Temperature (°C)	Density (kg/m³)	Vapor Pressure (N/m²)	Vapor Pressure (torr)
0	999.8395	613.2812	4.6
4	999.9720 (maximum density)	813.2642	6.1
10	999.7026	1226.562	9.2
15	999.1026	1706.522	12.8
20	998.2071	2333.135	17.5
22	997.7735	2639.776	19.8
25	997.0479	3173.064	23.8
30	995.6502	4239.640	31.8
40	992.2	7372.707	55.3
60	983.2	19918.31	149.4
80	971.8	47342.64	355.1
100	958.4	101324.7	760.0

TABLE B.7

Thermal Properties of Some Common Materials

Material	Density (kg/m³)	Heat Capacity C_P (kJ/kg·K)	Temperature Range (ΔT in °C)
Water	1000	4.190	0–100
Ethanol	780	2.460	−117–79
Glycerin	1260	2.420	17–290
Oil	910	1.800	−10–204
Synthetic oil	910	1.800	−10–400
Engine oil	888	1.880	0–160
Propanol	800	2.500	0–97
Brick	1600	0.840	20–70
Concrete	2240	1.130	20–70
Sandstone	2200	0.712	20–70
Granite	2650	0.900	20–70
Marble	2500	0.880	20–70
Clay sheet	1460	0.880	10–70

TABLE B.8

Electric Power Common Terms

Term	Definition	Standard Unit	Common Used Unit
Power density	Power per unit volume	W/m³	kW/m³
Specific power	Power per unit mass	W/kg	kW/kg
Electric power			
Output	Power × time	J	$kW_e h$
Rated power	Power output of a power plant at nominal operating conditions	MW	MW
Hate rate (HR)	Thermal BTUs required to generate 1 $kW_e h$ (3412 Btu = 1 $kW_t h$)	$W_t h/W_e h$	$kW_t h/kW_e h$
Capacity factor (CF)	Average power/rated power (per a specific time period)	—	—
Load factor (LF)	Average power/maximum power (per a specific time period)	—	—
Availability factor (AF)	Fraction of time that power generation system (Unit) is available (usually non-dispatchable units)	—	—

Index

Note: Page numbers in italic and bold refer to figures and tables, respectively.